VISUALIZING WITH

C A D

DANIELA BERTOL

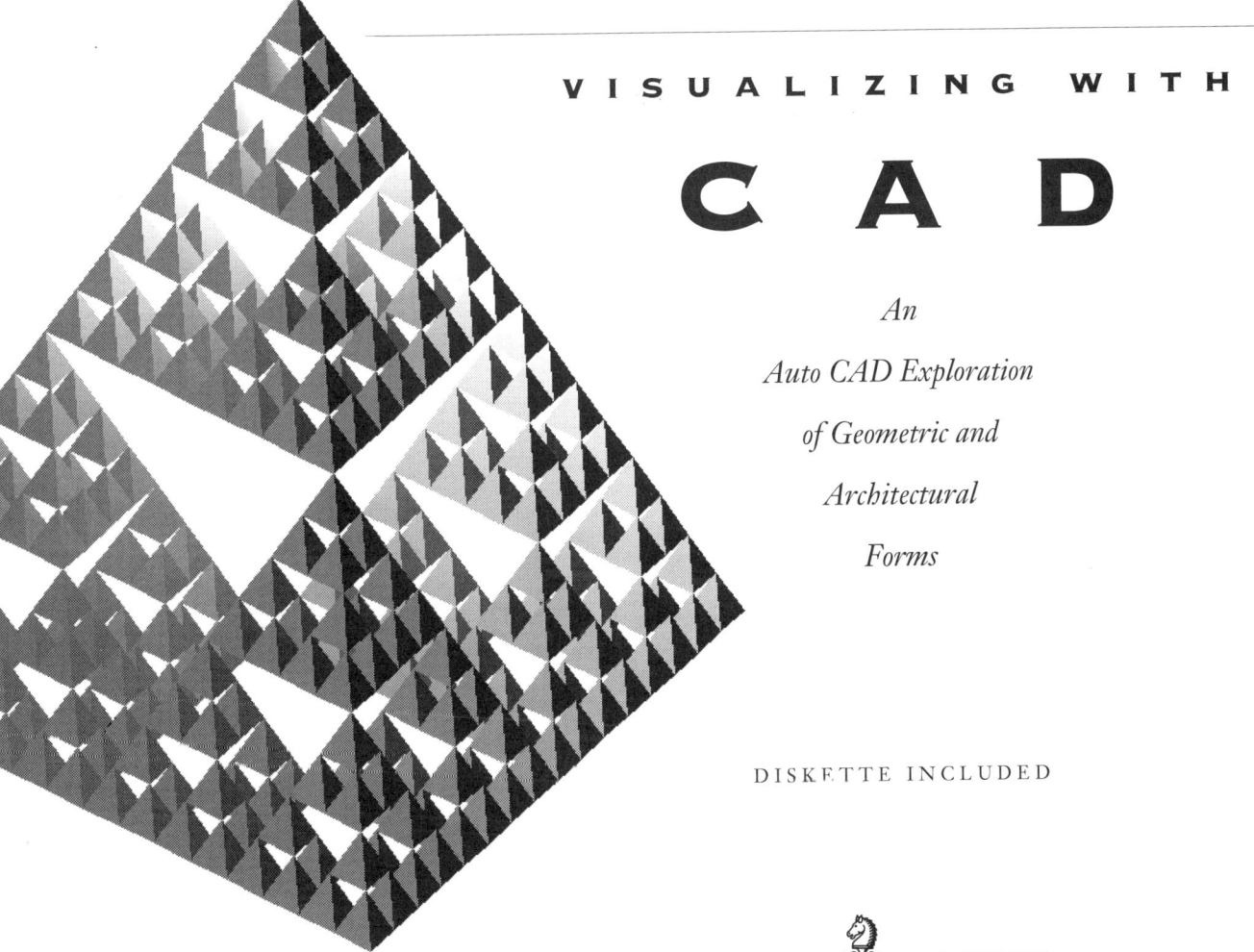

VISUALIZING WITH

CAD

An

Auto CAD Exploration

of Geometric and

Architectural

Forms

DISKETTE INCLUDED

SPRINGER-VERLAG

TELOS

THE ELECTRONIC LIBRARY OF SCIENCE

Publisher:	Allan M. Wylde
Publishing Assistant:	Kate Young
Promotions Managers:	Jacqueline Jeng / Paul Manning
Produced and Prepared by:	Jan Benes, Black Hole Publishing Service
Text and Cover Designer:	Janet Wood
Production Artist:	Jim Predny
Copy Editor:	Paul Green

Bertol, Daniela, 1958-
    Visualizing with CAD: why and how to generate forms from geometry
to architecture using AutoCAD / Daniela Bertol.
            p.   cm.
    Includes bibliographical references and index.
    ISBN 0-387-94275-0
    1. Computer graphics.   2. Visualization.   3. AutoCAD (Computer
file) 4. Computer-aided design.   I. Title.
T385.B467   1994
745.4—dc20
                                                                            94-17611
                                                                            CIP

Printed in the United States of America.

9 8 7 6 5 4 3 2 1

ISBN 0-387-94275-0 Springer-Verlag New York Berlin Heidelberg

THE
ELECTRONIC
LIBRARY
OF
SCIENCE

TELOS, The Electronic Library of Science, is an imprint of Springer-Verlag New York with publishing facilities in Santa Clara, California. Its publishing program encompasses the natural and physical sciences, computer science, economics, mathematics, and engineering. All TELOS publications have a computational orientation to them, as TELOS' primary publishing strategy is to wed the traditional print medium with the emerging new electronic media in order to provide the reader with a truly interactive multimedia information environment. To achieve this, every TELOS publication delivered on paper has an associated electronic component. This can take the form of book/diskette combinations, book/CD-ROM packages, books delivered via networks, electronic journals, newsletters, plus a multitude of other exciting possibilities. Since TELOS is not committed to any one technology, any delivery medium can be considered.

The range of TELOS publications extends from research level reference works through textbook materials for the higher education audience, practical handbooks for working professionals, as well as more broadly accessible science, computer science, and high technology trade publications. Many TELOS publications are interdisciplinary in nature, and most are targeted for

v

the individual buyer, which dictates that TELOS publications be priced accordingly.

Of the numerous definitions of the Greek word "telos," the one most representative of our publishing philosophy is "to turn," or "turning point." We perceive the establishment of the TELOS publishing program to be a significant step towards attaining a new plateau of high quality information packaging and dissemination in the interactive learning environment of the future. TELOS welcomes you to join us in the exploration and development of this frontier as a reader and user, an author, editor, consultant, strategic partner, or in whatever other capacity might be appropriate.

TELOS, The Electronic Library of Science
Springer-Verlag Publishers
3600 Pruneridge Avenue, Suite 200
Santa Clara, CA 95051

for Giustina

TABLE OF CONTENTS

IV

**CAD GENERATION AND
EVOLUTION FOR
GEOMETRIC FORMS**

ACKNOWLEDGEMENTS

A book of this type, in which a complete integration of images, words, graphics design and electronic files is essential to its continuity, requires a quite unusual effort. I would like to acknowledge a few people who helped in shaping the book to its final form.

First, I would like to thank Allan Wylde of Telos and Jan Benes, for making it happen. Because English is my second language, learned in adulthood, I am grateful to David Foell, who did the initial editing, for his support and his insight into architecture and his critical eye in graphics. Principal editing was performed, sensitively and knowledgeably, by Paul Green.

A special thanks to Luis Summers of the University of Colorado at Boulder, whose constructive criticism and generous advice greatly contributed to the contents of the book, which was reshaped completely from the draft he first reviewed.

I would like to express my gratitude to Autodesk, particularly to Robert Fassberg and Neele Johnston. Robert's knowledgeable assistance was valuable in the development of the tutorial section, as well as in the images production.

And thanks from my heart to Barbara Staples and Andrea Harrison for their support, encouragement, and faith.

I spent the first twenty six years of my life in Rome. I used to go for ice cream to a popular place near the Pantheon and I remember the excitement I felt, beyond the chocolate and whipped cream, when I entered this ancient Roman temple. After staring at the "shower" of light coming from the circular opening at the center of the dome, as strong as a spotlight, I remember being attracted almost hypnotically to the place below the opening. I remember counting the coffers carving the concave dome, composed in five rows of circular arrays, and could feel the power and protection created by the concave space.

I also recall going every Sunday to Piazza San Pietro. This Baroque square is well known for its colonnades, which have an oval shape defined by two interlocking circles. For each of these circles there is a mark, located approximately at its center, from which the four aligned rows of columns appear as one. Before entering the church, almost as a part of a ritual, I had to find the mark in the pavement of the oval square. I was amazed by how the rows of columns could appear and disappear according to my position in relation to the mark.

I remember, too, spending summer evenings sitting on the steps of the convex facade of Santa Maria della Pace. The church faces a small square, enclosed by the walls of the adjacent buildings. The perfect interpenetration between solids and voids, concavity and convexity, made this public space feel as protected as a living room.

These early experiences gave me the sense of architecture as the expression of a tangible geometry. Buildings and public spaces were "ideal" solids, defining a physical space where you could walk and touch a perfect harmony of numbers and proportions. I could feel how the power of these architectural and urban spaces arose from an underlying geometry, which made it possible to experience them directly. Later, as an architecture student, my fascination continued at a more intellectual level. I began to read with interest the treatises by Leon Battista Alberti, Luca Pacioli, Filarete, and Sebastiano Serlio, who emphasized the relation between architecture and geometry. These writings deal with architecture as well as with geometry, graphic type fonts, and perspective, all of which are integrated in a graphic layout that powerfully expresses the contents. The rich visual imagery and the outstanding logical continuity of these works, inspired me to write this book; I was driven by the fantasy of a contemporary treatise, but the main question was: What would differentiate a contemporary treatise from the Renaissance ones? I felt that, probably, the principal difference should be a consideration of the effect of the information revolution and computers on architecture, which is more prominent than the difference in construction technology and style.

If Serlio had written his eight books of architecture today, he probably would have added a ninth book on computers. And, given his prolific use of images to illustrate concepts, he probably would also have enclosed a computer diskette or CD ROM with the electronic models of his temples and columns. Today, computer visualization is probably as influential in architecture as perspective was in the Renaissance. The linking of geometry, CAD, and architecture in the same book is similar to the inclusion of geometry, perspective, building technology, and typologies and graphics in the treatises. The difference in the aesthetic of computer images versus traditional renderings inspired me to use both to express two different processes that achieve the same result: the representation of three-dimensional artifacts through two-dimensional images.

The analogies with the Renaissance treatises mentioned here and throughout the book may sound presumptuous. I hope that, in spite of its many limitations, this book has at least the merit of bringing these works to the reader's attention; even though their discussions of architectural typologies and construction technologies may be outdated, they can still excite us for their broad view of architecture and the universality of their approach. In the same spirit, I also hope that this book can inspire the reader to consider the integration of contemporary technology with historical foundations. I believe that only in understanding the past can we grow in the present.

Daniela Bertol
June, 1994
New York City

Computers and telecommunications have revolutionized traditional perceptions of space and time. The images we see and the sounds we hear often do not come from the physical world of our immediate surroundings but from a thousand miles away, thanks to the ability to transform information about physical objects and phenomena into electromagnetic media and then retrieve it as images and sounds representative of the original. The equation between physical reality and perception is no longer satisfied: what was once the realm of science fiction has become an integral part of our daily experience. The term *cyberspace*, likewise born in science fiction, has been increasingly used in the media to denote a digital world which exists in contrast to the physical one.

This transformed contemporary environment suggests the need for an investigation of physical forms in their relation to our modified visual perceptions. The disciplines dealing with exploration of forms, such as architecture and intuitive geometry, also need to address the implications of the new digital media. While in the past the means of communication in these disciplines were limited mainly to signs made of ink on paper (still the most common publishing medium) computers bring a new approach. The issues raised by this new way of dealing with visual forms in a digital world are the topic of this book. There are two main interlaced themes: the analysis of digital models seen from various perspectives, and the definition of geometric procedures and laws for the generation of spatial configurations to be used in the digital models. The viewpoint is that of architecture, the discipline concerned with forms from their ideation to their physical creation. Although the book is not strictly about architecture, the approach is that of an architect exploring space in terms of intuitive and visually constructible forms, as opposed to a strictly geometrical investigation of space. The digital universe we will be entering is that defined by computer-aided design (CAD) which is at the heart of the computer application for architectural and engineering design.

The aim of this book is not to create a "Theory of Everything" for CAD, but to suggest an approach to the exploration of forms using this remarkable tool. The focus is on geometry and architecture as "formal" expressions. Computers can enhance the creative process by placing clearly defined operations in the service of the user's intuition making possible a wide array of solutions to a given design problem. This creative potential is evident in the way that CAD makes it possible to construct **models of three-dimensional forms**.

Long before the advent of computers, physical models served as the tools for architecture and for mathematics and the physical and social sciences, as well. But with CAD it is possible to build visual models of three-dimensional forms. Such models are defined by a data structure in the computer memory that contains information about the geometric primitives constituting a particular form. In this way CAD, analogously to the visual perception process, provides two-dimensional images from an electronic model of a three-dimensional world. This **visualization of forms by means of two-dimensional projections** is the essence of CAD. Visualizing, however, should not be confused with viewing: in CAD, visualizing is synonymous with creating.

In the course of this book we will examine the geometrical principles that provide the foundations for the generation of dynamic models of three-dimensional forms. The term *dynamic* is used because, due to the data structure typical of CAD, the generated models can evolve into several different shapes. From this perspective, architectural forms, too, can be derived from an evolution of models of fundamental geometric forms. We will see how geometry and form are closely related and how any man-made form can be easily interpreted as a geometrical model. Furthermore, any CAD operation in itself can be defined in terms of geometry. We have made an effort to preserve a rigorous mathematical language in an attempt to link any specific CAD operation to its geometric background. Too often, architects and designers are not aware that a form can be axiomatically defined; this prevents them from taking advantage of geometry as an inspirational source, as well as a working aid.

The use of CAD in investigating man-made forms, from geometry to architecture, is analyzed from an interdisciplinary perspective, where concepts of geometry and linguistics are applied to issues of visual perception. We start with a discussion of theoretical principles and conclude by presenting final models of three-dimensional forms listing the basic CAD commands used for their generation. Many of the CAD models, illustrated on paper are also provided as electronic files. Although all the examples have been generated in AutoCAD, they can be of interest to other CAD users because of the universality of form generation algorithms and design visualization procedures.

The book is organized in six chapters. Chapter I discusses briefly the disciplines involved with computer-aided design, such as geometry and architecture; it also introduces general concepts such as space and form. A discussion of theoretical principles is important in order to provide solid foundations for the more specific applications described in the following chapters. Readers already familiar with these principles may elect to briefly review them here, or skip this beginning chapter altogether. Chapter II explains the methods used in CAD in contrast with those of traditional design, and introduces a few basic AutoCAD commands. This chapter represents a bridge between the interdisciplinary introductory chapter and the actual proposed CAD theory and applications which are discussed in the following chapters. In Chapter III, a theory of generation and visualization of forms is established through a series of operations based on primitives and relations borrowed from geometry. Chapter IV applies the established theory for the generation of models of geometric forms. All these forms, although they belong to the realm of geometry, can evolve into architectural models. The discussion then more specifically addresses architecture in Chapters V and VI. The geometric forms generated and visualized in Chapter V are the models of architectural elements commonly used in traditional buildings. In Chapter VI these basic elements are articulated to create architectural compositions according to the same type of rules and diagrams used for more generic forms. Following the same approach, models of famous historical buildings, from classic architecture to modern, are defined and visualized as a demonstration of our initial assumption that forms may be created by using geometry and logical deductive processes as an aid for creative intuition.

Some of the three-dimensional models described in Chapters IV, V, and VI are also provided as electronic AutoCAD (*.dwg) files in the enclosed diskette. The models included on the diskette are indicated by the icon

in the margin next to the paragraph discussing the model.

AutoLISP files are also included for some of the generated models. Each AutoLISP routine can be run from AutoCAD and the corresponding model will be automatically generated according to parameters defined by the user. The AutoLISP routine can also be listed and modified as a text file. The AutoLISP routine and electronic models provided here can be used only from the AutoCAD software. Additional information is provided in the Appendix.

Chapters II and III include a tutorial which takes the following format. Each section contains a general description of the topics indicated. The general definitions can be applied to a CAD software. At the end of each section, a brief AutoCAD implementation of the given definitions is presented. The tutorial segments are set off by a border and changes in font style so that readers who are not AutoCAD users can skip this section. The tutorial is provided for those readers who wish to implement the discussed models, as well as their own. These segments provide only the commands for the implementation of three-dimensional models, generated according to the theory illustrated in Chapter III.

I

A FEW CONSIDERATIONS ABOUT SPACE, ARCHITECTURE, DESIGN, AND COMPUTERS

Space and perception represent,

at the core of the subject, the fact of his

birth, the perceptual contribution of his

bodily being, a communication with

the world more ancient than thought

Maurice Merleau-Ponty

The focus of this book is the theory and practice of developing computer-aided design (CAD) models as a way to optimize the generation of form—a process that may evolve into building design. Generation of form, in turn, requires an understanding of space, the investigation of which can be approached from a variety of disciplines, using definitions and analysis from philosophy, psychology, mathematics, semiotics, and other areas. Although these fields may seem somewhat extraneous to the applications discussed in this book, a more theoretical approach reveals that CAD has a deeper relation to such disciplines than is suggested by its role as a mere tool.

WHAT IS SPACE?

Few words in our language have as many connotations as *space*. We can complain, on a crowded subway, of not having enough of it; we can also claim that it comfortably contains entire galaxies. As defined in different contexts, such as geometry, physics, or architecture, the term can have different meanings and properties. Physics, astronomy, philosophy, chemistry, mathematics, and architecture all deal with space, but each offers a different interpretation and analysis, although their formulations often find a common ground. For example, building design, which uses geometric models of architectural space, shares with physics a mathematical interpretation of space.

Unlike philosophy, physics, and geometry, however, in which space is often treated as an abstract concept and investigated by means of verbal descriptions or mathematical equations, architecture is concerned with the physical space of our everyday experience, and its transformation. The physical space of CAD itself might be the electronic two-dimensional array of pixels of the computer screen, but the space which CAD images allude to is physical three-dimensional space.

SPACE AS A PERCEPTUAL FIELD: FORMS

Regardless of the particular context, space can be defined as a field—a sort of "container"—in which we move and which we

3

perceive through all our senses. Objects in space have extension (that is, they "take up space"). Our own bodies extend and move in space; spatial existence, as the philosopher Maurice Merleau-Ponty wrote, is "the primary condition of all living experience." Further, our bodies have position and orientation with respect to other objects in space. Relative to such objects, we can move forward or backward, up or down, left or right. Thus, space itself, like the objects it contains, can be viewed in terms of three dimensions: We can define spatial extension as a physical quantity measured by width, length, and depth. As is shown in a later section concerning geometry, position in space and spatial extension are recurrent fundamental concepts that also have several applications in architecture and computer graphics.

The complementary views of space suggested here—in terms of both container and the objects contained—lead to the concept of form. *Form*, like *space*, has several meanings and is a subject for many disciplines, from biology to philosophy, from mathematics to psychology. Plato defined *form* (ειδος (eidos), synonymous with ιδεα (idea)) as the permanent ideal reality of things, as opposed to their temporary attributes. According to Aristotle, form (ειδος) is what makes things emerge from an undifferentiated primordial matter (υλη). In our discussion, *form* means physical form or shape as a perceivable entity, characterized by spatial extension and dimension, that creates distinctions in an otherwise amorphous field. According to our perception (which provides us with our first and most immediate knowledge of the world outside ourselves), the world is a three-dimensional space ordered and defined by constant or changing forms. Forms are really what make space perceptible; they provide a series of contrasts, like solid and void, inside and outside, figure and ground, positive and negative, and contrast between different colors. We can perceive all these oppositions because of **boundaries**, which create discontinuities in an otherwise uniform field. Therefore, space cannot be defined without the boundaries that give shape to forms.

An interesting definition of form (which finds application in

Chapter III) can be found in a book of mathematical logic by G. Spencer Brown: "A distinction is drawn by arranging a boundary with separate sides so that a point on one side cannot reach the other side without crossing the boundary... Call the space cloven by any distinction, together with the entire content of the space, the form of the distinction. Call the form of the first distinction the form." Even in this context the primary characteristics of form are given by boundaries, which determine distinctions.

Figure and Ground

Among the contrasts that form gives rise to, that of figure and ground (figure I-1) is one of the most fundamental for our perception: Boundaries define an object (figure) with respect to surrounding

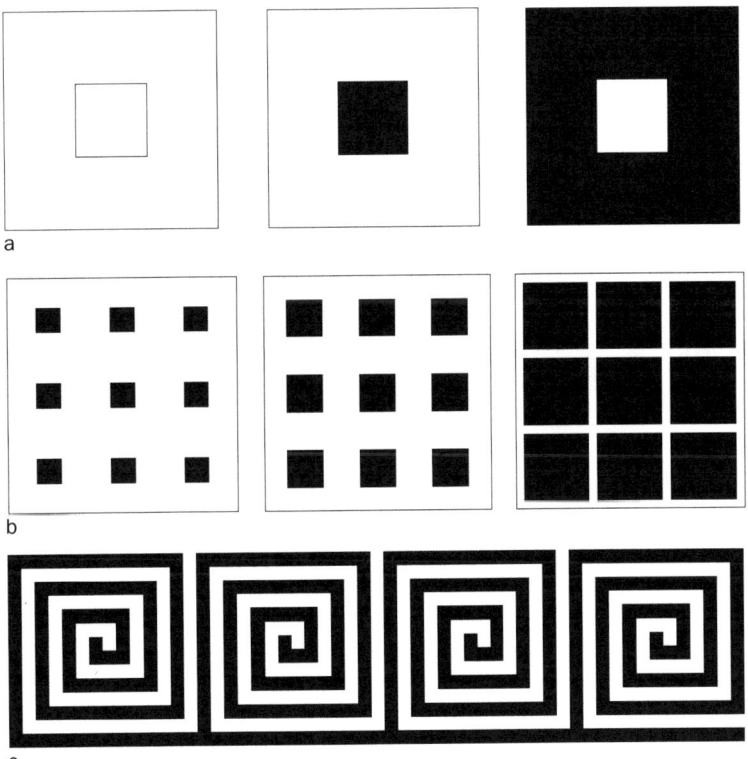

I-1 ►

Figure-ground perception and ambiguities.

a. The black square is recognized as a figure, while the white square can be perceived either as a figure or ground.

b. Small shapes are perceived as figures. When their size increases, shapes become ground (from Arnheim).

c. Another example of figure-ground reversal.

a

b

c

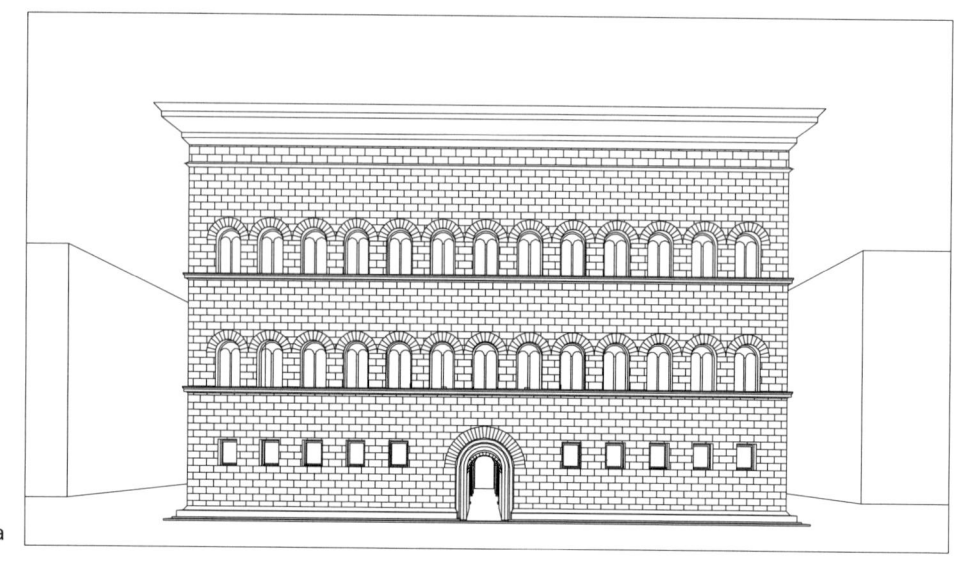

a

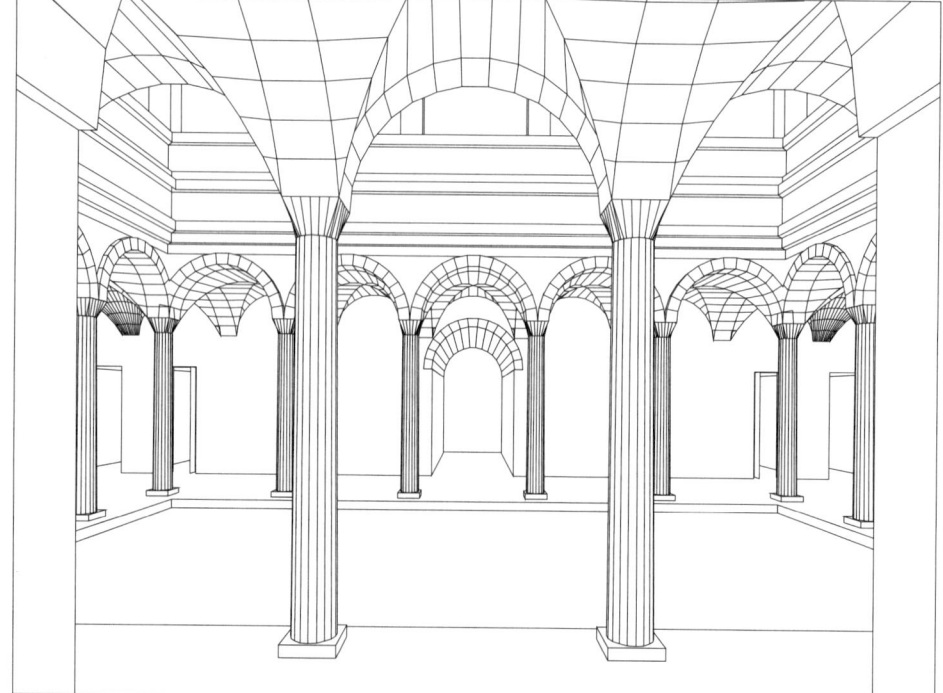

b

environment (ground); therefore a form becomes a figure distinct from the ground. In two-dimensional images, either on paper or on the computer monitor, figure and ground are distinguished by fields of different colors. In the context of architectural space, a built environment usually becomes figure, while the surrounding landscape is the ground [Norberg-Schulz 1980].

It may be erroneous, however, to consider a figure as a defined object against a boundless, amorphous field or ground: The figure-ground distinction does not necessarily represent a fixed relation, but mainly defines qualitatively different spatial attributes. What is figure in one context can become ground in another. In figure I-2, for example, which shows different views from a model of the Palazzo Strozzi (see Chapter VI), the building as a whole reads as a figure in the urban context (figure I-2a); viewed from the courtyard, however, the building reads as ground (figure I-2b). In fact, the figure-ground relation has a greater presence in the subject's perception than as a characteristic belonging to a physical object. Figure and ground are often ambiguously defined and can be alternatively perceived, particularly in two-dimensional imagery (figure I-3).

Solid and Void

In the perception of space, the relationship of solid and void is another fundamental dichotomy which configures form (figure I-4). The perception of solid (mass) and void (empty space) gives the most primitive information about space. This distinction is perceivable not just visually but through all our senses; sounds, for example, will be perceived differently depending on whether the listener is inside or outside an enclosure. Our body itself can be generically seen as a solid. If a solid occupies a position in space, no other solid can have the same position at the same time. In the same way that space can be viewed in terms of both container and extension, the perceptual field of space is articulated through a series of solids separated by voids. Often in architecture solid is defined as *positive* space and void as *negative* space; they are complementary to each other and one

a. A famous example of figure-ground ambiguity.

b. The ambiguity tends to disappear when one of the ambiguous elements is represented in perspective view.

a

b

could not exist without the other. Thus, room is defined both by the walls (solids) and the enclosure (void) which the walls create, and the room's spatial qualities are a result of both.

The meaning of *solid* in perception can be different from its meaning in physics, where it refers to a state of matter in which the binding forces are strong enough to restrict the movement of atoms, as a result of which solids have both definite volume and shape. In perception, a solid-void configuration can be seen as an on-off state, like the presence or absence of matter. Solid-void can also be seen in terms of density, in the same way that, in two-dimensional images, the perception of figure against background is due to the contrast of inked areas over the blank background.

The solid-void relationship can be found at the macrocosm level (planets, galaxies), or at the level of our everyday life experience down to a microscopic scale. It can also be hierarchical; an initially solid form might be further divided into solid-void relations, for example, at atomic and subatomic levels.

I-4 ▶

Voids are shaped by the enclosing solids.

a

b

VISUALIZING WITH CAD

A torus as a solid boundary, seen from.

a. Outside.   b. Inside.

Surfaces are the perceptual boundaries which separate a solid from the surrounding void. What we actually perceive of a solid object are just the surfaces bounding it. Anything beneath the most superficial layers is perceptually unknown. Previous knowledge or experience might suggest what a physical object is like in its totality, but the perceptual experience is only of its external surfaces. For example, a golf ball is perceived only as the external hard plastic facing which bounds it; one must section it to see the inner core of wound string and solid rubber, and even a sectional view will reveal only the surfaces delimiting the sectioned parts.

Walls, which are one of the primary solid elements of a building, can be considered as boundaries and they offer a clear example of how the same element can be considered both solid and void according to the level of observation. Walls at a first glance appear as solid, but upon detailed analysis their solidity is fragmented by the window openings. As we will see in the next chapters, the interpretation of forms in terms of solid and void is one of the first approaches in architectural design, where the solid element is often called *mass*. The rhythm of the composition is given by the proportion and alternation of the masses. It is the solid-void articulation which gives shape to a building and defines its primary visual and functional characteristics.

Inside and Outside

As we have already seen, forms are characterized by oppositions, in a sort of dialectical relation. Another fundamental relationship based on opposites is inside-outside, directly derived from that of solid and void. A closed solid boundary (as distinct from the surface boundaries of solids themselves) encloses a void space. The solid boundary becomes a limit between outside and inside space (figure I-5a,b). In this view, all man-made constructions, or "architecture," can be seen as enclosures or boundaries separating an inside from the outside, where functionally the inside represents a space sheltered from the outside elements. (Note that an inside requires a solid boundary, whereas solid elements, do not necessarily delimit an

VISUALIZING WITH CAD

A set of solids does not necessarily
create a physical enclosure.

A spiral extrusion creates ambiguous
inside-outside relations.

enclosure (figure I-6)). As with solid-void, the inside-outside opposition can be perceptually ambiguous: A spiral (or its extrusion as shown in figure I-7) offers a configuration of space where the inside is not completely bounded by the outside and the same part of space can be alternatively perceived as outside or inside, relative to different elements. In terms of the figure-ground relationship, an enclosure can be variously perceived: Seen from the outside, its boundaries represent a figure (as a building is seen as a figure against an urban ground), while from an inside view they become the ground (as, in figure I-2b, the interior walls become the ground against which the columns are seen as figures).

In the design of forms, where the focus is on the boundaries that distinguish an interior space from an exterior one, inside-outside, like solid-void, can be part of a hierarchical succession. The urban texture represents a clear example of how these relations become interconnected and exchangeable at different hierarchical levels, relative to different positions and scales of observation. From inside a building, the walls are perceived as solid (or "figure"), while the building is perceived as an interior space, or void (the "ground" for the building's interior elements). From the outside, the building is perceived as solid ("figure") in relation to the void of the street ("ground"), which at this scale represents the inside space, while the landscape surrounding the city becomes the outside. The next level of scale would be represented by an aerial view of the city, in which the city as a whole becomes the solid, or figure, in the void of the surrounding landscape, which is now the inside space, or ground. A clear visual reading of this succession is given by maps of different scales, from architectural to territorial. In maps the figure-ground opposition is representative of solid-void elements.

It is interesting to note that the inside-outside relation does not belong just to the perception of space, but is literally and metaphorically used as a dualism in other disciplines. The French philosopher Gaston Bachelard writes about outside and inside as a dialectic, which is reminiscent of the opposites of being and not being. There are several other philosophical and psychological associations: inside as a center and outside as the universe radiating from it, the "I" and the external world, the mind and the physical experience—these are only a few of the dialectic oppositions emanating from this dualism.

FORM AS AN ABSTRACT SYSTEM: ELEMENTS AND RELATIONS

The succession of forms constitutes our basic spatial perception, and the perceptual characteristics of forms are determined by boundaries. A form can also be interpreted as a whole made of parts. The psychologist Rudolph Arnheim distinguishes between parts in a quantitative sense ("any section of a whole") and "genuine parts," which he defines as "sections representing a segregated sub whole within the total context." A form cannot, however, be considered just as a collection of parts. The interpretation of form as a whole that is more the sum of its parts is typical of **gestalt theory**, from which many of the previously discussed ideas about perception arise.[1] The principle behind gestalt investigation of form is that the perception of an element is determined by its context. As Arnheim states, "the appearance of any part depends, to a greater or lesser extent, on the structure of the whole, and the whole, in turn, is influenced by the nature of its parts."

For our purposes, we can encompass the gestalt classification of parts and whole in a more general view of specific forms as systems of elements and relations linking the elements together. In the present context we focus only on the spatial properties of forms and ignore other properties, such as color or material characteristics, of the material the form is made of. In this case, relations among elements are given only by the spatial extension and position of each element. The relations involving **position** are those like above, below, left, right, back or in front, between, surrounding, outside and inside, often defined as topological relations [Piaget 1969]. In terms of extension, one form may, in relation to another, be equal, smaller or

Shape generated from
the fusion of other shapes.

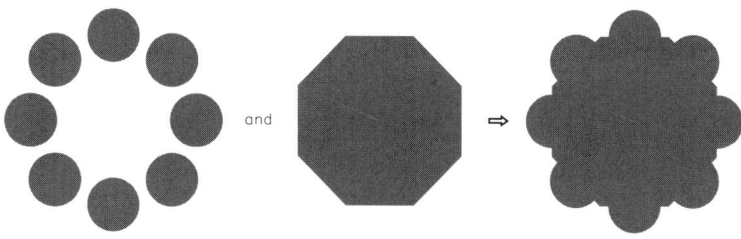

larger. Relations of both position and extension can be ordered and grouped according to the geometrical transformations of translation, rotation, reflection, and scaling (see Chapter III).

Sometimes an element can be generated by the fusion of two or more individually recognizable forms (figure I-8). The relations involved are then of a different type and can be defined according to the so-called Boolean operations of union, difference, and intersection (see Chapter II).

Each element can have a dual existence, as a whole by itself or as a simple part of a more complex system. Figure I-9 is an example of how a whole can be generated from parts. The simple elements grow in complexity as they are articulated in a series of relations. Once again, part and whole are not definite attributes but are relative to a context; an element that is a part at one scale can be a whole at a smaller scale, in a series of hierarchical relations.

Natural and Man-Made Forms

Viewed in the more abstract way suggested above rather than as collections of specific elements, forms can be classified in terms of the relations among their elements and the articulation and complexity of those relations. In this way we can unite two classes of forms that appear to be completely different: forms found in nature and man-made forms. Natural and artificial forms can both be viewed as sets of elements linked by relations such as above/below, in front/behind, and so on: The ceiling is above the floor, the tree trunk is above the roots. Natural forms, however, to a much greater extent than man-

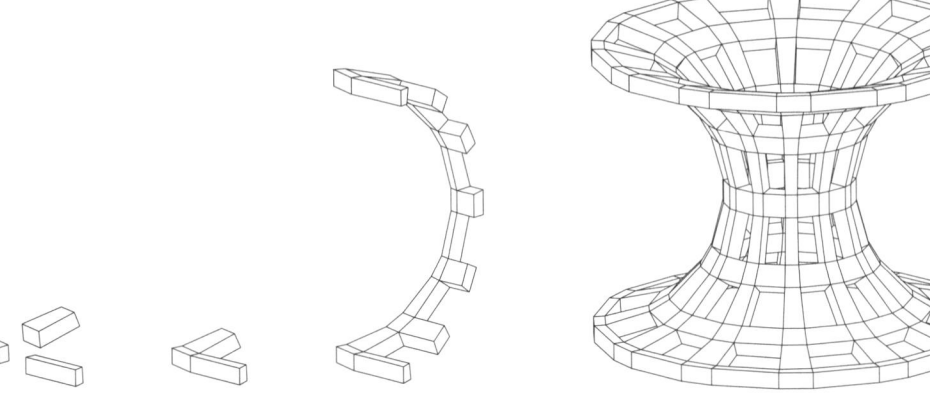

a

b

VISUALIZING WITH CAD

The whole is recognized as a hierarchical composition of the individual parts.

a. Individual parts.

made forms, are characterized by transformation in time as well as by spatial configuration. This is evident in the growth of organisms, the flow of water, the shapes of clouds and flames, to mention just a few examples of forms in continuous transformation. Although our concern is with man-made forms, which are very regular, and can be interpreted through geometric models, the methods of form generation methods require an awareness of the parallels between organic and inorganic forms.

Neither type of form is characterized by confusion and chance; both are ordered systems, recognizable and subject to classification. As with man-made forms, the relations between the elements of natural forms can usually be interpreted through their geometric properties. There is a branch of biology, morphology, which deals with the form and structure of animals and plants, without recourse to their function. Goethe used the word morphology specifically for botany, where he pursued a classification method based on forms and their transformation. Goethe's method was developed beyond botany by D'Arcy Thompson. His famous book *On Growth and Form*, published in 1917, despite its outdated mathematical method, represents the most outstanding analysis of biological forms and processes in terms of their mathematical and geometrical descriptions. The relation between forms in nature and geometric shapes is exemplified in figure I-10, taken from Ernst Haeckel's plates, often used by D'Arcy Thompson.[2] Forms in nature (to use D'Arcy Thompson's examples) such as "the waves of the sea, the little ripples on the shore, the sweeping curve of the sandy bay between the headlands, the outline of the hills, the shape of the clouds," and the human body itself, can still be interpreted through geometric models. Cells are shaped as minimal surfaces, shells have spiral configurations and some skeletons resemble tetrahedra and hexagonal meshes. The geometric properties of these and many other examples reflect that, again in D'Arcy Thompson's words "the form of an object is a diagram of forces."

VISUALIZING WITH CAD

Skeletons of Radiolaria resemble
complex geometric models
(reprinted with permission from
Dover Pictorial Archive Series).

As suggested by the brief digression on morphology above, geometry is fundamental to the study of form and space. The physical space that we experience is often different from geometrical space: Geometry is independent from empirical observations; geometric models of space arise instead from generalization and deduction.

The word *geometry* comes from the two Greek words: γεοσ (*geos*), which means earth, and μετρον (*metron*), meaning measure. Geometry was used by the ancient populations, Egyptians, Chinese, Babylonians, Romans, and Greeks, for practical applications in surveying, navigation, and astronomy. The practical applications of geometry were generalized and systematized by the Greeks, who developed logical reasons and relationships among them. Deductive methods made it possible to demonstrate propositions or theorems using logical reasoning rather than empirical arguments. In this way, geometry, which started from empirical observations, became a theoretical science, and its discoveries were eventually compiled and systemized in Euclid's *Elements*, around 300 B.C. This fundamental text has been in use for over 2000 years and is the basis of Euclidean geometry that we know from high school. Euclidean geometry is based on five postulates, which are specifically geometric, and five axioms, which are related to arithmetic concepts. All further definitions and theorems rest on these postulates and axioms or on theorems derived from them.

Beyond Euclidean Geometry

The truth of Euclid's postulates is considered to be self-evident without recourse to logical proof; their validity can also be found in analogies with objects in the physical world. The fifth postulate, however, known as the parallel postulate, does not rest on the same evidence as the others.[3] This postulate states that, "given a line *l* and a point *P* not on *l*, there exists exactly one line *m*, in the plane of *P*

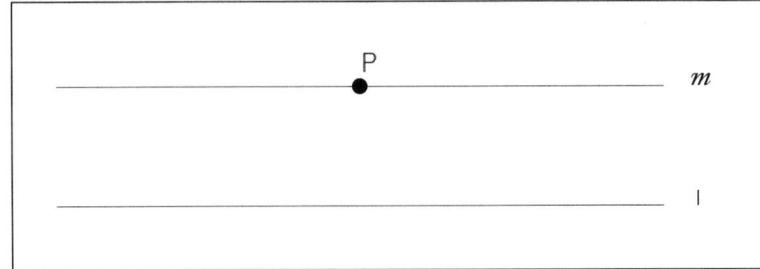

and *l*, that is parallel to *l* (figure I-11). Although geometers tried (without success) to derive the fifth postulate as a theorem from the other postulates, Euclidean geometry, with all its postulates and axioms, was considered irrefutable until the nineteenth century. Mathematicians then began to consider the possibility of replacing the fifth postulate, which is logically independent of the others. From this effort non-Euclidean geometries were born and, with them, questions about the validity of geometry and its correspondence to the physical world. In the hyperbolic geometry, developed by N. I. Lobachevski and F. Bolyai, numerous lines can be drawn through a point lying outside a given line, all of which are parallel to that line. In the other non-Euclidean geometry, the elliptic geometry, the mathematician G. F. B. Riemann replaced the fifth postulate with the statement that no such parallel line can be drawn through the given point [Coxeter 1967].

Non-Euclidean geometries may appear extraneous to the general discussion: The geometry used for computer-aided design is based on Euclidean space, and alternative versions of the fifth postulate are irrelevant. Nevertheless, non-Euclidean geometries need to be mentioned for several reasons. They are one of the most famous examples of how geometric ideas can be generalized, and thus they underscore the abstract nature of those ideas. Geometry is concerned with entities such as points, lines, and planes, and with concepts such as continuity, regularity, and infinity, which are idealizations of physical objects and their properties. Geometric models, even those starting from intuitive considerations, become ideal abstractions of natural forms. For instance, a dot drawn by pencil can be visualized as an extremely small object, but there is nothing in the physical world that corresponds to a geometric point, which has no dimension. Similarly, it is impossible to find objects in our experience that are infinite or (in the mathematical sense) continuous. The space of Euclidean geometry, however, while not perfectly corresponding to the physical space of experience, can still serve as a model for it, because geometric propositions are valid, not for their correspondence with real objects, but by virtue of their logical relationships.

Indeed, even non-Euclidean spaces, which are so contrary to our intuition, can themselves serve as models: Riemann elliptic geometry was found to describe the space that lies outside our experience or visualization better than Euclidean geometry. In physics, the general theory of relativity found a description of the curvature of physical space in Riemann geometry, which provides more interpretative models for curvature resulting from a gravitational field than does Euclidean geometry.

Analytic Geometry and Infinitesimal Calculus

Euclidean and non-Euclidean geometries do not represent the only geometric descriptions of space. In the seventeenth century, the philosopher and mathematician René Descartes developed **analytic geometry,** introduced a numerical and algebraic interpretation of space. According to analytic geometry, space can be characterized numerically by the use of coordinates. Initially, coordinates were applied only to the geometry of the plane, where a pair of numbers assigns a position to a point (figure I-12a). From the eighteenth century on, analytic geometry was also developed for three-dimensional space: a set of three numbers, or coordinates, can be assigned to each point, thus defining its position in space (figure I-12b). The coordinate definition is extremely useful in dealing with complex geometric configurations because it allows their components to be given algebraic

a. Analytical geometry in the plane.

b. Analytical geometry in three-dimensional space.

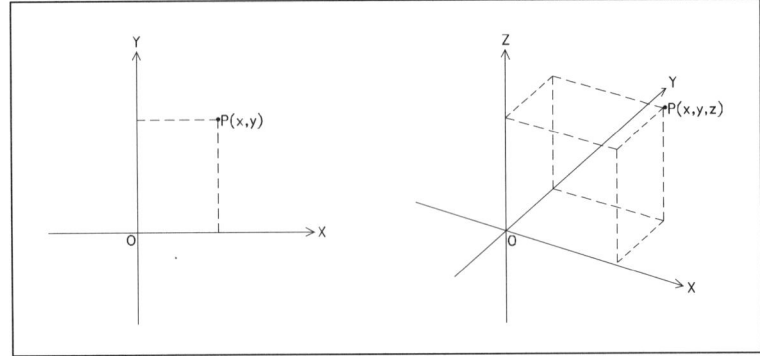

expression. A line, for example, can be defined by an equation. Analytic geometry also takes our geometric perception of space into the realm of numbers; it goes beyond geometrical intuition based on visual images and adopts numerical methods. Thus, for instance, a circle is represented not just by a shape, but by a mathematical expression. Still, the coordinate definition responds to one of our most basic empirical experiences of space, that of our bodies as objects characterized by a position relative to the position of other objects. Furthermore, analytic geometry offers a practical approach to geometrical problems: It is not by chance that all the procedures in computer graphics and computer-aided design use methods from analytic geometry.

The relation between geometry and algebra, initiated with analytic geometry, was further developed by Newton and Leibnitz in the seventeenth century with integral and differential calculus. Calculus deals with infinitesimal quantities and offers a numerical approach to otherwise unsolvable geometric problems. Problems common to physics and mathematics, such as the trajectory of a projectile, the area under a curve, the volume of a irregularly shaped solid, need calculus to be interpreted and solved.

More Geometries

In the last century, particularly since the development of non-Euclidean geometries, the abstract character of geometry became more pronounced. The validity of a particular geometry became a problem of an epistemological nature, and geometry became a science based more on logical relations than on an intuitive interpretation of space. In 1872, the mathematician Felix Klein proposed a classification from which emerged the validity of several different geometries, such as **metric geometry**, which includes Euclidean geometry, **projective geometry**, and analysis situs, or **topology** [Coxeter 1967].

Projective geometry is concerned with the properties of shapes projected on to a plane from a point representing a projection center. Although the shape is transformed through projection, features of the initial shape can still be recognized in the projected shape (figure I-13). Projective geometries deal with the properties of the shape which are unchanged after projective transformations. A perspective

I-13 ▶

Projective transformations.

a. Correspondence between sets of points of two lines.

b. Central projection of a cubic shape on a plane.

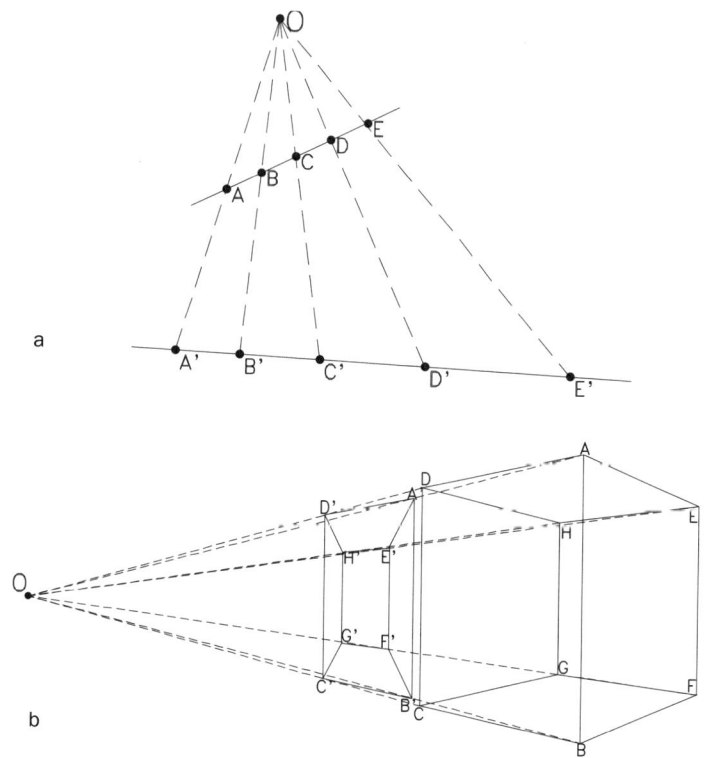

Topological transformations.

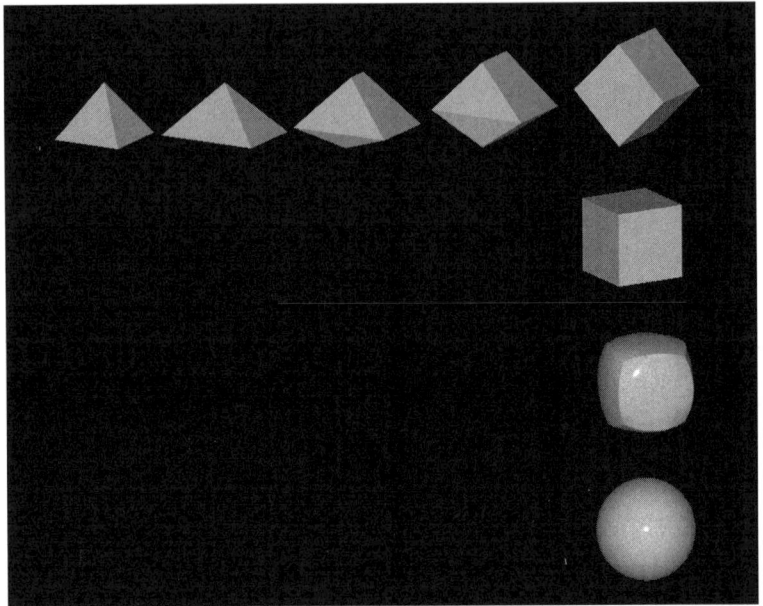

drawing is an example of a projective transformation. The study of perspective in art, formalized by Leon Battista Alberti in 1432, intuitively anticipated the development of projective geometry in the nineteenth century.

The most general and comprehensive type of geometric transformations are topological transformations. Topology—often called "rubber sheet" geometry—deals with the properties of a shape that remain unchanged after the shape is deformed, compressed, or stretched; an example is the deformation of a pyramid into a sphere (figure I-14). The properties that distinguish individual solids under metric and projective geometries are largely irrelevant; instead, different classes of shapes are distinguished by relations such as inside and outside or continuity and separation. Topology is also concerned with the qualities of juxtaposition and proximity. It is interesting to note that topological relations are the spatial properties that infants perceive first. The psychologist Jean Piaget asserts that a child's first spatial intuitions are of topological properties—such as whether a shape is open or closed, continuous or separate, two-dimensional or three-dimensional—rather than of metric Euclidean properties that depend on measurements or reference systems.

A final type of geometry worth mentioning is fractal geometry, a very popular subject in computer graphics. Fractal geometry was developed by the mathematician Benoit Mandelbrot, who in 1975 coined the word *fractal* to describe self-similar and irregularly fragmented shapes. Examples of fractal shapes are Sierpinski gaskets and carpets, Julia sets, and snowflake curves (figure I-15). Whereas figures from classical geometry have clearly identified dimensions—lines are one-dimensional, polygons are two-dimensional, solids are three-dimensional—fractal shapes usually have fractional dimension.

I-15 ▶

Fractal shapes.

a. Snowflake.

b. Sierpinski carpet.

a

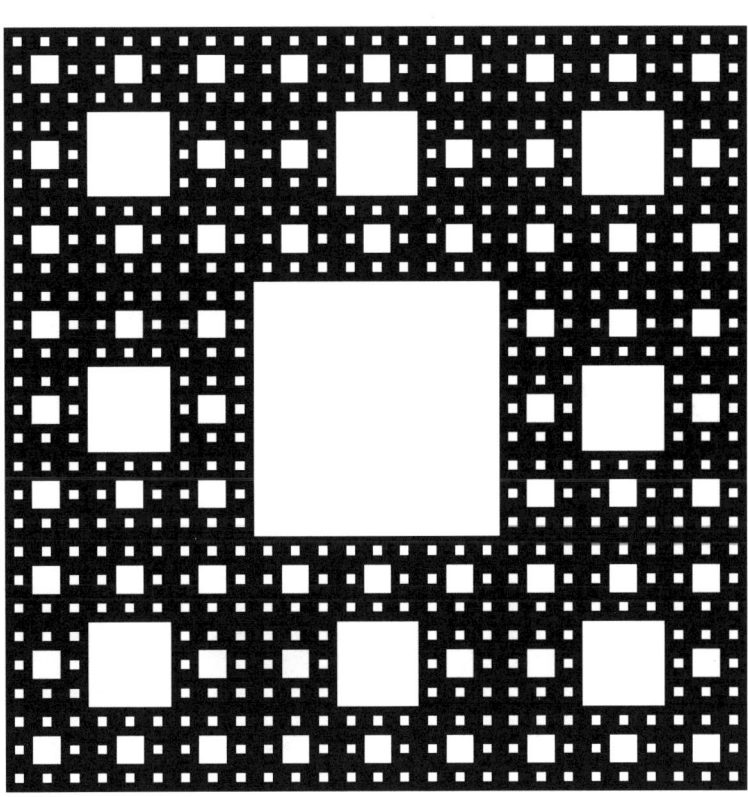

b

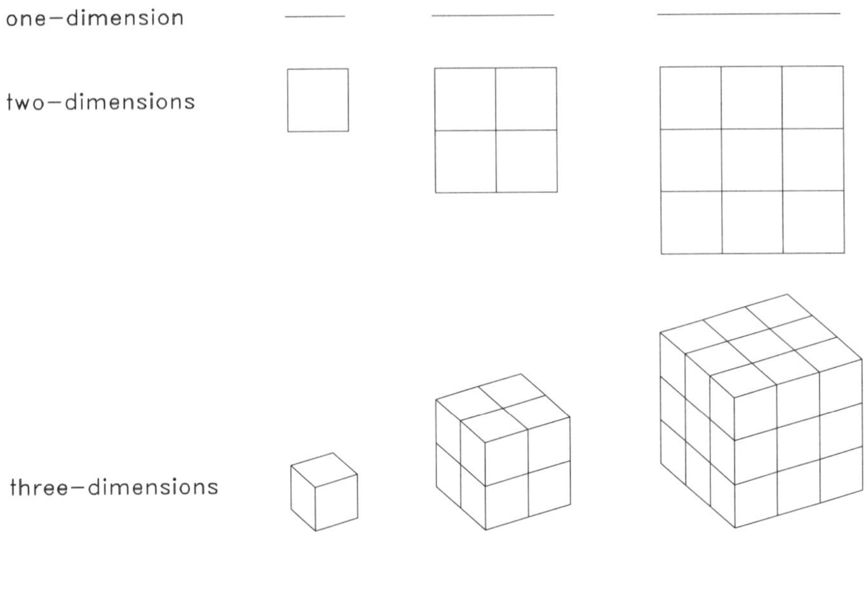

one—dimension

two—dimensions

three—dimensions

a

fractal dimension:
log3/log2

b

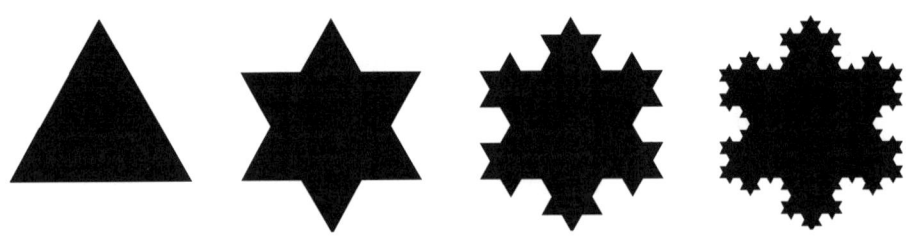

Exponential growth of a shape
according to its dimensions.

a. Line, square, cube.

b. Sierpinski gasket.

This arises from a curious fact: For familiar geometric figures, a change in scale resulting from a uniform change in the length of a figure's sides or edges can be expressed in terms of the figure's dimension (doubling the length of a square's sides or a cube's edges produces an increase of four times (2^2) the area of the square, eight times (2^3) the volume of the cube, and so on, where the exponents correspond to the dimension of the respective figures); by contrast, for fractal shapes such scaling factors usually involve fractional rather than integral exponents (figure I-16).

The other distinctive feature of fractals is self-similarity, the property that the parts of a figure have the same shape as the whole. This characteristic is evident in the visual representation of fractals (figure I-17).

Fractal models offer a better description of natural phenomena than most other geometries. Images of clouds, mountains, water, and other natural forms can be generated using fractal geometry. This geometry has become popular not just among mathematicians, but also among a broader audience, for the beauty and complexity of the pictures resulting from its underlying mathematics.

Applications in Design

Architects, artists, and designers should be aware that geometry is an essential tool for the generation of forms, especially in the schematic design phase, where forms are often represented as simple geometric models. Designers, whose interest in geometry is mainly pragmatic, will have recourse more to the intuitive empirical aspects of geometry than to its logical deductive aspects. Metric geometry offers the best interpretation of those properties of space relevant to the design of simple forms, while analytic geometry is used for its computational methods. From projective geometry are derived graphic representation methods such as perspective, by which three-dimensional forms are given two-dimensional representation.

The other lesson learned from the classification of geometries is that, notwithstanding their different descriptions of space, they are all, in essence, formalized systems of relations, based on primitive

Self-similarity is characteristic of
fractal shapes.

VISUALIZING WITH CAD

axioms, from which rules are derived for creating and recognizing shapes. This allows the creation, as is often required in computer graphics, of a purpose-oriented set of geometric rules.

SPACE AND DEPTH

Returning to a perceptual analysis of form and space, let us consider in more detail the visual perception of space. The three-dimensional space of our experience is received as an image projected on the two-dimensional surface of the retina. The stereoscopic effect, which causes the perception of depth, is given by our binocular vision. Of the three physical dimensions of space—width, height, and depth—depth is the most "subjective," since it is related more to the way visual perception works than to the physical reality of the objects perceived. Merleau-Ponty defines depth as "the most existential of all dimensions" since "it is not impressed upon the object itself, [but] quite clearly belongs to the perspective and not to things."

The two-dimensional projection of spatial objects that is the basis for perception differs from the three-dimensional reality of those objects. Perceptual space can therefore be different from physical space, as is evident from our most basic visual perceptions (figure I-18): Objects appear smaller at a distance, rectangles are perceived as trapezoids, angles change in apparent size, parallel lines meet in a point called the vanishing point, and parallel planes meet in a line—the horizon (familiar from everyday experience as the place where the sky meets the earth). Only the relations of objects parallel to the picture plane (i.e., the plane in which the two-dimensional representation is formed) are conserved. Nevertheless, we interpret forms according to their physical reality: We recognize that forms perceived as trapezoids are actually rectangles, and know that rails that appear to converge do not actually meet. Past experience, stereoscopic vision, the sense of touch and kinesthetic experiences, among other factors, allow us to recognize in our visual perceptions the physical reality of space [Arnheim 1974].

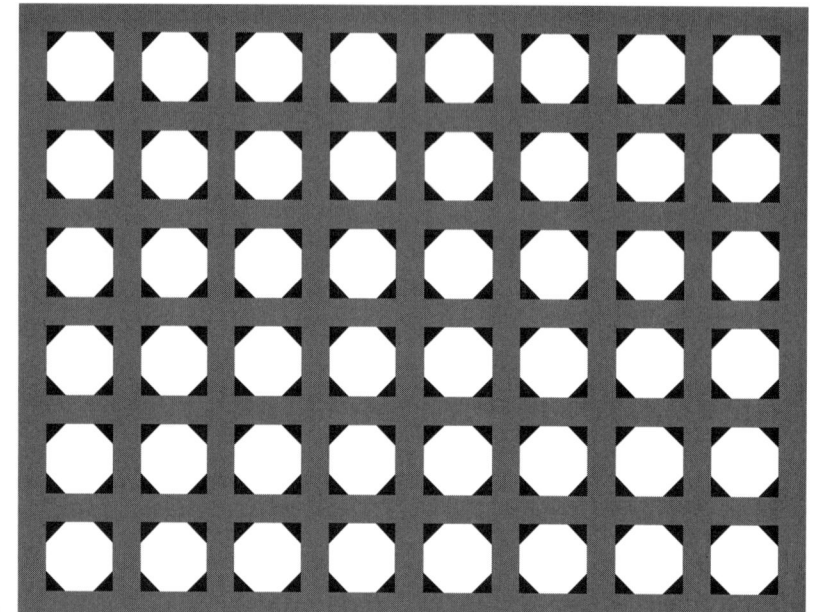

a

b

VISUALIZING WITH CAD

Two different categories of images.

a. Images representing two-dimensional shapes.

b. Images representing three-dimensional shapes.

Visualization

The relation between physical space and its visual perception is of fundamental importance for every type of visual representation, or **visualization**. Essentially visualizations are composed of the spatial boundaries of a form in terms of figure-ground, which provide information about visual characteristics such as contours, lines, area of colors, gradients, and so forth. In psychophysical terms, the visual characteristics of an object are only those given by the retinal stimulation which is induced by light radiation. Other physical qualities, such as density, weight, and viscosity, are not reflected in visualization, even though certain visual characteristics may allude to these nonvisual properties.

Visual representations are two-dimensional images, which can be classified in two ways. They can be regarded strictly as compositions, which take on meaning purely as two-dimensional systems of graphic signs, as in the examples in figure I-19a. As a subclass of all such images, we can consider all those that are two-dimensional representations of three-dimensional forms (figure I-19b). The perceiver, referring to the context of an image, interprets the image according to these two different possibilities. The representation of a three-dimensional object has therefore two semantic contents, consisting respectively of the two-dimensional composition and the three-dimensional object the image represents. Usually there is immediate correspondence between a two-dimensional image and the three-dimensional form the image represents.

In this second category of images, the problem of visualization recalls the geometrical interpretation of space previously investigated. Euclidean geometry, which is concerned with the metric (measurable) properties of objects, is not appropriate to describe the geometrical properties of vision, which can instead be expressed by projective geometry.

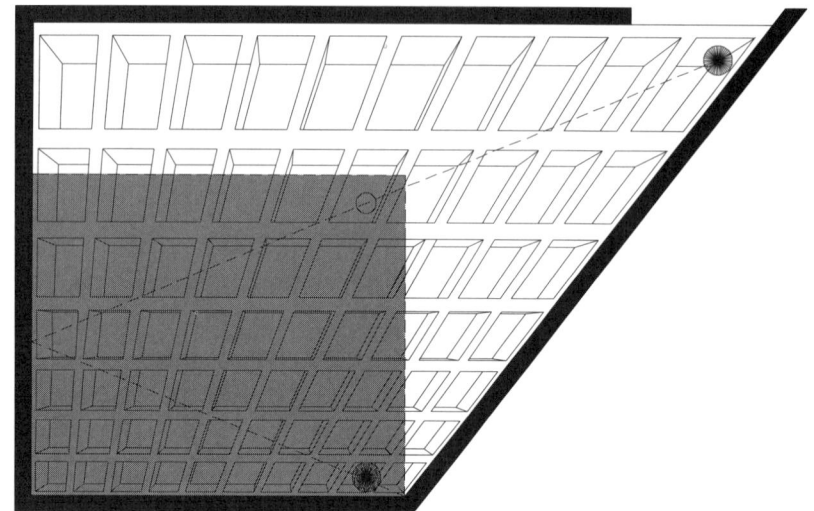

VISUALIZING WITH CAD

Visual Illusions

Optical illusions: the Ames room.

a. Floor plan showing the real trapezoidal shape and the rectangular room (shaded area) as perceived.

b. The perspective representation of the room from a determined viewpoint shows the illusion: two persons of similar height, standing at two opposite corners, are perceived to be of exaggeratedly different sizes.

Visual perception of space is often illusionary, in the sense that some physical forms can be mistaken for others. In *Visual Illusions*, M. Luckiesh writes: "In a broad sense, any visual perception which does not harmonize with physical measurements may be termed as an 'illusion.' "

Illusions figure greatly in the study of visual perception. Seeing is often deceptive; the perception of space we have through one of our senses has to be confirmed by the others, otherwise it generates illusions. Often, optical illusions are the result of our natural tendency to identify simple forms. If a perception can be the same for different physical spaces, the perceiver tends to believe the simplest one to be the object of the perception. A well-documented example is the Ames room (figure I-20), in which an irregularly shaped trapezoidal space is perceived, from a particular viewpoint, as a rectangular room. The Ames room demonstrates that perception of a three-dimensional space can be distorted through careful control of viewpoint and proportions.

Interest in visual illusions is not restricted to research in psychology and optics. Often illusions like trompe l'oeil imagery and false perspectives have been used in architecture, with the purpose of modifying an existing space through the use of images rather than three-dimensional construction. Baroque architecture offers several examples of two-dimensional imagery used to add a theatrical dimension. These topics will be considered again in the next chapter, dealing with representation methods in CAD.

CONTINUITY AND DISCRETENESS

The geometrical and perceptual properties of space reveal another dichotomy defined by the concepts of "continuity" and "discreteness," which represent two mathematical descriptions of the world. This dualism goes beyond geometry: From intuitive observation, any form, natural or artificial, can be defined in these terms, at least at a macroscopic level. *Continuous* entities are infinitely divisible; *discrete* are composed of separate elements. *Continuous* and *discrete* are not always (in their nonmathematical senses) antithetical terms, but can coexist at different descriptive levels.

The concepts of continuity and discreteness can be applied even to physical descriptions of the world. In spite of our perception of the natural world as continuous (exemplified, for example, by the flow of water, the sky around us, or the movement of fire), the microscopic level of atomic particles reveals a discrete reality, as represented by the discrete packets of energy in quantum physics or the discontinuities in the spectra of light. At the macroscopic level, however, where Newtonian mechanics are valid, phenomena are regarded as continuous. Almost every physical phenomenon in classical physics—*e.g.*, gravity, Newton's laws, heat, waves—can be described through mathematical methods, such as differential calculus, that are based on continuity.

Even though the notions of discreteness and continuity may seem unreconcilable, each can sometimes be used as an acceptable approximation of the other. Even in mathematics and physics, continuity problems, sometimes impossible to solve with differential calculus, can be solved approximately with discrete numerical methods.

The mapping of a form into a computer model often presents the fundamental problem of whether a continuous or a discrete method will be used to define a geometric model. The approach used in the next chapters for generation of form will be based on a discrete interpretation of space, and geometrically continuous shapes will be approximated by corresponding discrete models (figure I-21). This approach can be appropriate for several reasons. As previously dis-

I-21 ▶

Discrete approximation of
a continuous form.

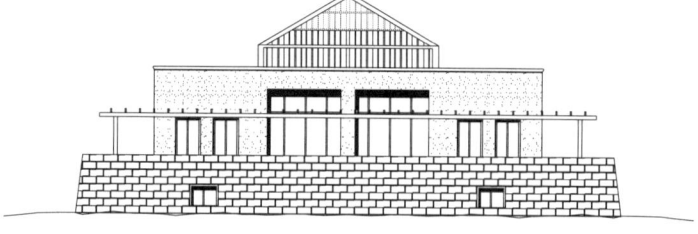

cussed, a form is defined by its boundaries and boundaries imply discontinuity. The definition of form in terms of its boundaries therefore suggests a discrete approach. Moreover, as discussed in Chapter IV, even forms that appear to be geometrically continuous can be approximated by a finite number of segments.

WHAT IS ARCHITECTURE?

Obviously, one of the primary applications of form generation is in architecture. Architecture is concerned with space and its molding, including its physical, existential, and functional aspects. While geometry investigates the properties of space through mathematical expressions, drawings, or verbal descriptions, and psychology investigates the perception of space, architecture not only deals with space in a theoretical way, but also involves its physical transformation.

The word *architecture* comes from the Greek words αρκε (arke), which means origin or beginning, and τεκτον (tekton), which means building. The Oxford Dictionary defines architecture as "The art or science of building or constructing edifices of any kind for human use." Another definition, often quoted in architectural history textbooks [Benevelo 1977], was given by William Morris in 1881: "Architecture embraces the consideration of the whole surrounding of the life of man: we cannot escape from it if we would, so long as we are part of civilization, for it means the molding and altering to human needs of the very face of the earth itself, except in the outermost desert." From these definitions—although extremely broad—it is clear that architecture is mainly concerned with man-made or artificial environments. More particularly, the architectural theoretician Christian Norberg-Schulz identifies three main elements of the language of architecture: *morphology*, which concerns the creation of forms; *topology*, which concerns the spatial attributes of proximity, continuity, and closure; and *typology*, which concerns the different categories of human settlement [Norberg-Schulz 1985].

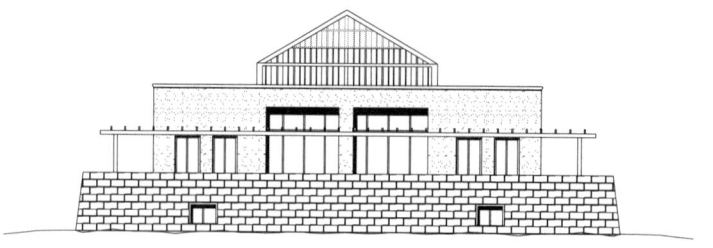

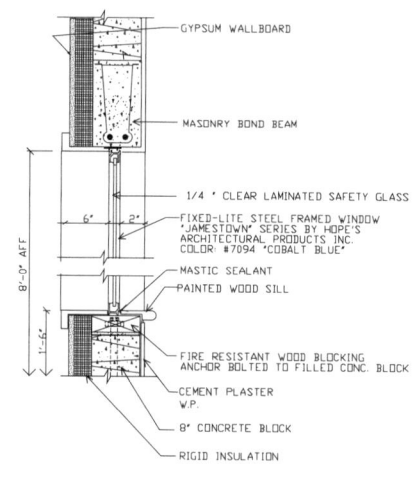

GYPSUM WALLBOARD

MASONRY BOND BEAM

1/4 " CLEAR LAMINATED SAFETY GLASS

FIXED-LITE STEEL FRAMED WINDOW
'JAMESTOWN' SERIES BY HOPE'S
ARCHITECTURAL PRODUCTS INC.
COLOR: #7094 "COBALT BLUE"

MASTIC SEALANT
PAINTED WOOD SILL

FIRE RESISTANT WOOD BLOCKING
ANCHOR BOLTED TO FILLED CONC. BLOCK

CEMENT PLASTER
W.P.

8" CONCRETE BLOCK

RIGID INSULATION

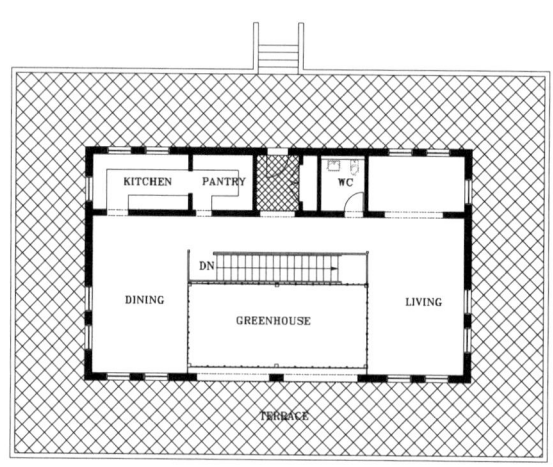

KITCHEN   PANTRY   WC

DINING   LIVING

DN

GREENHOUSE

TERRACE

VISUALIZING WITH CAD

Form, building technology, function
in the design of a house
(courtesy of David Foell).

Architecture deals with physical space in the creation of an artificial environment that satisfies certain specific functions, including psychological and social functions. One of the tasks of architecture is to endow a certain spatial organization with a character beyond its physical configuration. Norberg-Schulz writes: "Whereas 'space' denotes the three-dimensional organization of the elements which make up a place, 'character' denotes the general 'atmosphere' which is the most comprehensive property of any place" [Norberg-Schultz 1980].

Aesthetic judgments and feelings are associated with certain spatial memories and sensations. The forms perceived as boundaries between solid and void may have several psychological and semantic meanings. A wall at the most basic level is perceived as solid, geometrically defined as a parallelepiped bounded by rectangular faces. At different semantic levels, the same wall may be viewed as made of brick, or as part of a house, or as the wall separating the living room from the bedroom. Starting from the most basic perceptual level, at which geometric characteristics are the primary features, a space can become familiar to us as our home, or as the street and the city in which we live. The physical forms determining the space around us become identified with the values we attribute to them and the feelings they evoke. In this context space can be defined as a "place" that transcends the spatial properties which define its physical organization.

So far we have mainly considered just the perceptual characteristics of forms; beyond an abstract inquiry of this type, however, architecture involves the physical construction of spaces that serve social functions. To become built architecture, designed shapes have to satisfy requirements going beyond their formal configuration.

Form, Building Technology, and Function

The development of an architectural work involves three fundamental considerations: form, building materials and technology, and function (figure I-22). This classification dates back to Vitruvius, who, in the first century B.C., identified these elements as *venustas*, *firmitas*, and *utilitas*.[4] The architectural solution to the problem of human settlement comes from the interrelation of these elements. All three must be considered together; they can be individually analyzed at an initial stage, but the final architectural product is necessarily a result of their interaction. The relation between form and function has been one of the most debated issues in the history of architecture. Louis Sullivan's statement that "form follows function" has been one of the most recognizable catch phrases of the modern movement of architecture, where the architectural form is often nothing more than a container. More recently, however, form has been reconsidered and the relation between form and function has become more balanced.

In the present discussion, as part of a more general exploration of forms, architecture will be investigated as man-made transformation of the environment through the creation of three-dimensional forms, which is the aspect Norberg-Schulz defines as morphology. Traditional architectural elements will be analyzed together with forms of a more generic type since, in this conception, architecture is concerned not just with traditional buildings, but with every physically constructable or visualizable form. It should be emphasized that only the formal aspect of architecture will be considered, without any investigation of materials and functions. The only prerequisite is that a form can be geometrically described through a three-dimensional coordinate system.

Architectural Style

An architectural solution to a functional problem is a form realized with a certain technology and materials. The set of formal characteristics of an architectural product is what defines a style. Architectural styles are typical of an historical and geographical context as well as being the creations of individuals. For example, we speak of Gothic style or the style of Frank Lloyd Wright. A style can go beyond its historical and geographical context; for example, classical architecture can be found in several locations from various centuries.

The elements of style constitute what is often called architectural language. A style is determined by a set of formal rules, or grammar. The definition of the rules of a grammar and their permutations is one of the areas where computer-aided design offers a new and powerful approach [Kalay 1986].

Architecture and Form: Geometry

Architectural forms, with few exceptions, can be clearly identified by means of geometric coordinate systems and can be assimilated to geometric elements. For instance, walls and floors can be assimilated to planes, domes to spheres or surfaces of revolution, floor patterns to tessellations in the plane, and so on. The association with geometry can be more or less explicit.

Historically, in fact, architectural forms were not just interpreted through geometric models, but were explicitly inspired by geometry, and, more specifically, by Euclidean geometry. There are clear examples of this in the central symmetry of urban and architectural plans, the spherical surfaces of domes, the complex concave-convex volumes in Baroque architecture, and the canons of proportion of a facade (figure I-23).

Mathematics and geometry are related to architecture in two principal ways. First, they are a source of meaning for architecture [Perez-Gomez 1985]. This tendency can be found in architectural manuscripts of almost every historical period. Geometrical shapes and mathematical numerical relations have always represented an ideal static order which contrasted with a physical reality dominated by chaos and change. In particular, in Renaissance and Baroque architecture the use of geometric forms became a symbol of the power of human reason over nature. A philosophy of architecture based on mathematical principles emerged from the writings of Filarete, Hersey, Serlio, Scamozzi, and Alberti. The polygonal shapes of the grids of their ideal cities and the rotational symmetry and

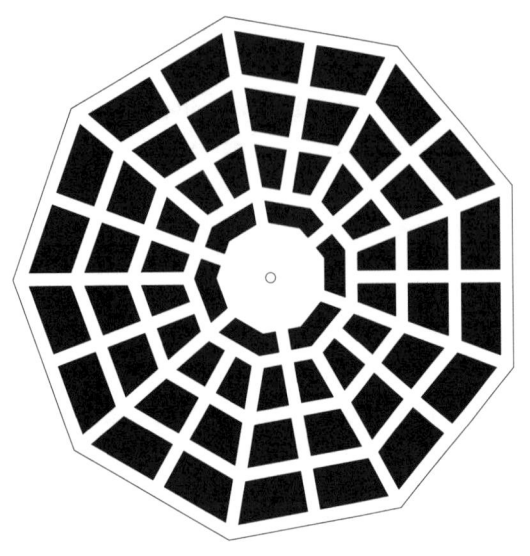

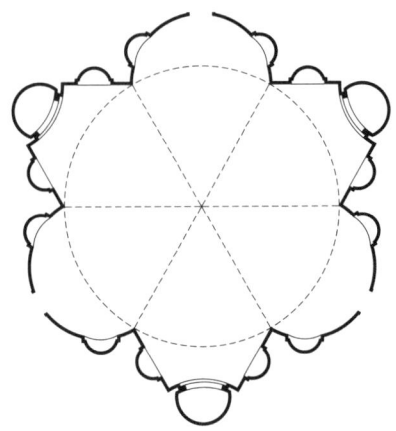

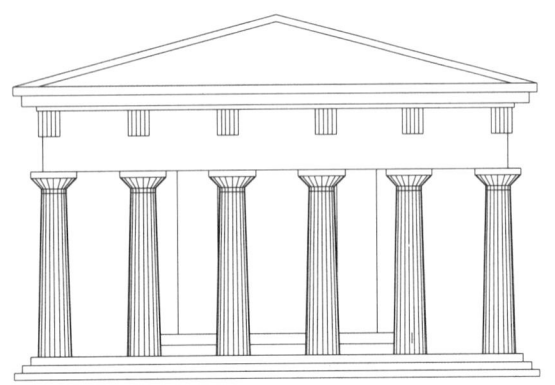

VISUALIZING WITH CAD

geometric operations in domes and vaults affirm the ideal perfection of Euclidean geometry.

The other relationship between architecture and mathematics is more pragmatic. Geometric laws have practical applications in surveying, measuring, and stereotomy. Mathematics is indispensable in structural analysis. Two-dimensional architectural representations are made possible by geometric projections; in particular, perspective representation uses descriptive geometry.

Thus, geometry in architectural design is mainly expressed in two ways: as inspiration for forms and for their representation on paper. This principle is central and will be illustrated in the chapters that follow.

WHAT IS DESIGN?

The human creation of a form is a design problem and the following considerations about design, while referring explicitly to architecture, can be applied to the creation of form in general.

Clearly there are conceptual differences between architecture, regarded as a physical configuration of space, and architectural design. Design is a way of developing and expressing ideas that precedes the realization of the artifact. The materials of design are not the physical materials which compose the architectural artifacts, but two-dimensional representations on paper, scale models, or data structures in the computer memory, media which use geometry to define the visual form of a work of architecture. Thus, design can be viewed as a bridge between a geometrical description of space and built architecture.

Even from a perceptual perspective, built architecture and design are distinct. As previously noted, forms are recognizable from the surfaces that bound them, and in the physical world surfaces are perceived tactically as well as visually. In a design representation on paper, the boundaries of a form are distinguished only visually. Design, as a process of creation and communication through a system of signs, must constantly relate the different signs which define a form to the physical materials of an artifact.

Design as Problem Solving

Design is a discipline pertinent not only to architecture but to many other fields, such as engineering, business, and medicine. Herbert Simon, one of the pioneers in the field of artificial intelligence, wrote: "Everyone designs who devises courses of action aimed at changing existing situations into preferred ones." Design is largely an exercise in problem solving; in architectural design, this involves the conception and representation of an artifact satisfying certain initial requirements and conditions. As suggested by Le Corbusier in the title of a well-known book, "creation is a patient search." Design is not necessarily an inspirational act; the search for form can be a combination of creative and deductive processes. Heuristic methods in design have an extensive literature [Rowe 1986, Simon 1969], and even a rigorous approach based on computational methods can be suitable for architectural design. The approach to design considered in this book relies on the definition of several

I-24 ►

A decision tree and a decision sequence.

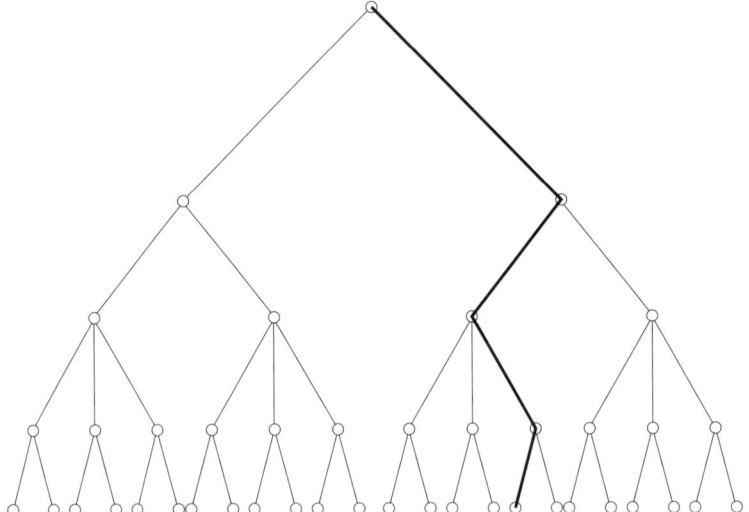

VISUALIZING WITH CAD

alternatives, which allows the possible evolution of one solution into another. A search conducted by various methods can lead to such alternative solutions; once these are defined, it becomes a decision-making problem to choose the best one for the specific situation (figure I-24).

Description and Simulation: Models

In addition to its problem-solving aspect, design can also be characterized by two different processes: description and simulation. In architectural design the descriptive phase usually consists of two-dimensional drawings, which provide all the information necessary to build the product. Architectural plans, elevations, and sections are necessary descriptive construction documents (figureI-25), but do not indicate how the built project will be visually perceived.

Simulation, as a means of imitating a real system and predicting its behavior, is an essential phase of an efficient design process. Models are used not just in architectural and engineering design, but also in physics, mathematics, and natural and social sciences. A model (figure I-26) is constructed to simulate both the visual aspects of a design and its structural and behavioral characteristics. An example of simulation in architectural design is given by a physical scale model. Even though a scale model may differ in materials and function from the real architectural artifact, formally it establishes a perfect correspondence. A traditional architectural model is a three-dimensional object, so it retains all the spatial characteristics of the configurations it represents except for size. Models have the disadvantage, however, of being costly and painstaking to construct, and do not always incorporate sufficient detail.

A much more efficient type of simulation is visualization, which provides an easily understandable graphic image of the model. In the design of forms, visualization consists of the two-dimensional representation of a three-dimensional object. Visualization as two-dimensional projection is the one of the main tasks of CAD, which provides—in a way analogous to the visual perception process—

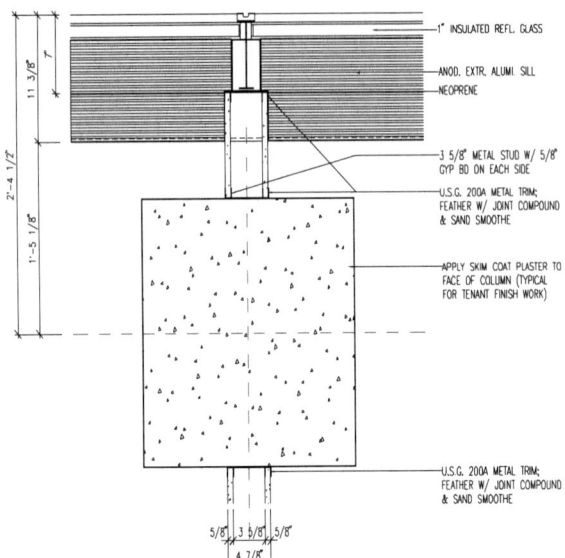

1" INSULATED REFL. GLASS

ANOD. EXTR. ALUM. SILL

NEOPRENE

3 5/8" METAL STUD W/ 5/8"
GYP BD ON EACH SIDE

U.S.G. 200A METAL TRIM;
FEATHER W/ JOINT COMPOUND
& SAND SMOOTHE

APPLY SKIM COAT PLASTER TO
FACE OF COLUMN (TYPICAL
FOR TENANT FINISH WORK)

U.S.G. 200A METAL TRIM;
FEATHER W/ JOINT COMPOUND
& SAND SMOOTHE

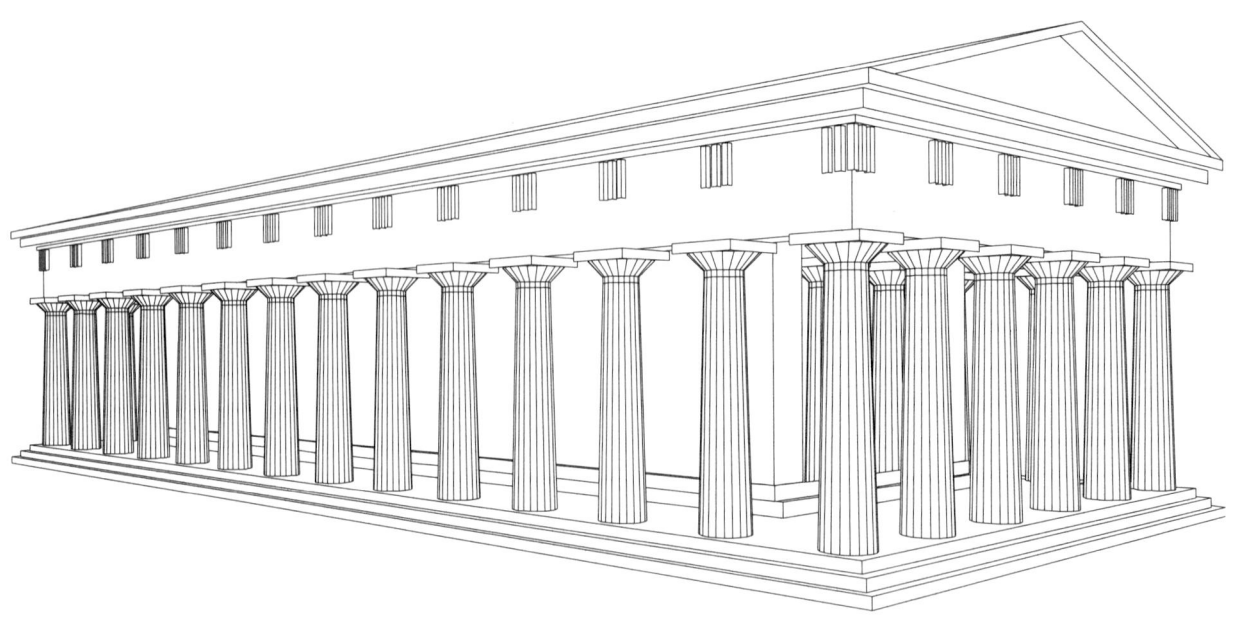

Working drawings provide description.

A model as visual simulation.

two-dimensional images from an electronic model of a three-dimensional world. The most faithful and powerful type of visualization is perspective projection; a series of perspective renderings can recreate the images we would experience if we were to walk through the built space.

Again, the distinction between the processes of definition and visualization should be stressed, even if in computer-aided design (and in traditional design as well) they overlap. The definition of a form is the description of its three-dimensionality in terms of plans, sections, and elevations, or even a set of numbers in the computer database. Visualization is the two-dimensional projection of the three-dimensional form, as it would be visually perceived by the human eye.

Language and Design

From the etymological roots of the word design—the Latin *de*, which means "out" or "from," and *signare* or *signum*, meaning "mark" or "sign"—it is clear that design deals with signs as a means to communicate a certain content. Thus, the relation between a design and the acutal construction can be described in terms of semiotics (that is, a doctrine of signs). Although the present discussion is concerned only with the very specific nature of signs in design, the study of semiotics in architecture [Preziosi 1979] is part of a more general investigation of nonverbal communication based on *visual* semiosis. Human language, however, can also be studied in semiotic theory, and parallels with language can be useful in a description of forms. A linguistic sign is composed of a *signifier* (the perceived sound of a word) and a *signified* (the meaning of the word) [de Saussure 1966]. A semiotic reading of the graphic signs—the signifiers—used to generate forms can be applied to their translations into models—the thing signified.

A language consists of an alphabet, a vocabulary, and a grammar. The alphabet is a set of the simplest, irreducible elements or signs

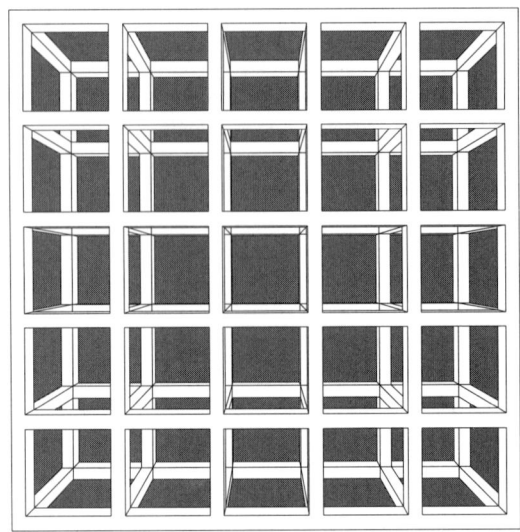

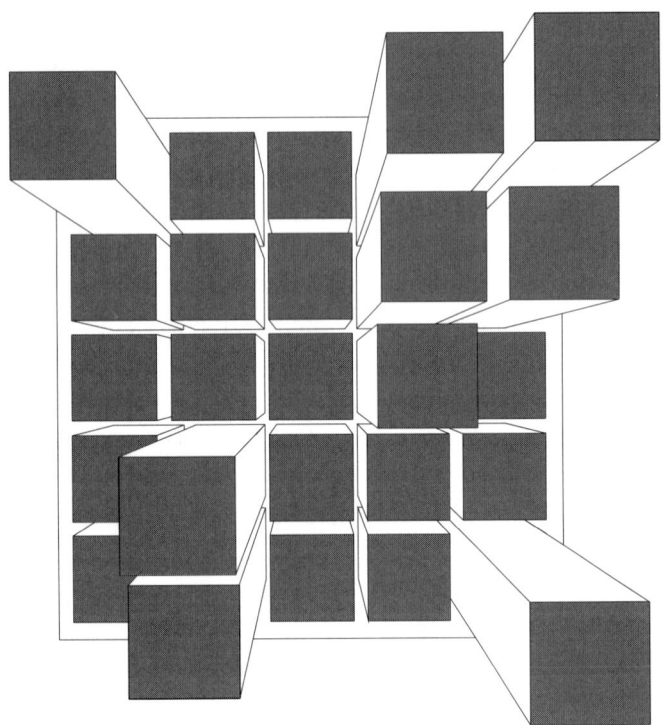

VISUALIZING WITH CAD

Semantic ambiguity: two-dimensional representations can have different three-dimensional interpretations.

[Eco 1986]. Some combinations of elements from the alphabet generate words in the vocabulary. The grammar associates words from the vocabulary according to a syntax, which is a set of rules, to produce a sentence. The syntax defines only the relations between words, independently from the meaning of the sentence. The relation between syntax and semantics, which generates questions such as "how can you construct a grammar with no appeal to meaning?" [Chomsky 1957], is largely discussed in linguistics, but it is also significant in form generation.

Ferdinand de Saussure, considered the founder of modern linguistics, recognized "the arbitrary nature of the sign" from the fact that there is no necessary natural association betweeen the signifier and the signified; for example, different words in different languages can have the same meaning. In physical architectural forms, the semantic content (as well as the syntactic arrangement) is usually easily recognizable, but unbuilt designs may be ambiguous, because a system of graphic signs can have many different possible meanings in physical construction. For instance, figure I-27, in plan view a square (signifier), has associated meanings going beyond its geometric definition: It can be read as a solid (*e.g.*, a box) or a void (*e.g.*, a hole). There are always several different meanings in the set of graphic signs which make a drawing. This ambiguity, however, can generate alternative design solutions.

Thus, the generation of forms can be explored in terms of linguistic analogies, especially in the case of computer-assisted design. The use of models—in which the elements of design compositions are viewed as alphabet and vocabulary while the relations between the elements are explained in terms of a grammar—can clarify the characteristics of an architectural style. The insight that a work of architecture can also be explained in terms of syntactic rules and semantic content can promote a better use of design media and offer further possibilities for exploration. This view provides a tool for analysis and can also help to generate design solutions.

Computer graphics can be generically described as the generation and manipulation—synthesis and processing—of images through a computer. The interest here is computer visualization of three-dimensional forms, which can be accomplished either by solid modeling or by computer-aided design (CAD). While the focus of this discussion is the use of CAD to generate forms, which is a specific application of computer graphics in the design process, it may be worth mentioning a few other recent developments.

Computer as Representation Tool: From Space to Cyberspace

Almost three decades have passed since the first breakthrough in computer graphics was introduced by Ivan Sutherland in 1962 at the Massachusetts Institute of Technology. With "Sketchpad," an interactive computer graphics system, the user could point at the screen with a "light pen" to define points, which were then connected in lines and polygons. The software and hardware for graphic representation have since developed at an exponential pace. We stand today at the gateway to a new level of computer simulation known as *virtual reality*. Virtual reality (VR) is a computer-simulated world, often referred to as *cyberspace* [Gibson 1984], in which the viewer experiences illusionary perceptual effects. These systems can be classified according to the interactive sensory devices used to experience the computer-generated world. Some popular systems use liquid crystal display video screens mounted into goggles together with sensor gloves. The images on the screens offer perspective representation from view points aligned with the viewer's eyes in order to achieve the stereoscopic effect necessary for immersion into the virtual world. This world can be computer generated or derived from video images of existing places (*telepresence*). The software is interactive, so the machine can also receive data through the sensorized clothing as it is manipulated by the participant. Because of the extreme realism of the simulation, this interaction with the virtual world produces illusionary effects—the experience of views, sounds, or even tactile sensations of the simulated environment, according to the head and body movements of the participant (the term *viewer* is no longer correct).

Virtual reality has several applications, in the military and medicine, from NASA space mission simulations to entertainment (where VR systems represent the most sophisticated arcade machines) [Rheingold 1991]. In architecture, VR simulations can offer the spatial experience of walking through a building before it is built. Beyond its more explicit applications, however, VR suggests a broader concept of space, based less on physical attributes than on information. Just as telecommunications have revolutionized our notion of distance, cyberspace changes the nature of spatial perception by dematerializing space itself. The idea of cyberspace in architecture is important not only for the realism of the representation, which offers an optimal simulation of an unbuilt architectural space, but mainly for offering an expanded vision. In cyberspace, architecture is no longer characterized by permanent construction, because, its configurations rely on an ephemeral environment created by electronic media with elements new to the architectural vocabulary.

Computer as Exploratory Tool

Since the 1960s, in the pioneering years of CAD, computers have been enthusiastically embraced. Nicholas Negroponte, from the Massachusetts Institute of Technology, in his book *The Architecture Machine*, discussed theories and ideas for the use of artificial intelligence in architecture, investigating the possible relationships between architect and machine, including that of a partnership, in which the machine and the architect would learn and grow from their interaction. The computer has been considered an integral part of the design process, and the discipline of architecture as a whole, revolutionizing its methods.

Academic research has progressed prolifically, establishing the theoretical foundations of CAD as well as advanced computational methods. One of the most interesting current areas of research in CAD is represented by the implementation of shape grammars. This approach, developed in the work of George Stiny, William Mitchell, Ulrich Flemming and others [McCullough 1990], was initiated both as a CAD theory and a general methodology to develop shapes. Shape grammars focus on the generation of new shapes from an initial set, using a given collection of rules which make a grammar. The established grammar can be usefully applied in the definition of an architectural style, since its rules define the style characteristics. Shape grammars are useful in generating new designs, but, they can also be applied to analyze and interpret existing works of architecture.

Knowledge base systems (KBSs) represent a prolific field in artificial intelligence that offers great potential for applications in CAD. By incorporating the knowledge about a certain architectural element or type necessary for the design process, knowledge bases would allow a CAD system to be used not just as a tool, but as a design assistant [Kalay 1986]. In the words of Richard Frost, "A 'knowledge base system' is a set of resources—hardware, software and possibly human—whose collective responsibilities include storing the knowledge base, maintaining security and integrity, and providing users with the required input/output routines, including deductive retrieval facilities, so that the knowledge base can be accessed as required." Such a system would "typically contain explicitly represented rules as well as simple facts." By combining specifically designed hardware, a database, and formal logic rules, a KBS would allow the designer optimal interaction with the computer.

In spite of these developments in research, the commercial applications used in the architectural office have mainly focused on CAD as a production drafting tool. At present, most designers use computers in various aspects of the profession, from drafting to data processing operations, but very few think of the computer as a design medium. In architectural practice, the computer is used mainly as an alternative to pencil and paper, which overlooks the fact that computer-aided design can also enhance the creative part of the design process. Even disregarding the most advanced applications, the way in which the computer makes geometry an accessible tool is itself a design aid. The sometimes complex computations involved in the geometric operations used to generate forms are performed accurately and rapidly by the computer. Thus, an understanding of geometry and the correct use of CAD procedures can make the design process more productive. Above all, however, CAD models provide an excellent means of visualization.

Visualization Through Computer Models

One of the most powerful means of computer generation and visualization of three-dimensional forms is provided by CAD models. As already emphasized, the construction of models contributes greatly to the development of a theory, in architecture and in other fields. With CAD it is possible to create models that incorporate different types of information about an object, from its spatial definition to its physical characteristics—information that is included in a data structure in the computer memory. Here, the emphasis will be on CAD visual models of three-dimensional forms and their spatial definition, where the computer database contains information about the geometric primitives (point, line, surface; see Chapter III) that constitute each form. The resulting visualization by means of projection provides two-dimensional images from an electronic model of a three-dimensional world (in a way analogous to the visual perception process).

The model of a form—taken from either the physical world or the mathematical universe—is constructed through a series of mappings. There are two processes in creating an electronic model. The imagined or existing form may be mapped into sets of points, which determine geometric entities such as lines, surfaces, solids, and meshes; conversely, a model can represent a form resulting from, say, algebraic equations, mathematical functions, or sets of logical rules. In both cases, a mapping is established between geometric primitives and the spatial definition of a form. Each geometric primitive is represented by means of three-dimensional coordinates

(x, y, z), and the set of all such representations composes a computer database. At this stage, the electronic model is defined by sets of numbers and cannot be visually perceived. The three-dimensional coordinates of the database, in order to be represented by the pixel of the screen or vectors plotted on paper, must be mapped once again into two-dimensional coordinates. Coordinates that define the point of view (POV) from which the model is visualized must also be provided, together with other information intrinsic to the viewing condition desired.

The mapping from three dimensions into two expresses the main characteristic of CAD electronic models: They are visual—that is, they consist of two-dimensional images—in contrast to traditional architectural models, which are three-dimensional physical objects. Unlike physical models, in which the dimensions and details different from those of the real form, electronic models can establish an optimal spatial correspondence. Also, the ability to change viewing conditions at will allows a visualization from the macroscopic to the microscopic level of detail.

As with physical forms, electronic models provide one of the most powerful kinds of visualization of geometric shapes. In fact, because geometric shapes can usually be defined through data points derived directly from mathematical functions of two variables—of the type $z = f(x, y)$—construction of an electronic model requires only that the set of coordinates defining the form be translated into a computer database. The accuracy of the representation is present in any of the different views. Even animated sequences that effectively describe the geometric shape can be obtained with no extra human effort.

THE PURPOSE OF THIS BOOK

In this chapter concepts such as **space** and **form** have been introduced in general terms, in the context of disciplines ranging from psychology to geometry, in the belief that a rigorous approach requires a consideration of the fundamental operations and defini-

Relations and processes in architectural design.

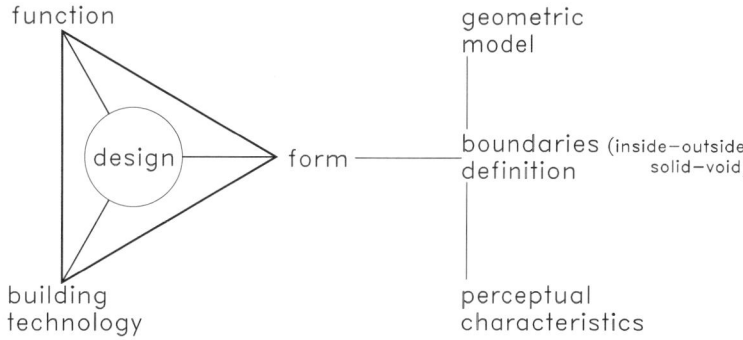

tions that underly the process of creating forms (figure I-28). In the course of this book, an approach to form generation and evolution will be developed, often following analogies with linguistic models, which provide interpretative categories. The generation of forms can be systematized as a set of rules and definitions, and the computer is part of the creative process, performing all the necessary geometric operations. This systemization starts from a perceptual analysis of form as a set of relations between boundaries that create a solid-void distinction. Any physical boundary of a form can be described in terms of its position and extension, defined by three-dimensional coordinates, and therefore can be interpreted through geometry. The use of geometric operations can also generate complexity from the elaboration of simple primitive elements.

Dynamic Models: From Geometry to Architecture

The geometric definition of form is therefore a necessary condition, not just for models of shapes belonging to the realm of geometry, but also for the articulation of more semantically complex models. Conversely, a static geometric definition of a model is not sufficient when we consider a different context, such as architecture. Even in the context of geometry, the model of a form sometimes needs a dynamic definition to evolve from a general case to more specific applications (figure I-29a). The data structures of computer-aided design systems,

a. A "dynamic model" and its geometric evolution.

b. Evolution in a different semantic context.

c. Evolution in a different semantic context.

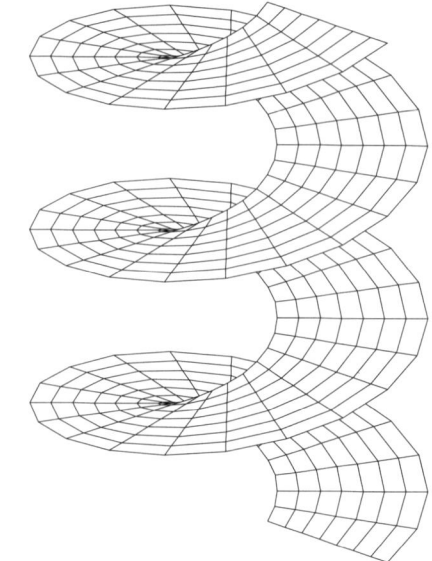

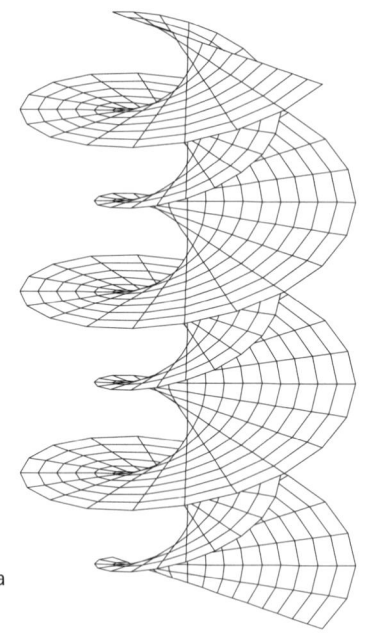

a

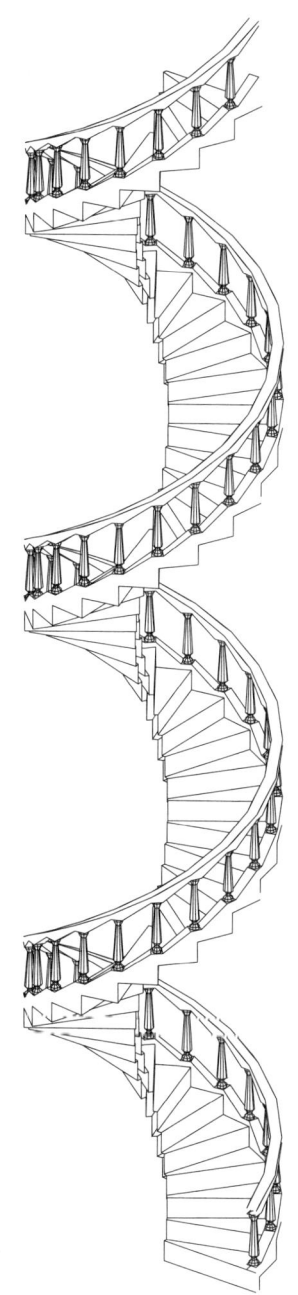

b

c

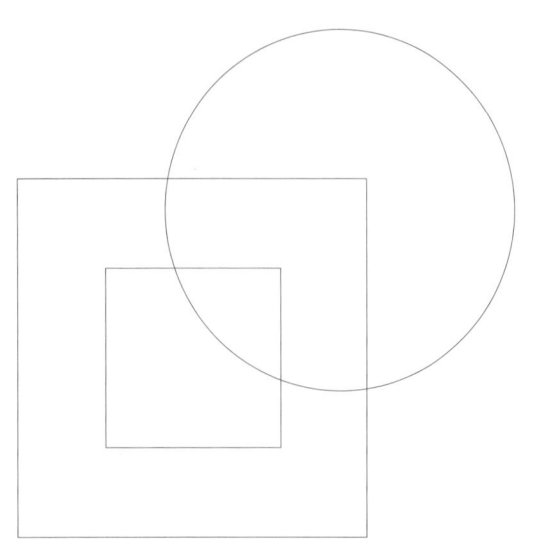

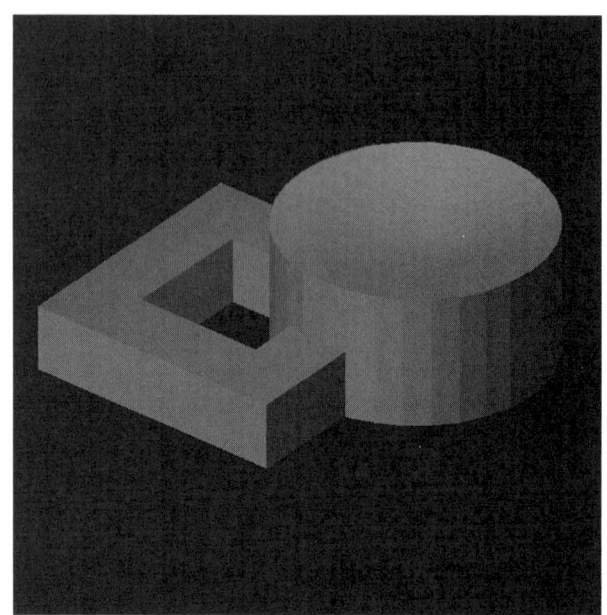

VISUALIZING WITH CAD

A "dynamic model" and design
alternatives.

as we will see in the following chapters, satisfy the requirement of a dynamically evolving model. Our initial geometric model of form can therefore be structured as a dynamic model, which may evolve from one semantic description to a different one (figure I-29b-c) and so represent different design alternatives (figure I-30).

It is important in this evolutionary process to distinguish between the semantic levels of geometric versus architectural form. In the transformation of a geometric model into architectural form, functional aspects must also be considered, with all the constraints that they will bring to the geometric form.

The present investigation of architectural models will be restricted to three-dimensional forms. The choice of basing design on three-dimensional models rather then on floor plan articulation—a method often followed in computer-aided design—is, once again, motivated by a perceptual approach. Additionally, architecture is three-dimensional; therefore a study of architectural forms should, from the very beginning, be concerned with three-dimensional models. Furthermore, the visual simulation of a three-dimensional model in itself can offer insights for design exploration, especially for models that are geometrically complex. Access to models of ready-made geometric configurations provides several design alternatives and becomes an aid in the design process.

Computer Model Implementation

This discussion will be oriented toward and implemented in the commercial CAD application AutoCAD.[5] For some of the most interesting operations and models, a listing of AutoCAD commands will follow the general description. An AutoLISP file will also be provided for some of the models described and illustrated in the book. The AutoCAD models will be structured in such a way that they can evolve according to parameters given by the reader.

The choice of a software such as AutoCAD, which is often used only for drafting, shows that an efficient approach to computer-aided design can be independent of the software used. What matters most is the methodology followed by the designer/user in defining design operations. The object here will be to show how even a commercial system, developed for general purposes, can be customized to offer a structured design language.

BIBLIOGRAPHY

Alberti, Leon Battista, *The Ten Books of Architecture*, Dover, New York, 1986

Arnheim, Rudolph, *Art and Visual Perception*, University of California Press, Berkeley and Los Angeles, 1974

Arnheim, Rudolph, *Visual Thinking*, University of California Press, London, 1969

Bachelard, Gaston, *The Poetics of Space*, Beacon Press, Boston, 1969

Benedikt, Michael (ed.), *Cyberspace: First Steps*, MIT Press, Cambridge, 1991

Benevolo, Leonardo, *History of Modern Architecture*, MIT Press, Cambridge, 1977

Bioy Casares, Adolfo, *L'Invenzione di Morel*, Tascabili Bompiani, Milano, 1989

Cassirer, Ernst, *The Problem of Knowledge*, Yale University Press, New Haven, 1950

Chomsky, Noam, *Syntactic Structures*, Mouton, The Hague-Paris, 1957

Coxeter, H. S. M. and S. L. Greitzer, *Geometry Revisited*, Random House, New York, 1967

D'Arcy Thompson, *On Growth and Form*, Cambridge University Press, Cambridge Eng., 1961

Denning, Peter J., Dennis, Jack B., and Qualitz, Joseph E., *Machine, Languages, and Computation*, Prentice Hall, Englewood Cliffs, N. J., 1978

Eco, Umberto, *Semiotics and the Philosophy of Language*, Indiana University Press, Bloomington, Ind., 1986

Euclid, *The Thirteen Books of Euclid's Elements*, Dover, New York, 1956

Filarete, *Trattato di Architettura*, Il Polifilo, Milano, 1972

Frost, Richard, *Introduction to Knowledge Base Systems*, Macmillan, New York, 1986

Gibson, William, *Neuromancer*, Ace Books, New York, 1984

Greenberg, Marvin Jay, *Euclidean and Non-Euclidean Geometries*, W. H. Freeman, New York, 1980

Haeckel, Ernst, *Art Forms in Nature*, Dover, New York, 1974

Hersey, G. L., *Pythagorean Palaces*, Cornell University Press, Ithaca and London, 1976

Hilbert, D. and S. Cohn-Vossen, *Geometry and Imagination*, Chelsea, New York, 1952

Kalay, Yehuda E., *Computability of Design*, John Wiley & Sons, New York, 1986

Le Corbusier, *Creation Is a Patient Search*, Frederick A. Praeger, New York, 1960

Luckiesh, M., *Visual Illusions*, Dover, New York, 1965

Mandelbrot, Benoit B., *The Fractal Geometry of Nature*, W. H. Freeman, San Francisco, 1977

McCullough, Malcolm, William J. Mitchell, and Patrick Purcell (eds.), *The Electronic Design Studio*, MIT Press, Cambridge, 1990

Merleau-Ponty, Maurice, *Phenomenology of Perception*, Routledge & Kegan Paul, London, 1961

Mitchell, William J., *The Logic of Architecture: Design, Computation, and Cognition*, MIT Press, Cambridge, 1990

Negroponte, Nicholas, *The Architecture Machine*, MIT Press, Cambridge, 1970

Norberg-Schulz, Christian, *The Concept of Dwelling*, Rizzoli, New York, 1985

Norberg-Schulz, Christian, *Genius Loci*, Rizzoli, New York, 1980

Norberg-Schulz, Christian, *Intentions in Architecture*, MIT Press, Cambridge, 1965

Perez-Gomez, Alberto, *Architecture and the Crisis of Modern Science*, MIT Press, Cambridge, 1985.

Piaget, Jean, *The Psycology of the Child*, Basic Books, New York 1969

Preziosi, Donald, *Architecture, Language and Meaning*, Mouton, The Hague, 1979

Rheingold, Howard, *Virtual Reality*, Simon & Schuster, New York, 1991

Rowe, Peter, *Design Thinking*, MIT Press, Cambridge, 1986

Saussure, Ferdinand de, Course in General Linguistics, McGraw-Hill, New York, 1966

Serlio, Sebastiano, *The Five Books of Architecture*, Dover, New York, 1982

Simon, Herbert A., *The Sciences of the Artificial*, MIT Press, Cambridge, 1969

Spencer Brown, G., *Laws of Form*, Julian Press, New York, 1972

Thom, Rene, *Structural Stability and Morohogenesis; an outline of a general theory of models*, W. A. Benjamin, Reading, Mass., 1975

Vitruvius, *The Ten Books on Architecture*, Dover, New York, 1960

NOTES

1   The German word *gestalt* means form or shape. From the beginning of this century, several experimental and theoretical studies in visual perception were grouped under this name.

2   Ernst Haeckel (1834-1919) was a biologist/philosopher and an exponent of Darwinism. His biological studies were illustrated by elaborate lithographic plates.

3   In its original form, Euclid's fifth postulate asserts that "If a straight line falling on two straight lines makes the interior angles on the same side less than two right angles, the two straight lines, if produced indefinitely, meet on that side on which the angles are less than two right angles."

4   See Vitruvius, *The Ten Books on Architecture, Book I, Chapter III*. This work on architecture provided inspiration for Leon Battista Alberti's treatise of the same name.

5   AutoCAD is a trademark of Autodesk.

II

COMPUTER-AIDED DESIGN

METHODOLOGY

Neither natural ability without instruction

nor instruction without natural ability can

make the perfect artist.

Vitruvius

Having examined the more theoretical foundations of form modeling in relation to computer-aided design, we now consider its technical aspects and analyze specific methods. As previously indicated, *form modeling* and *design* are terms used in several disciplines; but in this chapter the term *computer-aided design* will be used specifically with reference to architectural forms. Nevertheless, the generation of forms in other disciplines should not be considered an extraneous application, since any three-dimensional form can be constructed and visualized using the processes described here.

The specific CAD program chosen for our applications is AutoCAD, but almost all CAD software follows the same systematic approach. The nomenclature can vary from one software to another, but the methodology is shared and therefore the general definitions given here are valid for any CAD user or for anyone interested in the more general study of computer-aided generation of visual forms.

The objective of this chapter is to provide an overview of how a CAD system is structured and how its media are different from those of traditional design. This information is essential for readers unfamiliar with CAD, and valuable for those who have used CAD mainly as a drafting tool. The actual definition and implementation of CAD dynamic models will be introduced in Chapter III. This chapter supplies the terminology and tools to implement all the further developments.

NOTE TO THE READER

Chapters II and III contain a tutorial; each section discusses a particular topic and introduces general definitions that can be applied to any CAD software. At the end of each section there is a brief AutoCAD implementation of the given definitions; these tutorial segments are clearly set off, so that readers who are not AutoCAD users can skip them. Throughout the tutorial segments, different typefaces are used: Words in **boldface** are commands to be typed, those in *italics* are AutoCAD prompts, and words between <brackets> represent actions to be performed using the keyboard or pointing devices.

The tutorials are provided for those readers interested in using AutoCAD to implement the discussed models as well as their own. Usually, only those commands to be used in the implementation of three-dimensional models, generated according to the theory illustrated in Chapter III, will be shown. Unless otherwise specified, all discussion of AutoCAD refers to Release 12.

INTERACTIVE CAD SYSTEMS

The use of CAD in the design of an artifact of any type involves two processes. The first is the creation of a data structure containing the information necessary to describe the object; the second is the visual display of this information. Because of the interactivity of the CAD software, visual elements are drawn as soon as the information that describes them is given.

Like most other computer graphics software, CAD is an interactive program, structured in the form of languages, that is implemented through **commands** given by the user. Interactive computer graphics have some hardware requirements. The basic hardware configuration consists of the computing unit, a memory storage device, a video display screen (the output device on which the graphics and text are displayed), and a keyboard (the simplest device to enter programs and data; a light pen, mouse, joystick, or graphics tablet or digitizer can also be used). Optional hard copy output devices include various types of printers, plotters, and film recorders.

Often software commands are organized in lists called **menus**, displayed in portions of the screen. The way in which commands are organized in menus usually gives the user a first understanding of how the software works and what it is intended to achieve. Commands are words which usually correspond semantically to words from common language (with the verb implicitly given by the menu from which the command originated). For example, the command **line** taken from the **draw** menu will create an entity in the data structure corresponding to a geometric line segment. Menus will be discussed in more detail later in this chapter.

The definitions and tutorials that follow do not require a specific hardware configuration, only the basic keyboard and video display screen.

REPRESENTATIONS AND PHASES OF ARCHITECTURAL DESIGN

Architectural design uses graphics representations to communicate ideas (although such images are subject to different interpretations, both as representations of objects and as pure images. Drawings are by definition two-dimensional representations, but they can simulate three-dimensionality. This is illustrated in figure II-1, which shows examples of two-dimensional architectural representations. There are two ways to represent three-dimensional forms. The first type of representation is given by orthographic projections—plan, section and elevation (figure II-1). These representations are diagrammatic images: They contain a description of a physical form—often with references to dimensions and other quantitative information—but they do not offer any visual simulation of how the real form looks. By contrast, representations of the second type—perspective and axonometric projections (figure II-13d,e)—and in particular perspective, which is much more realistic than axonometric projection, do offer a visual simulation. Orthographic projections and perspective views therefore complement each other and are equally important for a complete design. The former provide a quantitative description necessary for physical construction and the latter give a qualitative preview of the design, based on visual simulation.

These considerations are generic and can refer to CAD as well as to traditional design. In both cases the design output is a drawing,

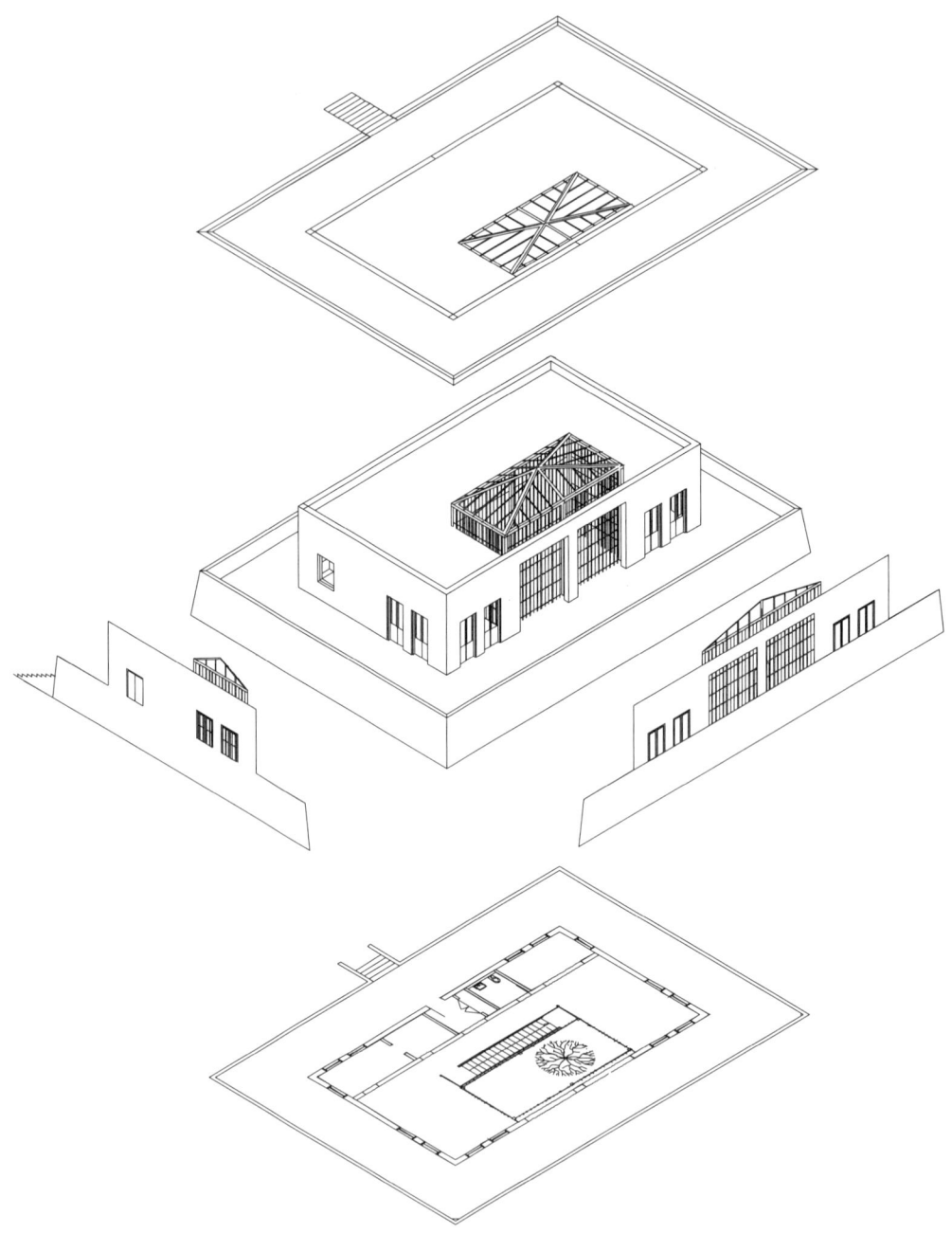

Two-dimensional representations
in architecture.

realized with different media. In traditional design, a drawing is realized in paper and pencil, while in CAD the image is electronic, even though it can eventually be printed on paper. The end output may be the same, but the process behind it is different.

Architectural design is usually not completed in just one phase. According to the type and size of the project, there are different descriptive methods and presentation tools. Usually there are three types of graphics involved (figure II-2): sketches, presentation renderings (included in the so-called design development), and working drawings, which represent a phase completely different from the others in that they allow a design on paper to be realized in construction. Design as exploration is not pursued in this last phase, although architectural construction details may still be designed. The use of CAD can overcome the need for different types of graphics, which is due to the impossibility of presenting all information at once through a physical medium such as paper. Electronic storage of information electronically allows the development of a unified model that contains all the data pertinent to each phase. In this way, the overlapping of information can be eliminated, making the whole process more efficient.

COMPUTER-AIDED DESIGN AND COMPUTER-AIDED DRAFTING

The separate phases in architectural design also usually reflect another division, that between design and drafting. Both syntactically and semantically, design and drafting represent completely different processes. While design is an exploratory process that results in a three-dimensional artifact, drafting is confined to the two-dimensional world of drawings, and takes place when the exploration is over and decisions have been made. The graphic signs used in drafting consist mainly of symbolic conventions and diagrams used to represent elements in the physical world: A circle may represent a column, or two parallel lines may signify a wall. In contrast, the design process

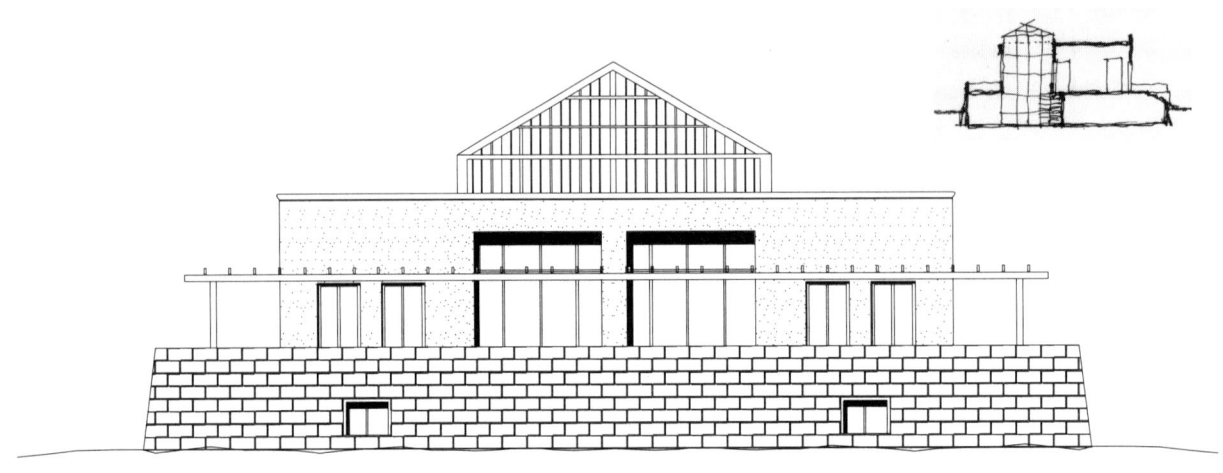

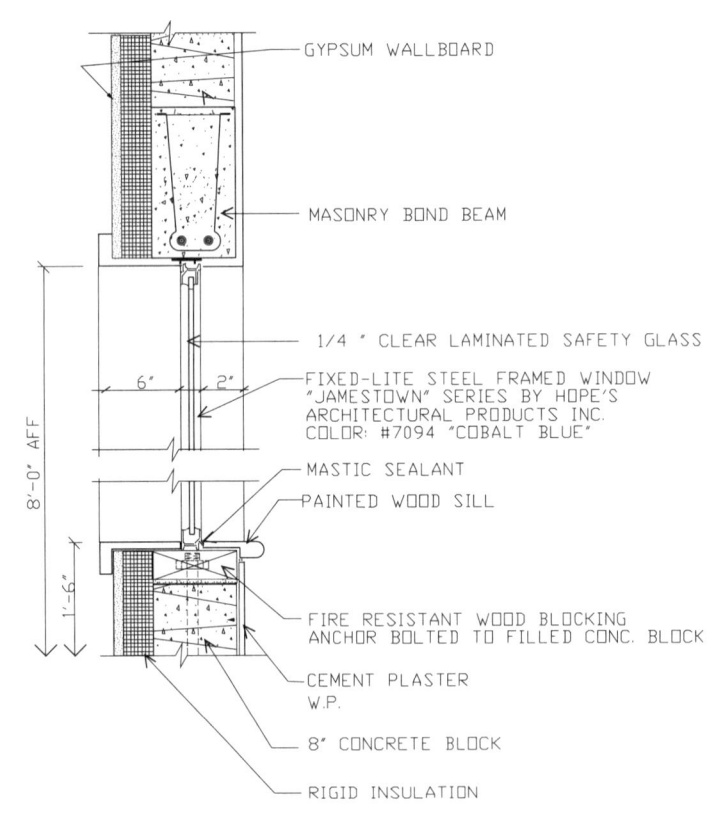

GYPSUM WALLBOARD

MASONRY BOND BEAM

1/4 " CLEAR LAMINATED SAFETY GLASS

FIXED-LITE STEEL FRAMED WINDOW
"JAMESTOWN" SERIES BY HOPE'S
ARCHITECTURAL PRODUCTS INC.
COLOR: #7094 "COBALT BLUE"

MASTIC SEALANT

PAINTED WOOD SILL

FIRE RESISTANT WOOD BLOCKING
ANCHOR BOLTED TO FILLED CONC. BLOCK

CEMENT PLASTER
W.P.

8" CONCRETE BLOCK

RIGID INSULATION

6"   2"

8'-0" AFF

1'-6"

requires a type of representation that can produce a simulative model in order to communicate completely. Therefore the graphics used in design should provide both two-dimensional and three-dimensional visualization.

CAD is often identified with drafting systems. In CAD, a drafting system operates only in two-dimensions; it is just a computer-aided version of manual drafting. The graphics information is organized in extremely structured and efficient way, however. Features such as text manipulation, graphics symbols templates, and separation of different categories of information in layers make computer-aided drafting a very powerful tool, faster and more accurate than manual drafting. The initial graphics input is the same for CAD as for manual drafting; that is, the basic construction depends on the same geometric definitions—a line is defined by two points, a polygon by three or more lines, and so on. CAD, however, has transformation capabilities such as rotation, mirror, scaling, and stretching that make it much more efficient than manual drafting. Nevertheless, in a two-dimensional context, automated drafting systems are not very different from manual drafting.

Although many CAD systems were born as two-dimensional drafting systems, they later acquired three-dimensional capabilities. In the transition from two to three dimensions, the designer should acknowledge the difference between the two types of representations and learn how to use one of CAD's most amazing features: the automatic generation of several views from the same three-dimensional model.

WIREFRAME, SURFACE, AND SOLID MODELING

Three-dimensional CAD systems can be classified according the type of primitive elements used in the generation and visualization of a three-dimensional model. The simplest type is the wireframe model (figure II-3a), where a form is described in terms of lines in three-dimensions. This model contains only information about the linear

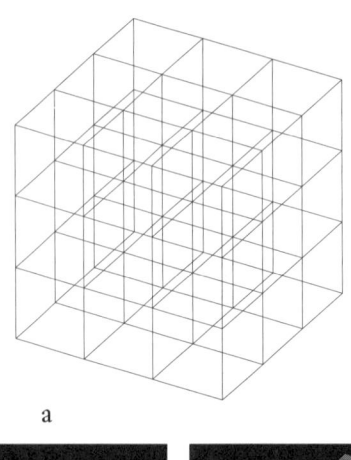

a

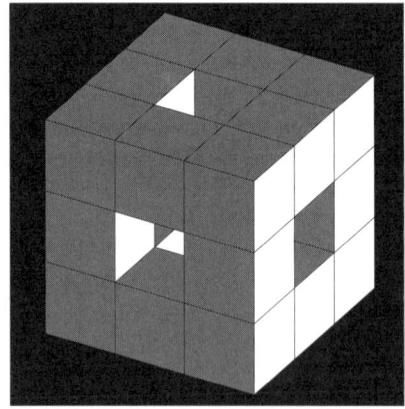

b

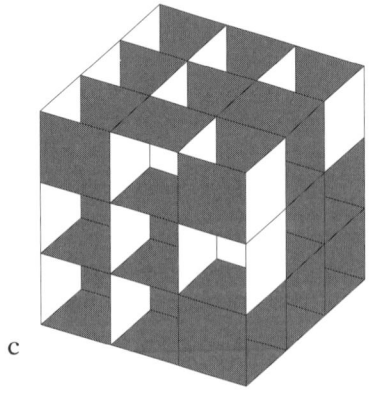

c

VISUALIZING WITH CAD

boundaries of the form; it does not describe the form's spatial characteristics in terms of solid-void boundaries. Thus, it is fully descriptive only for forms made of wire; to fully understand displayed forms of any other type, the viewer would require additional information not provided by the visualization. For instance, in the wireframe set of cubes represented in figure II-3a, several interpretations are possible regarding its foreground and background and the faces of the cube (figure II-3b,c).

The next category of three-dimensional system is visually represented by the polygon-based model in figure II-4, where a three-dimensional model is described in terms of the polygons defining its boundaries. The model so generated is fully descriptive of the spatial characteristics of the form. As stated in Chapter I, a solid can be completely defined by the surfaces bounding it. The polygons, being opaque, block out the hidden lines beyond them, producing a comprehensive visual simulation that clearly defines foreground and background.

A third type of system is represented by solid modeling (figure II-5), where the so-called primitive elements are represented by solid shapes such as cubes, spheres, pyramids, wedges, tori, and so forth. The geometry used to manipulate these shapes is called **constructive solid geometry**. The Boolean operations of union, intersection, and

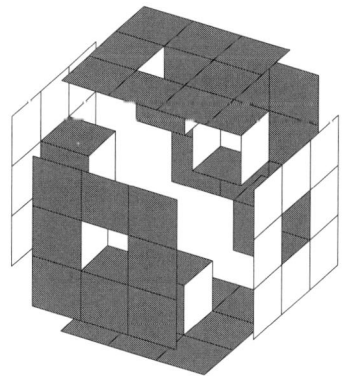

Solid model.

subtraction allow the generation of unlimited shapes through the combination and modification of two primitives, similar to the physical sculpting of forms by means of gluing and carving. This type of modeling is most representative of physical reality. The visual simulation it offers is similar to the polygon-based model, but it has the additional advantage of production use in a CAD or CAM (computer-aided manufacture) environment.

The CAD system referred to in the definitions and examples in this book is based on polygon modeling, but some examples of solid modeling will also be offered, primarily in cases where the exploration of forms is more purposefully achieved by using the Boolean operations.

In early releases, AutoCAD was a drafting system with only two-dimensional (2D) capabilities. Starting from release 2.5, 3D features have been gradually introduced. The first 3D constructions were just extrusions of 2D entities, based on a two-dimensional coordinate system. Later, three-dimensional coordinate systems were introduced, together with commands for the construction of quite complex surfaces. In releases 11 and 12, AutoCAD offers the optional program AME (Advanced Modeling Extension), which uses solids instead of surfaces as primitives. AME has complete solid modeling features, providing Boolean operations and even the possibility of assigning materials to shapes and calculating physical attributes such as mass, center of gravity, and moment of inertia, which can be used for structural analysis.

VISUALIZING WITH CAD

MENUS

Menus can enormously facilitate the use of a CAD interface. In menus, the software commands are subdivided in a list, created according to type of action they represent. Menus can also be customized according to the type of application they serve.

AutoCAD has a default menu that is automatically loaded (unless the user chooses otherwise) at the beginning of a new session. Pull-down menus began appearing in AutoCAD Release 9, and are visible when the pointing device is moved to the top of the screen. The pull-down menus—**file**, **assist**, **draw**, **construct**, **modify**, **view**, **settings**, **render**, **model**—are located at the top of screen. The name of the menu is explanatory of the action. The menus are based on a hierarchical structure in a top-down fashion. The first level can be represented by commands to be invoked or can contain more subdivision. For instance, the top-down sequence to draw a sphere is the following:

draw-3Dsurface-3Dobject-sphere

The reader is invited to practice navigating the pull-down areas. Observation of how the menus are structured can provide a good tutorial experience.

USE OF FULL SCALE: FROM SCHEMATIC DESIGN TO CONSTRUCTION DOCUMENTS

A necessary operation in architecture is the scaling of a proposed design to fit a sheet of paper. Different scales are used according to the level of detail desired. For example, architectural plans and elevations read well at a scale of 1/4"=1'0" or 1/8"=1'0", furnishing plans often are drawn at 1/4"=1'0", while a wall detail may require a 1"=1' scale. In manual drafting, problems of accuracy arise when a form is scaled down in size, because measurements cannot be scaled with extreme accuracy on paper and a change of scale requires a new drawing.

With CAD, the use of full scale is possible. Every distance between elements can therefore be electronically measured with extreme accuracy. Another useful feature is the ability to plot drawings at different scales. This flexibility allows the representation to grow in complexity without the visual confusion contributed by information that is relevant only at a different scale.

In the **settings** menu we find scale and units. The initial choice of a type of unit can be changed at a later time.

PAPER SPACE; MODEL SPACE

Another interesting feature of the latest releases of AutoCAD is the differentiation of the model/drawing environment to model space and paper space. Model space represents the three-dimensional design/mod- eling universe, where entities are represented at a full scale. Paper space is the drawing universe and provides the output on paper of the model created in model space. Paper space is the real size of paper, and includes annotations, borders, and other entities that are not representational but serve only as a communication tool. Model space is the default environment when we start a session; for our purposes, we will be dealing only with model space.

COORDINATE SYSTEMS

A basic understanding of coordinate systems is a necessary foundation for the use of CAD. The universe of our representation corresponds to the mathematical representation of three-dimensional space. While a paper drawing is limited by the size of the sheet in relation to the chosen scale, in CAD the limit for the space of construction of our model is extremely large, measured by the number of digits allowed rather than inches.

Coordinate systems in CAD have the same characteristics and properties as those in geometry. **Cartesian** coordinates (figure II-6a) are most often used in CAD, as well as in geometry. In a Cartesian system, a point P in three-dimensional space is defined by a set of three numbers (x, y, z), or coordinates, which specify respectively the distances along three mutually perpendicular axes at which P lies from the origin (the point at which the axes intersect, having coordinates (0,0,0)). The origin in CAD is usually represented by the lower left corner of the screen or viewport.

a. Cartesian coordinates.

b. Polar coordinates.

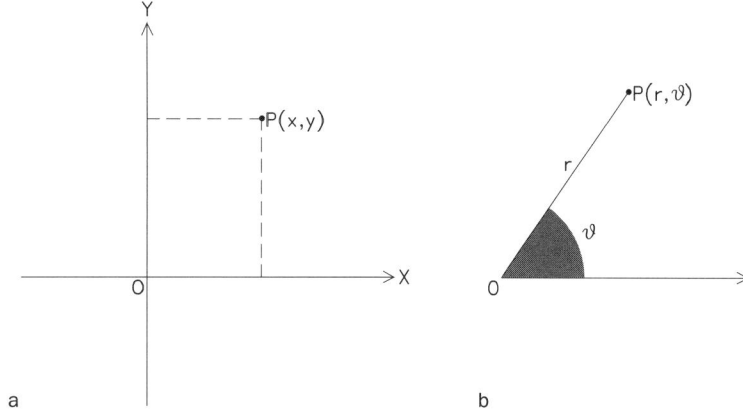

a

b

Polar coordinates (r,θ) of a point P (figure II-6b) are given by the distance r from the origin O and the angle θ (measured counterclockwise) formed by the line OP and a horizontal axis through O. **Relative** coordinates, in AutoCAD, give the position of a point with respect to the last-entered point (that is, the last-entered point is treated as if it were the origin).

In AutoCAD the default plane is the *x-y* plane, identified with the screen with origin in the lower left corner. The definition of an element (taken from the **draw** menu) by means of Cartesian coordinates is accomplished as follows:

Command: **line**
from point: **5,15**
to point: **55,65**
to point: **30,65**
to point: <enter>

The same lines can also be constructed using polar coordinates:

Command: **line**
from point: **5,15**
to point: **@70.57<45**
to point: **@25<180**
to point: <enter>

or relative coordinates:

Command: **line**
from point: **5,15**
to point: **@50,50**
to point: **@-25,0**
to point: <enter>

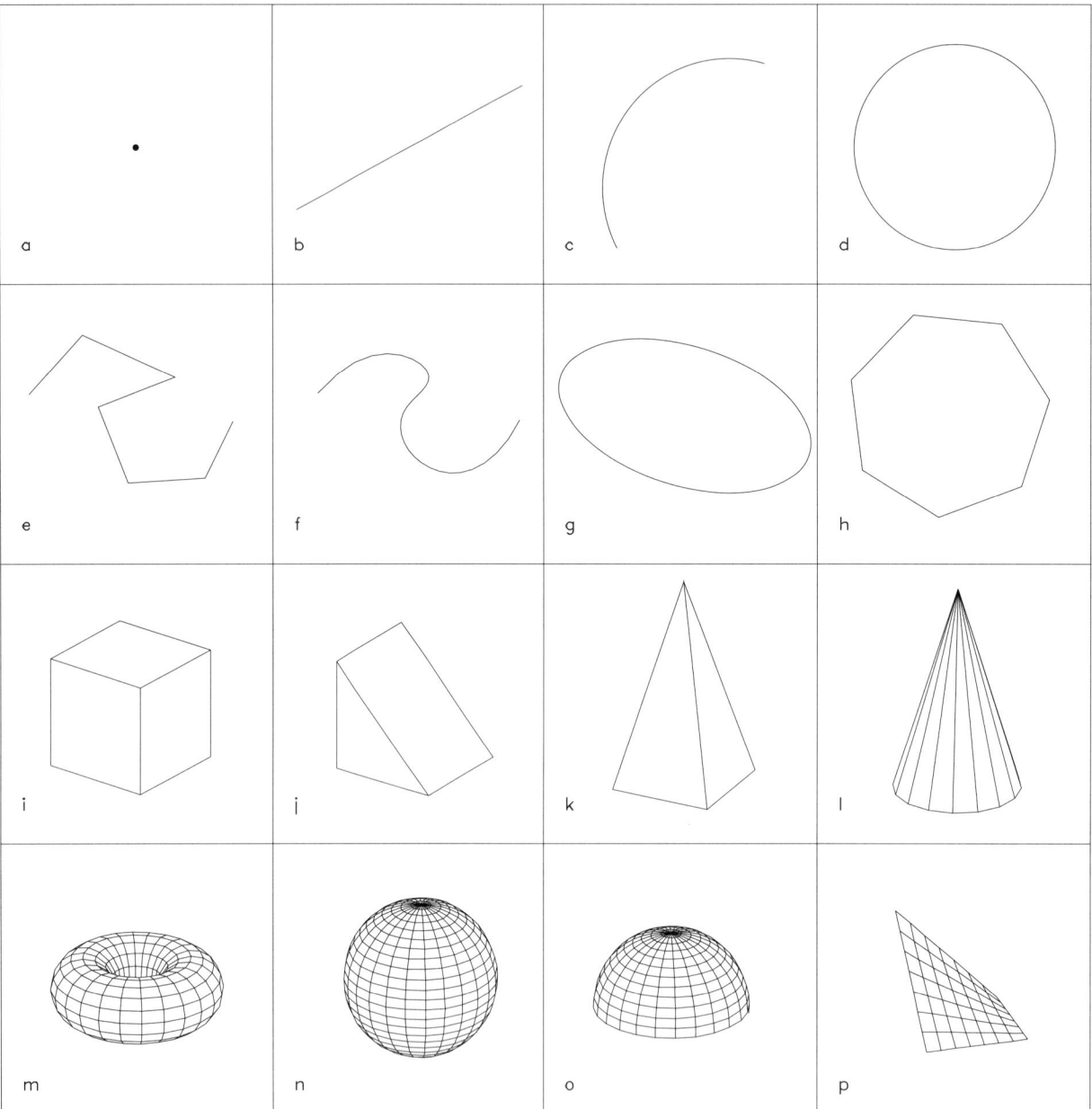

VISUALIZING WITH CAD

The **user coordinate system (ucs)** can easily be changed. The new user coordinate system can be a translation of the **original world coordinate system (wcs)** or, more radically, a rotation of all three axes through any generic angle. A new *x-y* plane is defined, replacing the old one. To work in a different coordinate system can be useful since there are many commands that can be used in the *x-y* plane but not in space. The command to change coordinate systems is:

Command: **ucs**

*Origin/Zaxis/3point/Entity/View/X/Y/Z/Prev/
      Restore/Save/Del/?/<World>:* **3p**

Origin point <0,0,0>: <enter>

*Point on positive portion of the X-axis <0'-1",
      0-0",0'-0">:* <enter>

*Point on positive portion of the UCS X-Y Plane
      <0'-0",0'-1",0'-0">:* **0,0,1**

To return to the original world coordinate system:

Command: **ucs**

*Origin/Zaxis/3point/Entity/View/X/Y/Z/Prev/
      Restore/Save/Del/?/<World>:* <enter>

GRAPHICS PRIMITIVES

CAD offers a series of graphics primitives, some of which correspond to basic geometric elements. In an analogy with language, they represent the alphabet, from which it is possible to construct, say, a noun— in our context, a form. A discussion about graphics and geometric primitives in relation to spatial form, although it pertains to CAD, also involves more general concepts from geometry and logic (see Chapter III). At the moment, however, it suffices to mention the existence of several CAD commands, the names of which correspond to the graphics entities they define as database and drawing, according to the user's given parameters.

The CAD graphics primitives most often used are (figure II-7): **point**, **line**, **circle**, **ellipse**, **polyline** (a set of lines, which may be interpolated into a curve), **polygon**, **face**, and **solid**. Often, three-dimensional shapes, such as a box, sphere, torus, or cylinder, which are not primitives in a geometric sense (since they are composed of simpler geometric primitives), are nevertheless referred to in CAD and solid modeling as primitives.

Starting with this section, all descriptions of CAD features will focus

◄　II-7

Graphics primitives

Two-dimensional AutoCAD primitives:
a. Point　b. Line　c. Arc　d. Circle
e. Polyline　f. Bezier　g. Ellipse
h. Polygon.

Three-dimensional AutoCAD primitives:
i. Cube　j. Wedge　k. Pyramid　l. Cone
m. Torus　n. Sphere　o. Dome
p. Mesh.

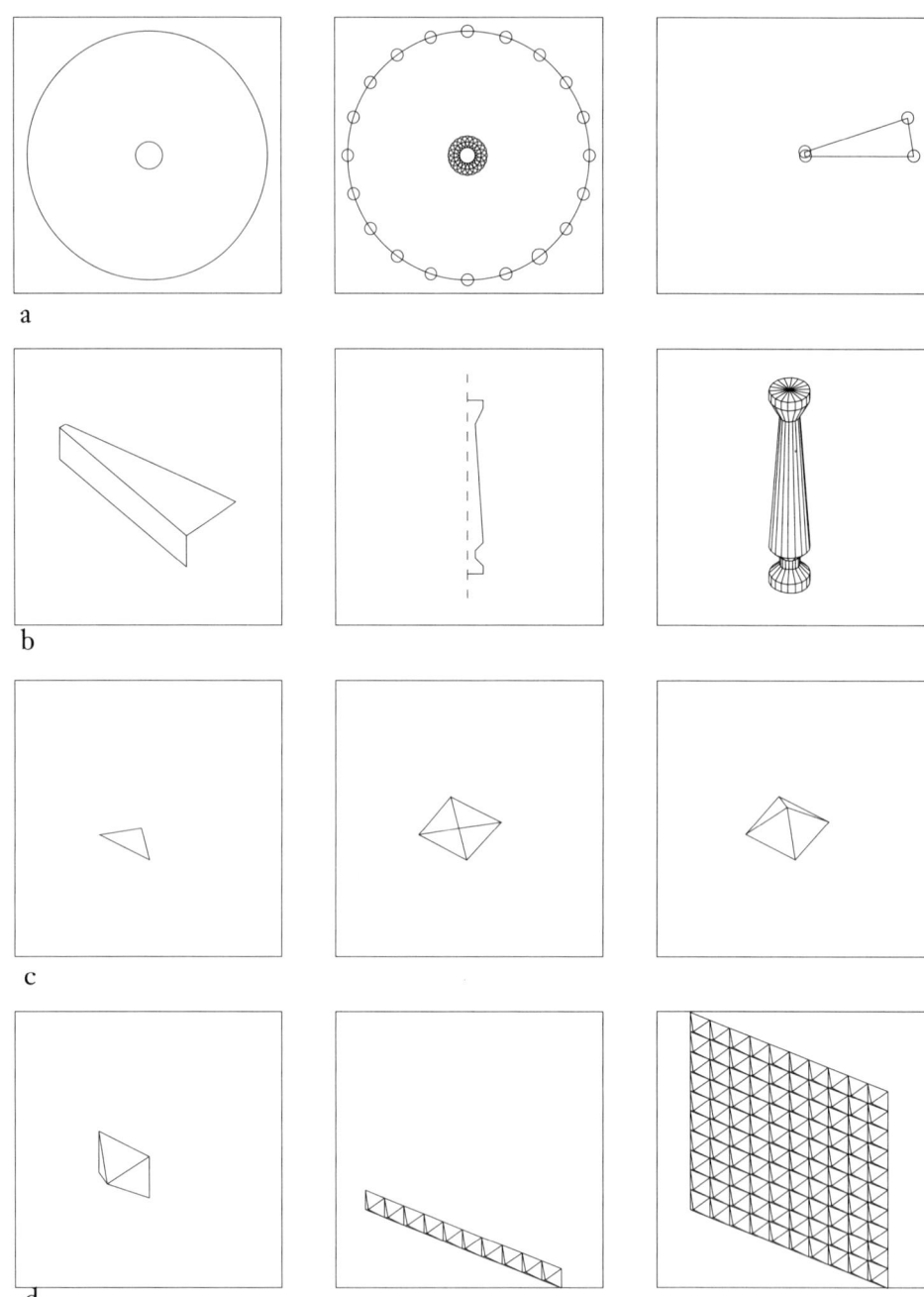

a

b

c

d

VISUALIZING WITH CAD

Model of a spiral stair.

a. b. Primitives. c. d. Editing.

on the construction of a model of a spiral stair, a good initial example of generation of a CAD dynamic model. The CAD model is implemented in AutoCAD and shown in detail in the tutorial sections. The sequence of illustrations follows the temporal sequence of construction of the model. Figure II-8a,b shows how the basic elements of the stair are constructed through the implementation of graphics primitives.

We are about to start the construction of a model of a spiral stair. The sequence presented in this section is illustrated in figure II-8a. The definition of the surfaces defining each of stairs is implemented as follows:

To draw each circle, from the draw menu we choose the command "circle":

> *Command:* **circle**
> *3P/2P/TTR/<Center point>:* **0,0**
> *Diameter<Radius>:* **0.5**
>
> *Command:* **circle**
> *3P/2P/TTR/<Center point>:* **0,0**
> *Diameter<Radius>:* **4.5**

To divide one circle into 20 equal segments delimited by nodes, from the **construct** menu we choose the command "divide":

> *Command:* **divide**
> *Select object to divide:* <pick one of the circles>
> *<Number of segments>/Block:* **20**

Repeat for the other circle. We can proceed to draw the surface of the step determined by four nodes as shown in figure II-8a3.

> *Command:* **3dface**
> *First point:* <pick node 1>
> *Second point:* <pick node 2>

> *Third point:* <pick node 3>
> *Fourth point:* <pick node 4>
> *Third point:* <enter>

To draw a revolution surface, representing the baluster (figure II-8b2), the second "pline" is selected from the "draw" menu:

> *Command:* **pline**
> *from point:* **0,0**
> *Arc/Close/Halfwidth/Length/Undo/Width/*
> *<Endpoint of line>:* **0,0.1**
> *Arc/Close/Halfwidth/Length/Undo/Width/*
> *<Endpoint of line>:* **0.2,0.1**
> *Arc/Close/Halfwidth/Length/Undo/Width/*
> *<Endpoint of line>:* **0.1,0.2**
> *Arc/Close/Halfwidth/Length/Undo/Width/*
> *<Endpoint of line>:* **0.1,0.3**
> *Arc/Close/Halfwidth/Length/Undo/Width/*
> *<Endpoint of line>:* **0.2,0.4**
> *Arc/Close/Halfwidth/Length/Undo/Width/*
> *<Endpoint of line>:* **0.1,1.9**
> *Arc/Close/Halfwidth/Length/Undo/Width/*
> *<Endpoint of line>:* **0.2,2.1**
> *Arc/Close/Halfwidth/Length/Undo/Width/*
> *<Endpoint of line>:* **0,2.1**
> *Arc/Close/Halfwidth/Length/Undo/Width/*
> *<Endpoint of line>:* <enter>

To draw the axis of revolution (figure II-8b2), the command "line" is selected from the "draw" menu:

Command: **line**
from point: **0,-3**
to point: **0,3**
to point: <enter>

From the draw menu we invoice the command to draw a revolution surface (figure II-8b3):

Command: **revsurf**
Select path curve: <pick the profile>
Select axis of revolution: <select the line>

LAYERS

Layering is the CAD structure that separates information related to different aspects of a project. Each entity in the CAD model belongs to a layer characterized by such attributes as name, color, and line-type, which correspond to intrinsic aspects of the entity, for instance, all the steel structural columns belong to a layer called column_2x12 and are displayed in red. When we enter a new entity, it is automatically assigned to the current layer, but can be moved to a different one later. Segregation of information can be based on such variables as size, function, and material. For example, in the spiral stair model, layers are created and named according to the different physical elements of the model—wall, step, railing.

Layers are also analogous to acetate overlay sheets in that they can be turned on or off as they are needed in a specific drawing. Therefore, several different paper drawings can be output from an electronic drawing, simply by turning layers on and off. Figure II-9 shows how the database representing a floor plan can be organized in four different sets of layers to produce four different construction drawings.

The use of layers to keep information separate has obvious advantages in drafting, and also has application in three-dimensional design. Sometimes a drawing can be thought of as a combination of three-dimensional elements as well as two-dimensional entities, such as text and graphics symbols. If the designer wants to generate a

Layers create different construction documents from the same floor plan (Courtesy of Trimbach Interior Design, Inc.).

a. Construction plan.

b. Reflected ceiling plan.

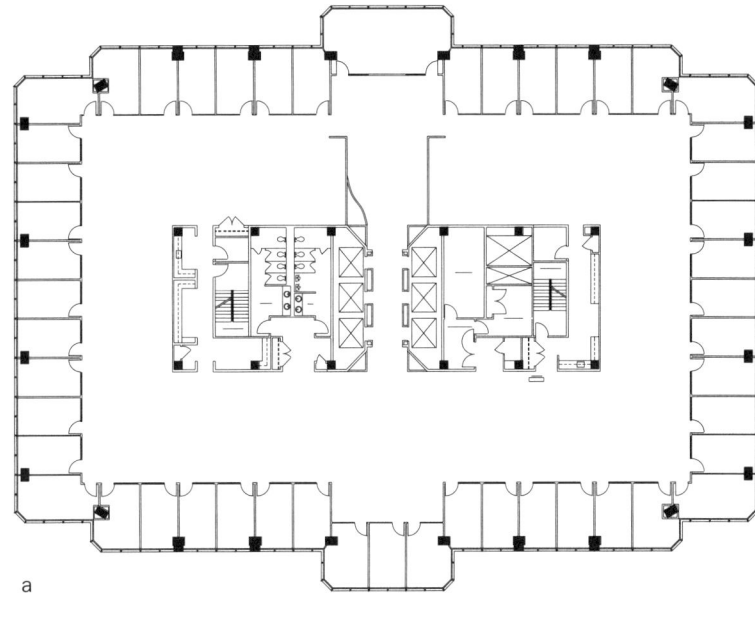

a

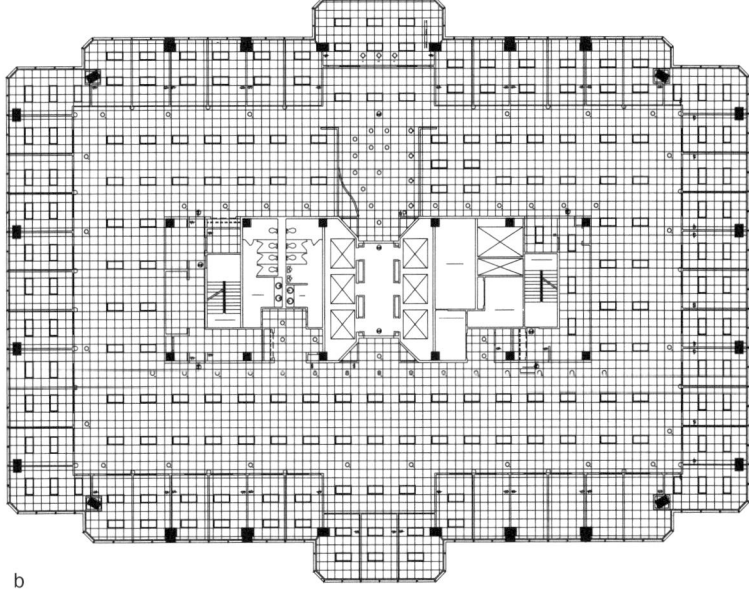

b

Layers create different construction
documents from the same floor plan
(Courtesy of Trimbach Interior
Design, Inc.).

c. Furniture plan.

d. Power and communication plan.

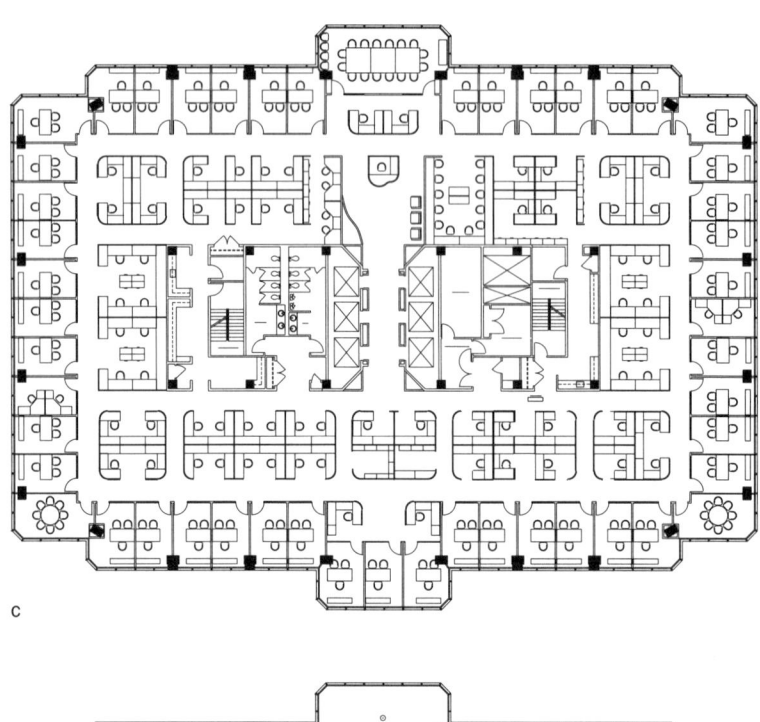

c

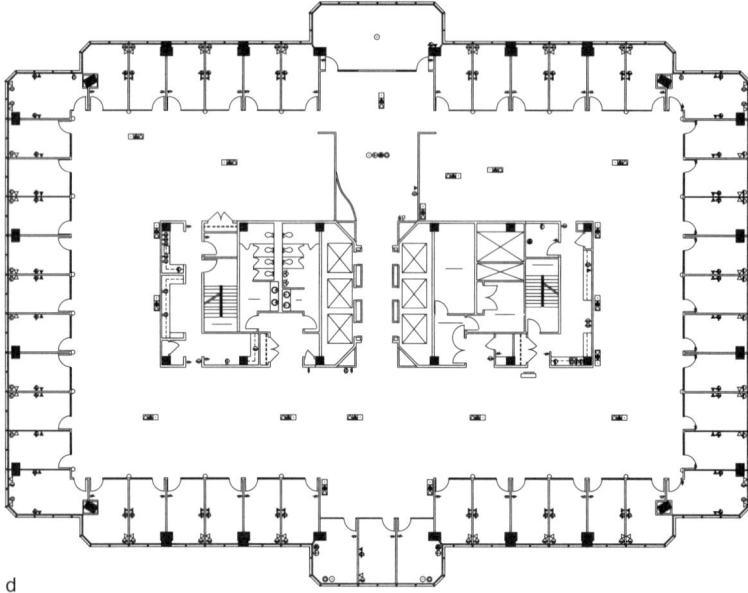

d

three-dimensional model of a building and at the same time use that information for drafting construction documents, separate layers are required for the elements belonging to the model and those needed for the elaboration of working drawings.

In AutoCAD the default layer is 0, and unless the layer command is invoked, all entities are by default assigned to layer 0. The creation of new layers is quite straightforward:

Command: **layer**

?/Make/Set/New/ON/OFF/Color/Ltype/Freeze/
  Thaw/LOck/Unlock: **make**

New current layer <0>: **railing**

?/Make/Set/New/ON/OFF/Color/Ltype/Freeze/
  Thaw/LOck/Unlock: **color Color: 1**

Layer name(s) for color 1 (red) <railing>: <enter>

?/Make/Set/New/ON/OFF/Color/Ltype/Freeze/
  Thaw/LOck/Unlock: **make**

New current layer <railing>: **step**

?/Make/Set/New/ON/OFF/Color/Ltype/Freeze/
  Thaw/LOck/Unlock: **color Color: 2**

Layer name(s) for color 2 (yellow) <step>: <enter>

?/Make/Set/New/ON/OFF/Color/Ltype/Freeze/
  Thaw/LOck/Unlock: <enter>

To change existing elements to a different layer:

Command: **change**

Select objects: <pick the baluster>

Select objects: <enter>

Properties/<Change point>: **properties**

Change what property(Color/Elev/Layer/Ltype/
  Thickness)?: **layer**

New Layer: **railing**

Change what property(Color/Elev/Layer/Ltype/
  Thickness)?: <enter>

EDITING

Editing is one of the features that most distinguishes CAD drafting from manual drafting. While the only editing tool in manual drafting is the eraser, in CAD there are several possible ways to transform a primitive or a set of primitives. The editing features allow such changes as stretching of graphic entities (figure II-8c) and more complex array replications (figure II-8d) of elements in the drawing.

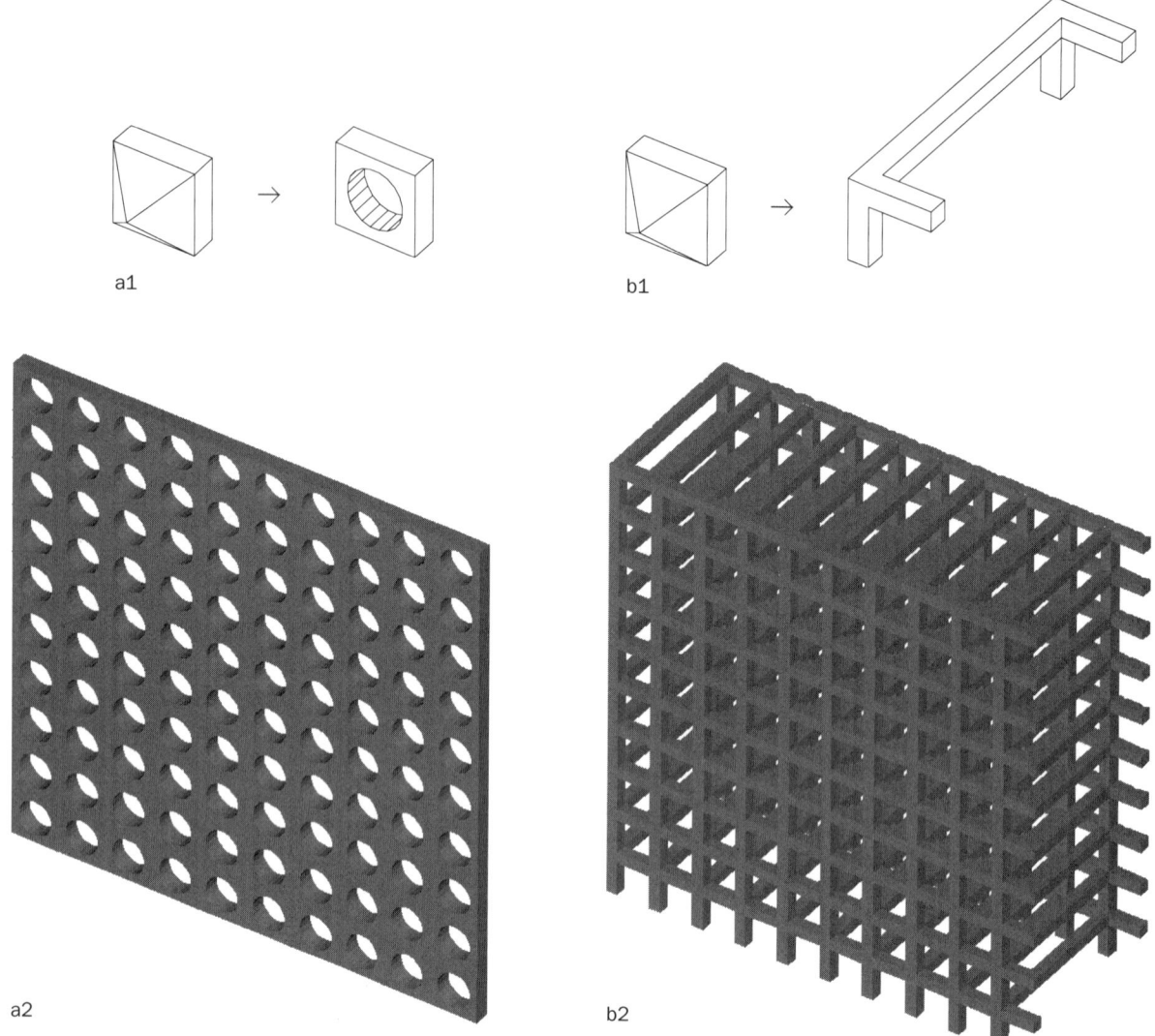

a1

b1

a2

b2

VISUALIZING WITH CAD

We will use editing commands to generate the model of a wall, made of diamond shaped stones. The Auto-CAD editing commands are subdivided in two different menus—**construct** and **modify**

To draw a stone, as shown in figure II-8c:

> *Command:* **3dface**
> *First point:* **-0.75,-0.75**
> *Second point:* **0.75,-0.75**
> *Third point:* **0,0.75**
> *Fourth point:* <enter>
> *Third point:* <enter>

> *Command:* **array**
> *Select objects:* **L**
> *Select objects:* <return>
> *Rectangular or Polar array (R/P):* **p**
> *Center point of array:* **0,0**
> *Number of items:* **4**
> *Angle to fill (+=ccw, -=ccw) <360>:* **360**

> *Command:* **stretch**
> *Select objects to stretch by window or polygon.*

> *Select objects:* **c** <draw a window to include the common vertices of the four faces>
> *Base point or displacement:* **0,0,0**
> *Second point of displacement:* **0,0,0.5**
> *Base point or displacement:* **0,0,0**

> *Command:* **rotate3d**
> *Select objects:* <pick the four faces>
> *Select objects:* <enter>
> *Axis by Entity/Last/View/X-axis/Y-axis/Z-axis/ <2 points>:* **0,0**
> *2nd point on axis:* **5,0**
> *<Rotation angle>/Reference angle:* **90**

> *Command:* **array**
> *Select objects:* **L**
> *Select objects:* <return>
> *Rectangular or Polar array (R/P):* **r**
> *Number of rows <1>:* <enter>
> *Number of columns <1>:* **10**
> *Unit cell or distance between columns:* **1.5**

Symbols.

SYMBOLS

Symbols are one of the most amazing features of CAD. A symbol (figure II-8d1) is a set of elements of any nature (points, lines, faces, text, or other symbols), grouped together with a name given by the user. After a symbol is defined, it can be inserted repeatedly in different positions. The content of the symbol can be replaced with a different set of graphic primitives (figure II-10a1,b1); all the symbols in the data structure bearing the same name as the transformed one will be automatically updated with the new information. For those familiar with programming languages, symbols may be thought of as similar to variables. A certain procedure, having variables as its argument, can be repeated in an unlimited number of contexts defined by different variables, generating completely different results. In a similar way, completely different forms (figure II-10a2,b2) can be generated according to the content of the symbols and their spatial position.

Symbols will be discussed again in Chapter III from a more general perspective, since they represent the fundamental CAD data structure which allows the evolution of a model.

The four faces making the geometric model of a diamond stone can be grouped into a symbol. The definition and insertion of a symbol is comprised of the following steps:

Command: **block**
Block name (or ?): **diamond**
Insertion point: **0,0,0**
Select objects: <pick the four "3dfaces" >
Select objects: <enter>

Command: **insert**
Block name (or ?): **diamond**
Insertion point: **0,0,0**
X scale factor <1>/Corner/XYZ: <enter>
Y scale factor (default=X): <enter>
Rotation angle <O>: <enter>

We can in a similar fashion create symbols for the step and for the baluster.

HIERARCHICAL DATA STRUCTURE

Symbols represent one example of a feature often used in CAD: the hierarchical organization of a data structure. A data structure in CAD applications contains information about the coordinates and nature of the elements of a form. Any type of form can be thought of as organized according to such a hierarchy (figure II-11). For example, a building consists of different groups of elements such as floor, exterior and interior walls, windows, stairs, reflected ceiling, structure, mechanical features, furniture, and so on.

Grouping elements by means of symbols makes the hierarchical structure clearly recognizable and the number of nested symbols can identify the hierarchical level. It is important for purposes of data

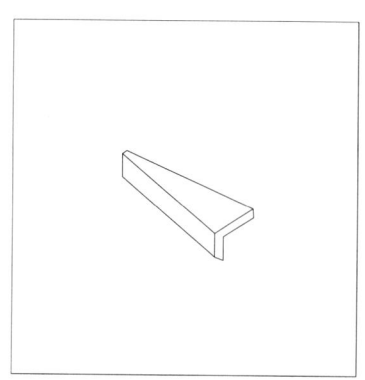

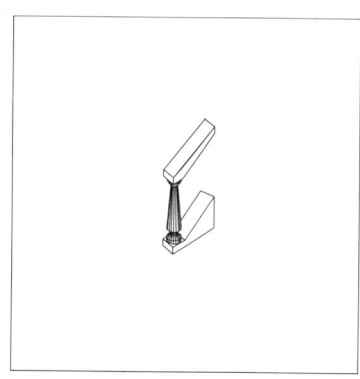

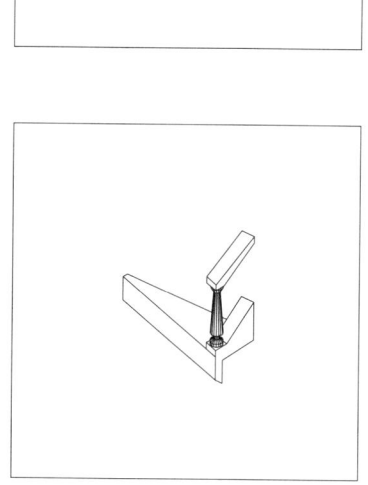

II-11

Model of a spiral stair:
hierarchy of elements.

organization to define the hierarchical level of an element and its relation to other elements.

NAMES

As we have already seen in the AutoCAD implementations, entities such as symbols and layers must be named by the user, and the names assigned to these graphics entities are extremely important. Names— the means by which we identify things in language—can give us a sense of the type of element, its hierarchical level, whether it is two-dimensional or three-dimensional, and the type of view.

For each project, names should be assigned in a consistent way. Names can be universal, but each project has its own organization. It

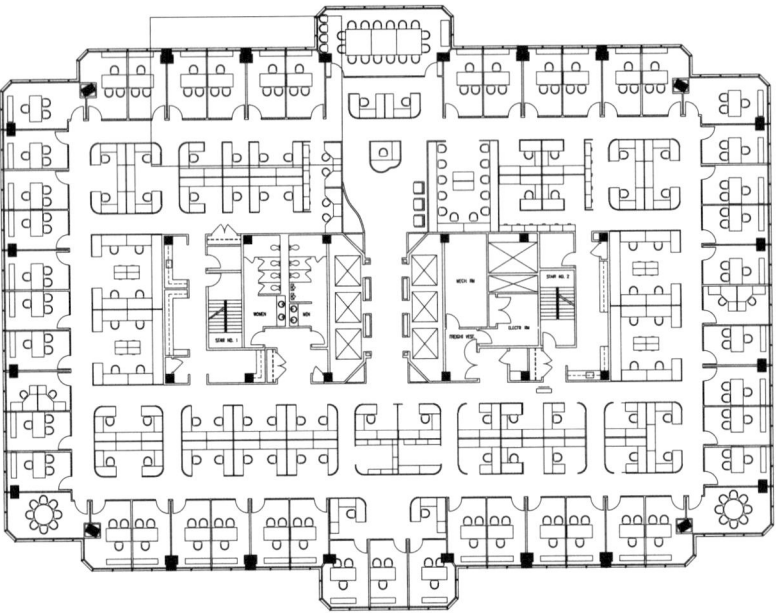

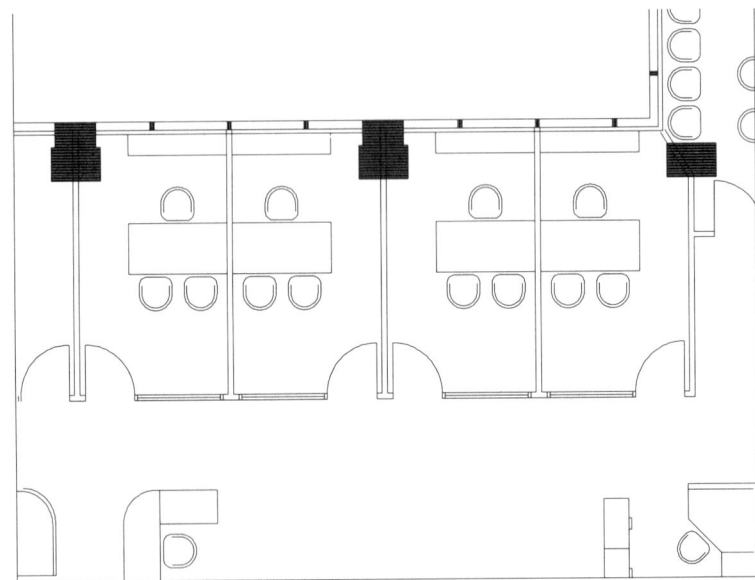

Viewing in a two-dimensional world: zoom
(Courtesy of Trimbach Interior
Design, Inc.).

VISUALIZING WITH CAD

is important to give a suffix indicating the level of information. For instance, elements defining a column should belong to the layer *column*, and if there are different types of columns in the same model, the name of the layer should contain more information about the characteristics of the column (*e.g.*, column_12x18_steel).

VIEWING AS PROJECTION

Viewing in CAD means the establishment of a correspondence, or mapping, between the graphic elements, stored as an array of numbers in the computer memory, and the grid of pixels of a raster display screen.

There is a major difference between viewing in a two-dimensional world and viewing in a three-dimensional world. Representations of a two-dimensional world are just transformations of the coordinates of the constructed form into the points of a viewport on the screen.

The command mostly used for viewing in a two-dimensional world is:

Command: **zoom**
All/Center/Dynamic/Extents/Left/Previous?Vmax/Windows/<Scale(X/XP): <select the desired view>

To view a three-dimensional form on the screen (or on a sheet of paper), it must be projected onto a two-dimensional plane by a series of operations from projective geometry. The three-dimensional coordinates of the points defining the model must be transformed into two-dimensional coordinates, which is achieved through one of two types of projections—**parallel** or **central**. Both are used for graphics representations. A projection maps a three-dimensional configuration into a two-dimensional configuration (image) in the picture plane. In a parallel projection, the projection lines connecting each point of the three-dimensional configuration to its image are all parallel. Parallel projections can be classified according to the angle that the projection lines make with the picture plane: If these lines are perpendicular to the picture plane, the projection is orthographic; otherwise it is oblique. If the projection lines are not parallel, but instead radiate from a single point (called the *center of projection*), this is known as a central projection, or perspective.

Orthographic Projections

Since orthographic projection lines are perpendicular to the picture plane, if the picture plane is itself perpendicular to one of the three coordinate axes, any line of the initial three-dimensional configuration that is parallel to that axis (*i.e.*, perpendicular to the picture plane) is projected as a point, and the plane surfaces perpendicular to the picture plane are projected as lines. For elements parallel to the picture plane, measurements and other geometric relations are perfectly preserved. Orthographic parallel projections are the most frequently used in architectural design. They are familiar to us as **plan** and **elevation** (figure II-13a,b). A **section** view represents what would be seen if the picture plane intersected the three-dimensional object (figure II-13c). A plan also can be thought of as a section, where the picture plane is parallel to the *x-y* plane and cuts through the model. The different orthogonal views are classified as two-dimensional projections parallel, respectively, to the *x-y*, *x-z*, or *y-z* planes. More specifically, the plan is the view parallel to the *x-y* (horizontal) plane, while section and elevation views are parallel to the vertical planes. Orthographic representations are diagrammatic images, since they contain a description of a physical form, often with references to dimensions and other quantitative information, but do not offer any visual simulation of the real form.

While the nomenclature used in the construction of a data structure is very similar in all CAD software, the viewing procedures can be very different; the following AutoCAD implementations may not be completely transferable to other software.

II-13

Model of a spiral stair.

a. Plan. b. Elevation.

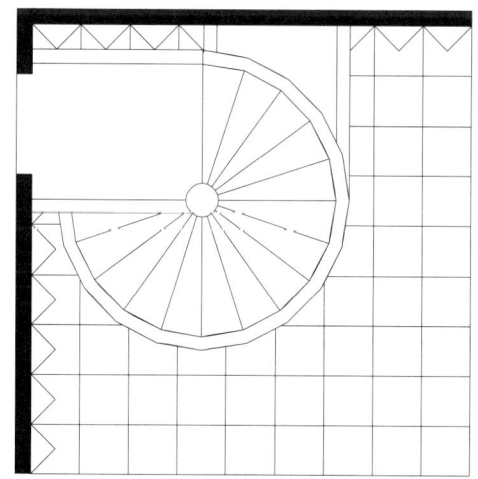

a

b

Model of a spiral stair.

c. Section.

In AutoCAD, the default view—the view in which elements are visualized as soon as drawing is started—is the plan view. If we want to invoke the plan view from a different type of view:

Command: **plan**
<Current UCS>/UCS/World:

A more general command for any type of view obtained from orthographic projections:

Command: **vpoint**
Rotate/<View point><1.0000,1.0000,1.0000>:
    0,0,1

The elevation view parallel to the *x-z* plane is obtained in this way:

Command: **vpoint**
Rotate/<View point><1.0000,1.0000,1.0000>:
    0,1,0

For the elevation view parallel to the *y-z* plane:

Command: **vpoint**
Rotate/<View point><1.0000,1.0000,1.0000>:
    1,0,0

Axonometric Projections

An axonometric projection is a type of orthographic projection in which the picture plane is perpendicular to the direction of projection but not to the direction of any of the three coordinate axes. In axonometric representations (figure II-13d), none of the angles are preserved and lines parallel to any of the axes are proportional in length to their projected images. In **isometric** projections, the lines of projection form equal angles with each of the three axes; measurements along lines parallel to any of the axes have the same proportionality constant.

To obtain an isometric view:

Command: **vpoint**
Rotate/<View point><1.0000,1.0000,1.0000>: **1,1,1**

Oblique Projections

In oblique projections, the direction of projection is not perpendicular to the picture plane (which is itself taken to be perpendicular to one of the three axes). The result is that the elements parallel to the

Model of a spiral stair.

d. Axonometric.

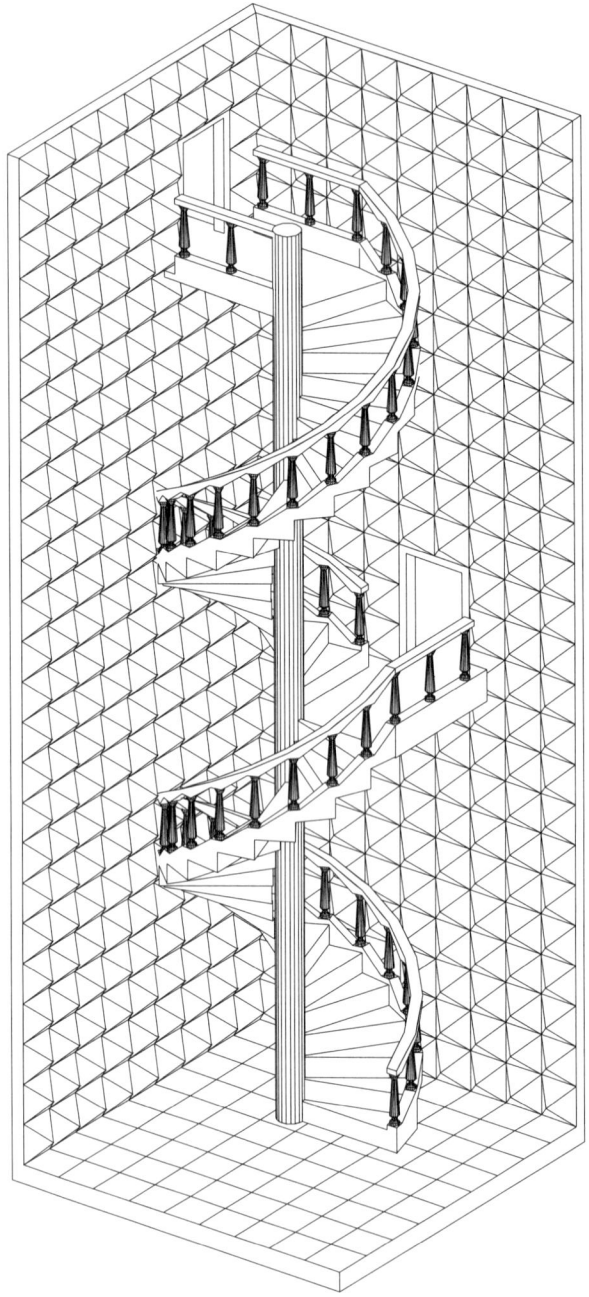

picture plane conserve their metric and angular characteristics, while the others preserve only certain geometric attributes, such as parallelism of lines and proportions, such that they can still be measured. The measures of the projections of the nonparallel elements will not be the same as the measures in the three-dimensional configuration, but will be proportional to them.

Perspective

Perspective (figure II-13e,f) preserves metric and other geometric characteristics, such as angles and parallelism, only for the elements lying in a plane parallel to the picture plane. All parallel lines not parallel to the picture plane converge to a point, and all parallel planes not parallel to the picture plane converge to a line. The converging point is called the **vanishing point** and the converging line for all *horizontal* planes is defined as the **horizon**. The center of projection is the **point of view**. Perspective projection represents the geometry of our visual perception, where the eye is the center of projection and the direction in which we look is perpendicular to the picture plane. For this reason, perspective is the most realistic type of representation. Perspective offers a visual simulation, a preview of the experience of looking at the designed three-dimensional forms once they are built. This allows a qualitative approach to design based on spatial characteristics.

The user can define several different parameters to obtain a perspective view. An analogy can be established between the command and the use of the optical components of a camera, such as lens, target point, eye point. This is shown also in the nomenclature of the parameters of the command dview (dynamic view), which is invoked to generate a perspective.

The most accurate way to generate a one-point perspective is to input the coordinates of the viewpoint and target point:

Command: **dview**
Select objects: <pick the desired entities>
Camera/Target/Distance/POints/PAn/Zoom/
   Twist/Clip/Hide/Off/Undo/eXit: **PO**

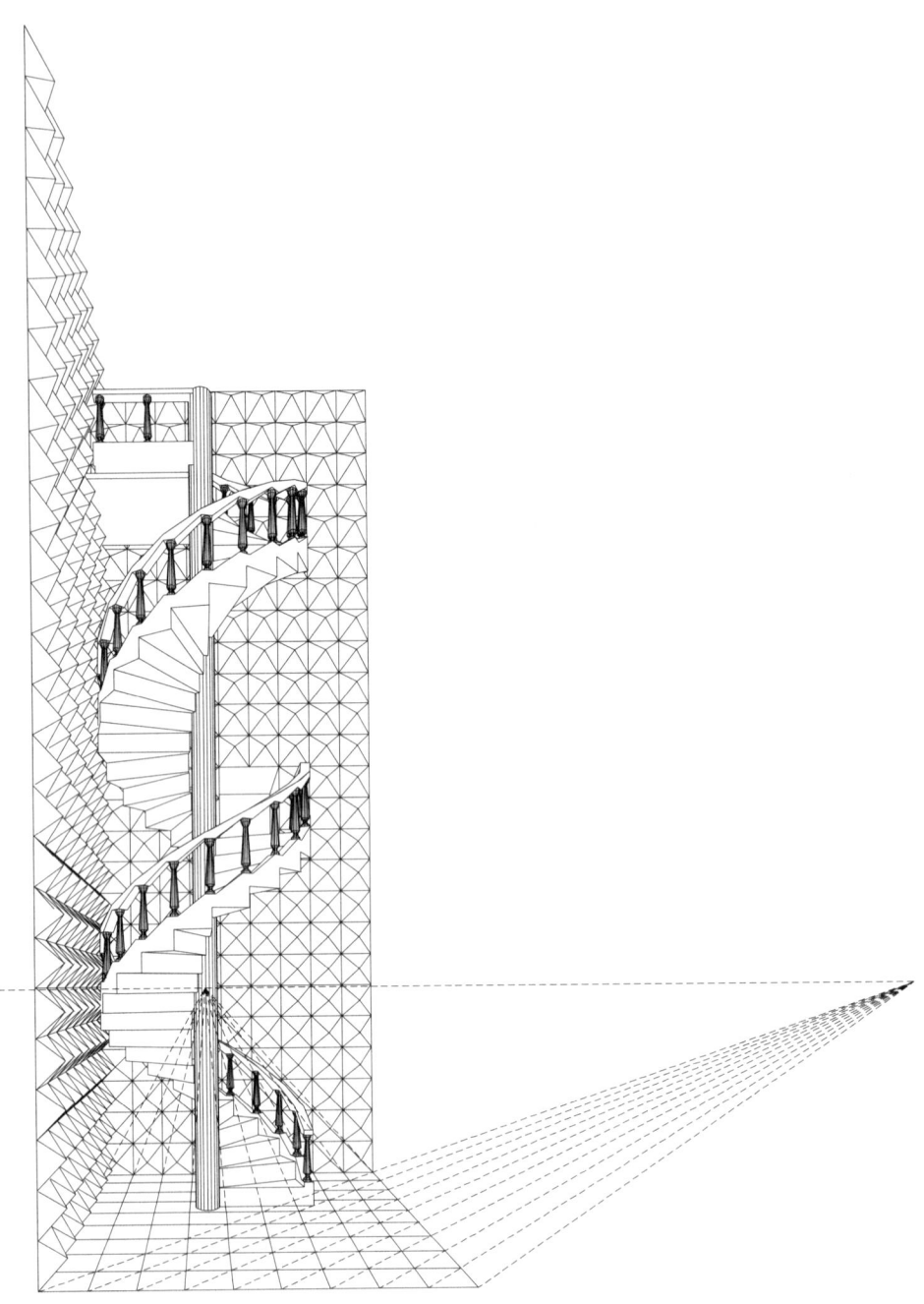

VISUALIZING WITH CAD

Model of a spiral stair.

e. Perspective (hidden line).

f. Perspective (rendering).

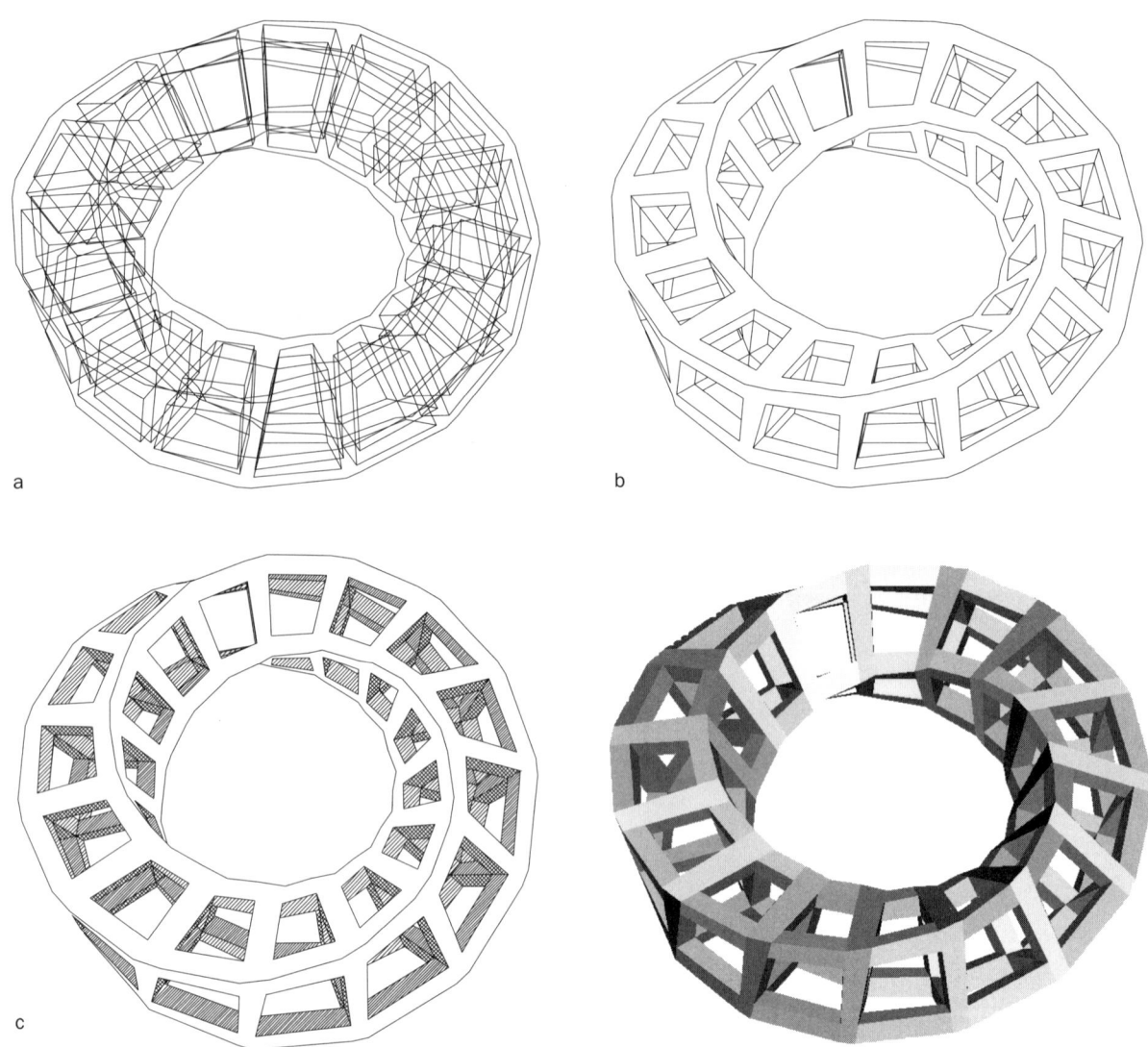

a

b

c

VISUALIZING WITH CAD

RENDERINGS

Renderings are manipulations of the views obtained as geometric projections of a constructed computer model. They do not add to the geometrical definition of a given form, but they can offer a qualitative type of information.

◄ **II-14**

Renderings.

a. Wireframe.   b. Hidden line.

c. Pen and ink.   d. Shaded.

Renderings can be classified in the same way as the different types of three-dimensional computer graphics systems. The most basic type of representations are **wireframe** (figure II-14a) and **hidden line** (figure II-14b). Hidden line renderings afford the clearest visual simulation of an object's spatial form and boundaries. In a hidden line drawing, hatching can be added to emphasize the different planes to which the surfaces belong and to simulate shading. This type of rendering is often referred to as pen and ink (figure II-14c).

The wireframe representation is the default. To obtain a hidden line representation:

Command: **hide**

Shaded models (figure II-14d) further suggest the perceptual characteristics of an object, according to the attributes of its surfaces and one or more light sources. Figures II-15a-d show the same geometric model (a sphere) with different surface chacteristics; in figure II-15a, the sphere is made of reflective surfaces, while in figure II-15b, the surfaces are transparent. Other surface characteristics are bumps (figure II-15c) or texture mapping (figure II-15d).

Shaded models and different type of surfaces.

a. Reflective surfaces.

b. Transparent surfaces.

a

b

Shaded models and different
type of surfaces.

c. Bumps.   d. Texture mapping.

c

d

Starting with Release 12, AutoCAD has a rendering feature. To render a scene, just invoke the command:

Command: **render**

From the **render** menu more commands can be selected to change the other parameters, such as **light**, **view**, **finishes**, **shading**. Photorealistic effects can be achieved from AutoCAD® using AutoVision™.

A surface model constructed using AutoCAD® can be translated into a format that can be used by several third party rendering programs. The model of a sphere shown in figure II-15 has been rendered using Autodesk 3D Studio; this is not just a rendering program, but a modeling and animation software as well. An AutoCAD model can be imported into 3D Studio as a dxf file.

Command: *dxfout*
(enter filename)

Enter decimal places of accuracy
(0 to 16)/Entities/Binary <6>: <enter>

CUSTOMIZING

Macros provide a powerful tool to customize the use of a generic CAD software for specific purposes. Macros range in complexity from a simple sequence of software commands for executing a given task to more complex routines constructed according to the rules of a programming language. Macros become useful when there is a recurrent procedure to be executed with varying parameters.

Some of the models implemented in the next chapters are accompanied by a macro (AutoLISP routine) contained in the enclosed diskette (see Appendix).

In AutoCAD, the most efficient way to create macros is to use AutoLISP. AutoLISP is a derivation of the LISP programming language. A discussion of AutoLISP is too specific for this context. The interested reader can refer to the bibliography.

BIBLIOGRAPHY

AutoCAD® Release 12 - Advanced Modeling Extension®
Release 2.1 Reference

AutoCAD® Release 12 - AutoLISP Reference

AutoCAD® Release 12 - Command Reference

Ching, Francis, *Architectural Graphics*, Van Nostrand Reinhold,
New York, 1985

Foley, James D., *Computer Graphics Principles and Practice*,
Addison-Wesley, Reading, 1992

Jacobs, Stephen Paul, *The CAD Design Studio*, McGraw-Hill, 1991

Kalay, Yehuda E., *Computability of Design*, John Wiley & Sons,
New York, 1986

McCullough, M., W. J. Mitchell, and P. Purcell (eds.), *The Electronic
Design Studio*, MIT Press, Cambridge, 1991

Mitchell, William J., *Digital Design Media*, Van Nostrand Reinhold,
New York, 1991

III

Ubi materia, ibi geometria

Johannes Kepler

This chapter investigates the creation of CAD models of three-dimensional forms according to the foundations established in Chapter I. The syntax used in generating models of architectural forms is the same as that used in creating geometrical forms, especially in computer-aided design, where any architectural configuration is defined initially by its geometric interpretation. Forms in architecture and in geometry clearly have two different semantic contents. As already emphasized, while geometric forms exist in "intellectual space" and are subject only to logical relations, architectural forms deal with the constraints of the physical world as well as functional requirements. Nevertheless, at the syntactic level, architectural and geometric forms can both be investigated in terms of their descriptions as sets of points, lines, and surfaces in three-dimensional space, and how they determine a series of perceptual dualistic relations, particularly those of solid-void and inside-outside.

FORM AS A SET OF ATOMS AND RELATIONS IN THREE-DIMENSIONAL SPACE

A form can be considered an aggregate of primitive elements, or atoms (figure III-1). Atoms (the word, as used here, does not have the same meaning as in physics) cannot be decomposed into simpler elements; they are the simplest elements that have geometric meaning. The configuration, or geometric relation, of several atoms identitfies them as a set, and differentiates each such set of atoms. Atom set A can be changed into atom set B through geometric operations called **transformations**. A series of transformations also determines the complexity of a form.

A form can also consist of just one primitive, or can be composed of parts or subshapes more complex than primitives. By decomposition of all its subshapes, a form can always be reduced to an elemental level consisting of primitives.

As before, in the discusion that follows, a three-dimensional coordinate system will be used to define a form in terms both of its spatial

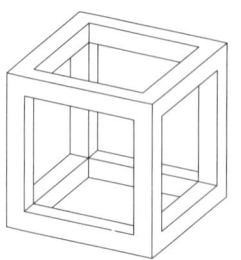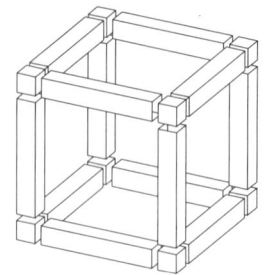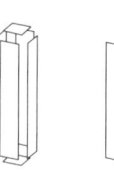

A surface-bounded form as an aggregate
of simple elements.

extension and its **position**. Some constructions may require the use
of only two coordinates, but it should emphasized that all operations
can be transferred to three-dimensional space.

PHYSICAL CHARACTERISTICS AND
DESIGN DEFINITIONS

As discussed in Chapter I, there is a difference between objects in the
physical universe and their forms in the design universe. Not all the
visual characteristics of physical shapes can be perceived in their
models. For instance, in the physical world, surfaces are characterized
not just by their spatial extension, but also by texture and color, which
are functions of the materials they are made of and the effect of light.
The geometric definition of spatial characteristics is the first necessary condition for the generation of a model in CAD. Perceptual
aspects such as solid-void and inside-outside are determined by geometric boundaries and represent the basic physical characteristics of
objects in the design universe.

Usually in CAD the geometry of forms can be associated with different colors. Some CAD and solid-modeling systems also offer
material attributes such as lighting effects (shininess, transparency,
reflectivity), texture patterns, and bumps (see Chapter II). The form,
if modeled as a set of "solids" (see *Solids or Volumes*, below) can be
assigned properties related to its mass, as well as material properties

such as density, elasticity, stress, and thermal characteristics [AutoCAD®, Advanced Modeling Extension®, 3D Studio™].

GEOMETRIC PRIMITIVES AND BOUNDARIES

The term *primitive* as used here is not to be confused with the primitives that often appear in CAD system menus as part of constructive solid geometry (which include three-dimensional forms such as cubes, spheres, pyramids, wedges, and tori). These shapes are not really primitives according to the geometric meaning of the term, which will be followed here.

Using language as an analogy to describe form generation, geometric primitives can be compared to an alphabet; they are the atoms in the universe of computer-aided design. As in Euclidean geometry, **point**, **line**, and **surface** represent the primitive elements, undefined in terms of other geometric elements and operations, but understood in an intuitive way. All further constructions will use these elements. Form, as emphasized in Chapter I, is *the intersection of regions of space delimited by boundaries*. Therefore a solid, for example, as an enclosed portion of space, can be defined by the surfaces which bound it. Each primitive must be defined in terms of the coordinates of its position in space and also by its boundaries. (Note that lines are bounded by points and surfaces are bounded by lines.)

Points

A point P (figure III-2a), because it has no extension, can be considered zero-dimensional; it is just a position in space, defined by a set of three coordinates (x, y, z). A point does not have any physical value, except as a reference.

In AutoCAD:
Command: **point**
Point: **4,3,2**

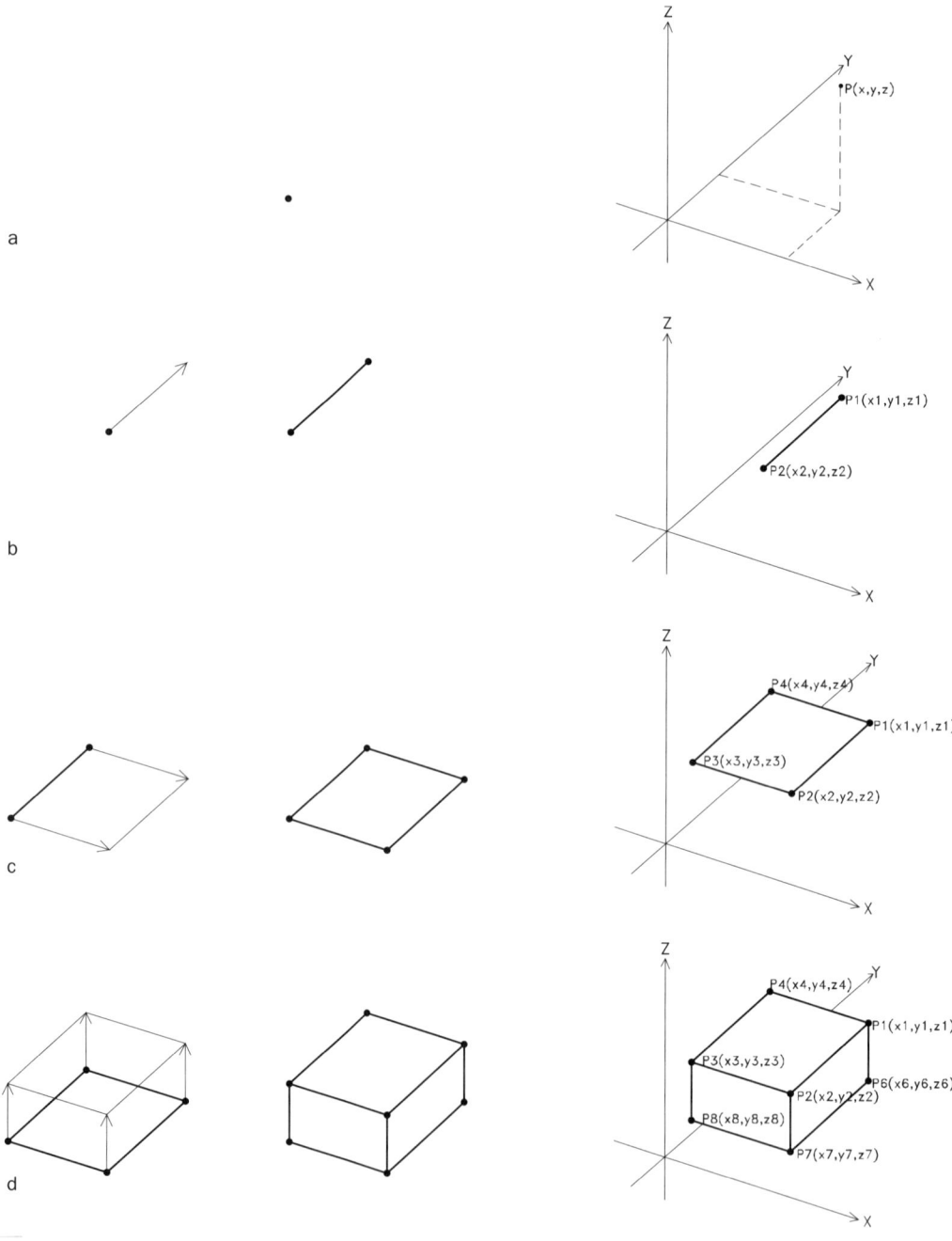

a

b

c

d

VISUALIZING WITH CAD

◄ III-2

Geometric primitives:
a. Point. b. Line. c. Surface.
d. Volume.

In other systems, although terms such as *node* or *vertex* can be used in place of the term *point*, the implementation is still the same.

Points have no boundaries—or rather a point is a boundary of itself—but they can become the boundaries for all other primitives.

Lines

A line (figure III-2b) has extension just in one dimension; it has length, but no thickness or width. It can be defined by any two points, $P_1(x_1, y_1, z_1)$ and $P_2(x_2, y_2, z_2)$, that lie on the line. While a line as a geometric entity is unlimited in its extent, in reality we deal just with line segments; this is also true in CAD, where we deal only with segments, defined by their end points. Lines can also be classified as straight or curved. Here, however, the word *line* will indicate just a segment of a straight line.

A line is bounded by two points at its extremities. A set of lines can become the boundary of a surface.

In AutoCAD a line is constructed by defining the coordinates of its endpoints. Cartesian coordinates can be combined with a polar coordinate system, where the origin is represented by the last defined endpoint:

Command: **line**
From point: **4,3,2**
To point: **9,8,5**
To point: @**9<<72**
To point: <enter>

Associated with lines are geometrical characteristics such as perpendicularity, parallelism, and bisection. CAD systems usually provide commands to create lines according to these geometrical relations.

The coordinates of the endpoints of a line (as well as the other primitives) can be defined in conjunction with other geometric constraints. These are defined in the AutoCAD **osnap** modes.

> *Command:* **line**
> *From point:* **4,3,2**
> *To point:* **per**

> *Perpendicular to:* <pick the desired entity>
> *To point:* **mid**
> *midpoint of:* <pick the desired entity>
> *To point:* <enter>

Surfaces

A surface (figure III-2c) extends in two dimensions; it has length and width but no thickness. In this context, surfaces are portions of plane surfaces, with boundaries represented by coplanar lines, and therefore by the points bounding the lines. Surfaces that do not lie in a single plane cannot be considered primitives, since they can be decomposed into planar surfaces. Surfaces are boundaries for a volume, which is an enclosed portion of space.

In AutoCAD:

> *Command:* **3dface**
> *From point:* **-5,-5,-5**
> *Second point:* **5,-5,-5**
> *Third point:* **5,5,5**
> *Fourth point:* **-5,5,5**
> *Third point:* <enter>

Surfaces are called **polygons** in several CAD systems. The polygons in the AutoCAD command should not be mistaken for regular polygons in Euclidean geometry. In AutoCAD, the command **polygon** does not generate an enclosed surface, but only a perimeter made of lines of the same length.

Solids or Volumes

The last element, the solid (figure III-2d), has different characteristics from the others and may not be considered a primitive in our discussion of the perception of surfaces. A solid is extended in three dimensions: length, width, and depth (thickness). It is an enclosed portion, or volume, of three-dimensional space, which is surfaces. Therefore, the boundary primitives defining a volume can be reduced to the lines and points that determine its surface boundaries. Since a solid may be decomposed into simpler surface boundaries, it is no longer a primitive. In the generation of three-dimensional forms we are really concerned more with the boundaries than with the enclosed volume. At this semantic level, since forms will be treated just as visual models rather than physically built objects, the solidity of forms—the matter enclosed in the surface boundaries—is of no concern. (This is consistent with the perceptual phenomena discussed in Chapter I: We visually perceive only the surfaces bounding an object.)

Although solids are not primitives in the geometric sense that we are using here, the primitives of solid modeling usually refer to geometric solids, such as parallelepipeds, cylinders, cones, spheres, and wedges. Even if we disregard this approach in the construction of dynamic models, this view of primitives can provide a hint for the design of certain models.

RELATIONS

As discussed in Chapter I, a form, considered as a whole, is defined not just by a sum of parts, but also by the relations between them. In the same way, in a CAD model, a form is composed of parts and each part is a set of primitives and relations. The basic topological relations among parts, and primitives as well, are **separation**, **touching**, and **interpenetration** (figure III-3). Two parts are separated when they have no common boundaries; they touch when they share, partially or totally, one or more boundaries; and they interpenetrate if they share an enclosure. The position of two separated or touching elements or primitives can be interpreted by geometric transformations while the interpenetration can be defined by Boolean operations.

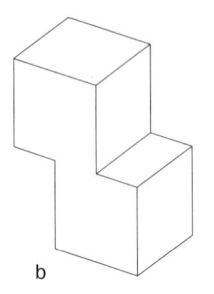

a

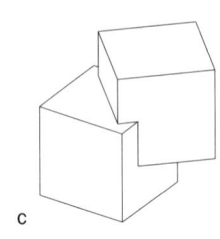

b

c

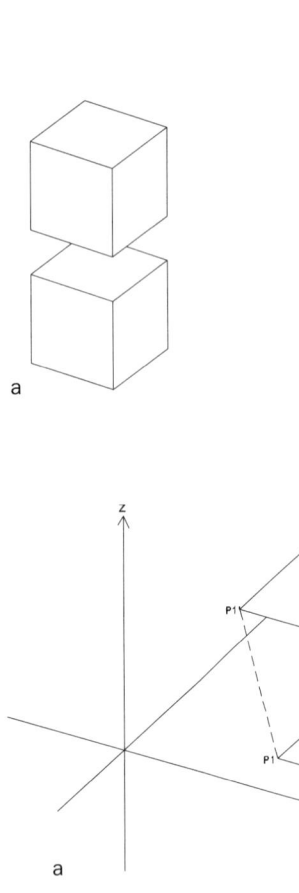

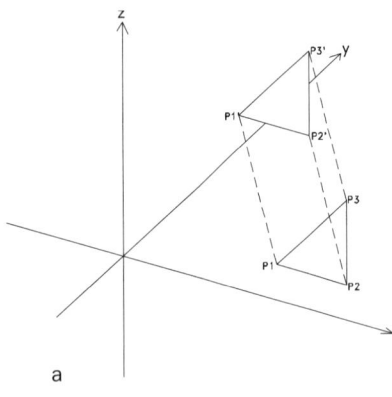

a

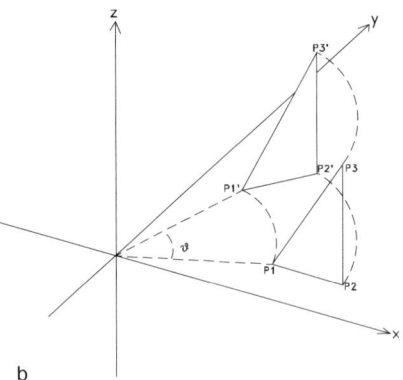

b

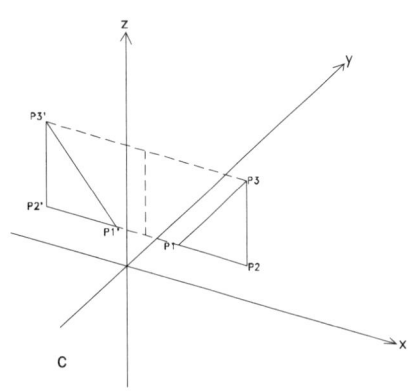

c

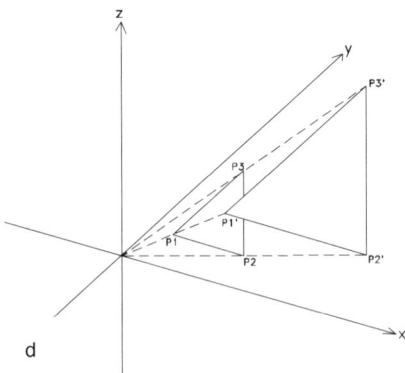

d

VISUALIZING WITH CAD

GEOMETRIC TRANSFORMATIONS

The basic topological relations.

a. Separation.   b. Touching.
c. Interpenetration.

Geometric transformations.

Orthogonal Transformation:
a. Translation.   b. Rotation.
c. Reflection.

Similarity transformation:
d. Scaling.

Geometric transformations represent one of the most common mathematical operations in computer graphics. They allow the evolution of a complex shape from a set of the primitive elements discussed above.

Transformations are a fundamental topic in modern geometry. In the Erlangen Program (the inaugural speech at the University of Erlangen, source of the modern classification of geometry), Felix Klein classified different geometries according to the group of transformations under which the properties of a given manifold (the higher-dimensional analogue of a surface) are unchanged [Coxeter 1967]. Readers interested in the developments in geometry arising from the study of transformations can refer to the extensive literature on the topic [Coxeter 1973, Yaglom 1962, Modenov 1965].

In computer-aided design, the operations of transformation and replication are often used simultaneously, so that the initial element remains in the initial position.

The following discussion of transformations will focus on the geometric meaning of these relations. The transformed elements can be primitives—points, lines, and surfaces—or more complex shapes. The designer will be surprised to discover how operations and terminology used implicitly in the graphics and design process are given rigorous definition in more specialized areas of geometry.

Orthogonal Transformations

Orthogonal transformations change the position of an initial configuration without changing its shape (or dimensions). Therefore orthogonal transformation displaces a rigid body from one position in space to another position.

The following discussion of transformations will deal mainly with points, the most basic primitive. Transformations are used to create a more complex shape from an initial set of primitives, but they can also be applied in the definition of primitives themselves.

In fact, transformations of an initial set of one or more points generate the boundaries for higher-dimensional primitives such as lines and surfaces.

Translation

A translation (figure III-4a) is a displacement of a configuration of points to a new position, where for each point P of the initial configuration there is a correspondent P'. The distance PP' is always the same for each point of the configuration. We can substitute lines and faces for points while the relation remains the same. In terms of coordinates, for each point $P(x, y, z)$ the translated point $P'(x', y', z')$ is given by:

$$x' = x + a$$

$$y' = y + b$$

$$z' = z + c$$

The AutoCAD command is:

Command: **move**
Select objects: <pick the desired entities>
Base point or displacement: **5,10,15**
Second point or displacement: **@21<30**

The replication and translation of an element is given by just one command:

Command: **copy**
Select objects: <pick the desired entities>
Base point or displacement: **5,10,15**
Second point or displacement: **@21<30**

ROTATION

The counterclockwise rotation of a configuration (figure III-4b) through an angle θ about the z-axis takes each point $P(x, y, z)$ of the configuration into a new point $P'(x', y', z')$. The new position of $P'(x', y', z')$ is calculated as:

$$x' = x \cos \theta - y \sin \theta$$

$$y' = x \sin \theta + y \cos \theta$$

$$z' = z$$

Rotations about the other axes are described in the same way; about the x-axis, for instance, x is fixed and the values of y and z allowed to change as functions of θ as above.

In AutoCAD the implementation of rotations is different from the other transformations, which are automatically executed in three-dimensional space. Rotation instead happens implicitly in the x-y plane, about an axis parallel to the z-axis:

> *Command:* **rotate**
> *Select objects:* <pick the desired entities>
> *Select objects:* <enter>
> *Base point:* **4,6,9**
> *<Rotation angle>/Reference:* **-36**

The implementation of a rotation about an arbitrary axis in space can be obtained using a specific command. To rotate entities about an axis parallel to the y-axis:

> *Command:* **rotate3d**
> *Select objects:* <pick the desired entities>
> *Select objects:* <enter>
> *Axis by Entity/Last/View/Xaxis/yaxis/zaxis/*
>   *<2points>:* **15,0,0**
> *Second point on axis:* **15,5,0**
> *<Rotation angle>/Reference:* **60**

The replication and rotation of an element can be obtained by using a single command. In the x-y plane:

> *Command:* **array**
> *Select objects:* <pick the desired entities>
> *Select objects:* <enter>
> *Rectangular or Polar array (R/P):* **p**
> *Center point of array:* **5,2**
> *Number of items:* **5**
> *Angle to fill (+=ccw, -=cw) <360>:* **75**

For an array replication in three-dimensional space:

> *Command:* **3darray**
> *Select objects:* <pick the desired entities>
> *Select objects:* <enter>
> *Rectangular or Polar array (R/P):* **p**
> *Center point of array:* **5,2**
> *Number of items:* **5**
> *Angle to fill (+=ccw, -=cw) <360>:* **75**

REFLECTION

Reflection (figure III-4c) is a transformation about a line called the **mirror**. In a given configuration, the reflection of each point $P(x, y, z)$ not belonging to the mirror is the point $P'(x', y', z')$ located on the line perpendicular to the mirror such that PP' is bisected by the mirror. If we assume the y-axis to be the mirror, $P'(x', y', z')$ is determined as:

$$x' = -x$$

$$y' = y$$

$$z' = z$$

For a mirror defined by any given line in space, we can use the same relations, changing only the coordinate system in such a way that the mirror is identified with the y-axis.

In AutoCAD a reflection transformation in the x-y plane can be implemented as follows:

> Command: **mirror**
> Select objects: <pick the desired entities>
> Select objects: <enter>
> First point of mirror line: **5,0**
> Second point: @**5<30**
> Delete old objects? <N>: <enter>

If the mirror lies in an arbitrarily defined plane (a rotation of the x-y plane about the y-axis in the following example):

Command: **mirror3d**
Select objects: <pick the desired entities>
Select objects: <enter>
Plane by Entity/Last?zaxis?View/XY/YZ/ZX/
   <3points>: <enter>
First point on plane: **0,0,0**
Second point on plane: **0,5,0**
Third point on plane: **10,10,5**
Delete old objects? <N>: <enter>

Reflection transformations create **symmetry**, which is one of the most important relations in architecture as well as in geometric and organic forms. There are several types of symmetry relations; the simplest is bilateral symmetry, which is present even in the human body, the left side of which is the mirror image of the right. Rotations also can be interpreted as a symmetric movement. In fact, a series of successive reflections gives the same configuration that would be obtained by a rotation about the axis passing through the point of intersection of all the mirrors.

Similarity Transformation: Scaling

In computer graphics, scaling (figure III-4d) is treated similarly to translation and rotation, while in geometry it is considered a different type of transformation. Recall that translation, rotation, and reflection, all defined as orthogonal transformations, leave both the shape and the size of the initial configuration unchanged. Scaling, by contrast, does not change the object's shape but instead transforms its size. In geometry this is referred to as a similarity transformation.

Each point $P(x, y, z)$ of a given configuration can be scaled by a scaling factor s about the origin. The scaled point $P'(x', y', z')$ can be calculated as:

$$x' = xs$$

$$y' = ys$$

$$z' = zs$$

It is also possible to scale an element by different scaling factors for each axis:

$$x' = xs_x$$

$$y' = ys_y$$

$$z' = zs_z$$

In AutoCAD, only a global scaling about all the three axes is possible:

   Command: **scale**
   Select objects: <pick the desired entities>
   Select objects: <enter>
   Base point: **5,2,8**
   <Scale factor>/Reference: **1.25**

The implementation of a scaling transformation about only one of three axes is not immediate, since there is no command which directly performs it. We have first to create a block of the elements we want to scale by different factors, then insert the block. Thus, we will be prompted by the computer to supply the scale factors about x, y, and z.

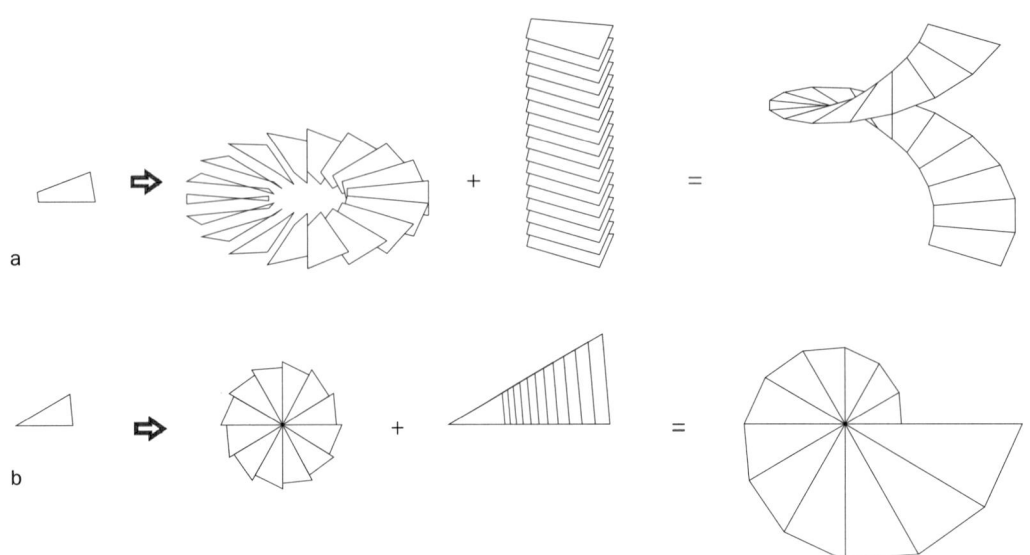

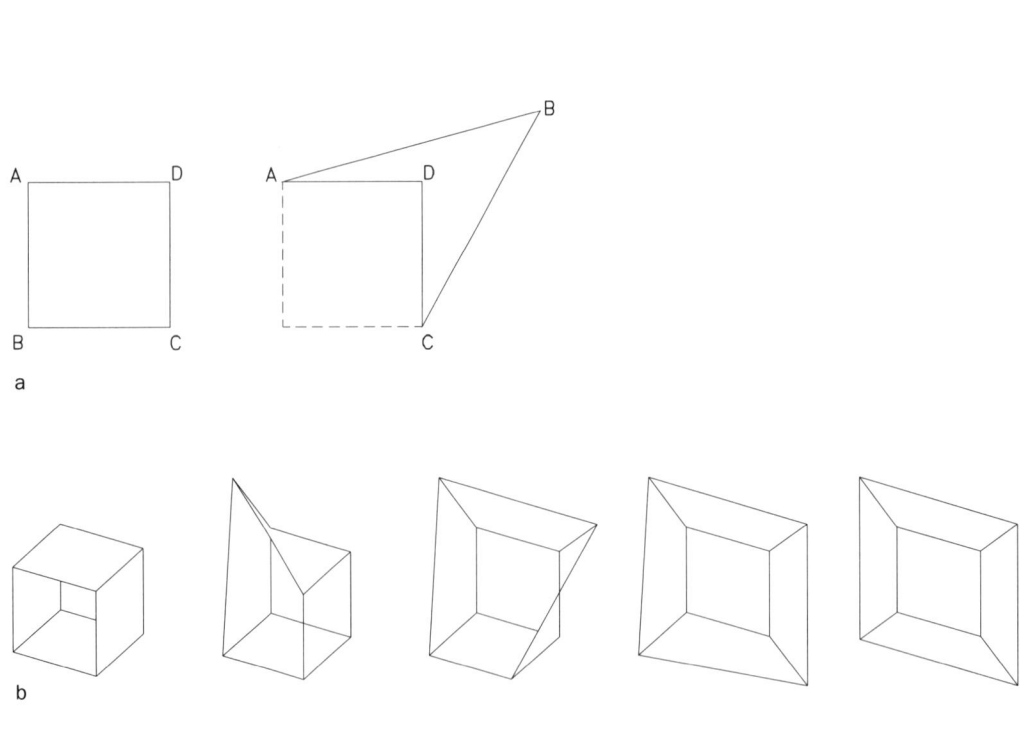

VISUALIZING WITH CAD

Combined transformations.

a. Twist.   b. Spiral.

Transformations beyond metric
properties: stretching.

a. In two-dimensions.

b. In three-dimensions.

COMBINED TRANSFORMATIONS

The transformations described above represent the possible simple movements in a plane or in three-dimensional space. Interesting results are obtained by combining two or more simple transformations into a complex movement.

Twist

A twist is generated by the product of a **rotation** and a **translation** along the axis of rotation (figure III-5a). A point moving in three-dimensional space according to a continuous twist traces a curve called a **helix**, which in plan projection appears as a circle. The helix will be discussed in the paragraph on helicoids (Chapter IV).

Equiangular Spiral

The combination of **rotation** with **scaling** generates relations among elements similar to those represented by an equiangular spiral, a curve traced by a point moving continuously according to this combination of movements (figure III-5b).

To create a spiral made of discrete elements, the following commands are repeated the desired number of times, using the most recently generated elements in each new iteration:

Command: **array**
Select objects: <pick the desired entities>
Select objects: <enter>
Rectangular or Polar array (R/P): **p**
Center point of array: **5,2**
Number of items: **2**
Angle to fill (+=ccw, -=cw) <360>: **15**

Command: **scale**
Select objects: <pick the desired entities>
Select objects: <enter>
Base point: **5,2**
<Scale factor>/Reference: **2.5**

TRANSFORMATIONS BEYOND METRIC PROPERTIES: STRETCHING

Stretching is a transformation of a completely different type from the others thus far described. Translation, rotation, and reflection conserve both the metric and angular properties of a shape; scaling, though it does not preserve metric relations, nevertheless maintains a morphological similarity, by preserving the angles between elements. Stretching—sometimes called one-way stretching—instead changes the position only of some points of a configuration, leaving the others fixed (figure III-6a). The result is that, if the original set of points defines lines or surfaces, these elements become completely distorted under the transformation, losing their original metric and angular properties. In a certain way, the stretching can be considered a topological transformation, in which the configuration to be transformed takes on rubber-like qualities.

Stretching also allows some operations of projective geometry, such as reducing an *n*-dimensional configuration of points to one of *n*-1 dimensions. The example of figure III-6b shows how the three-dimensional configuration of faces defining a cube (with one face missing) can be reduced to a two-dimensional configuration through a series of consecutive stretching transformations.

The AutoCAD implementation is quite simple:

Command: **stretch**
Select objects to be stretched by window:
   <pick the desired entities using a window>

Select objects: <enter>
Base Point: **5,8,12**
New point: **@5<24**

POINTS AND CONNECTIVITY

The definitions of geometric primitives and transformations need to be clarified to be complete and consistent with a more rigorous geometric interpretation. Recall that there are four types of primitives—

point, line, solid, and surface (and for our purposes we exclude solids)—that are the most elemental geometric constituents of form. Now that the concept and applications of transformations have been introduced, a further distinction must be made between points and the other primitives. Although they are all regarded as indecomposable elements of a spatial configuration—that is, lines and surfaces cannot be decomposed into points—points are nevertheless simpler than the other elements in the sense that a configuration of points can determine lines and surfaces. Still, even though lines and surfaces (that is, their boundaries) can be defined in terms of points, they must be considered primitives because of their unique geometric and perceptual characteristics.

If the primitive elements are considered not just in terms of their geometric definitions, but in regard to how they can generate perceivable forms, surfaces are the only elements that can characterize a form in physical three-dimensional space. Points can mark a location in space, or position, but are not perceivable; lines may enclose a portion of a plane, but they can not enclose a three-dimensional space. (Solids, although the exclusive constituents of the material world, are redundant elements in a geometric description, since they can be perfectly defined by surfaces.) The method of constructing three-dimensional geometric forms, therefore, is mainly a matter of defining boundary surfaces.

A surface defined by points may be obtained through transformations—by translation, rotation, scaling, or reflection—of an initial point. A configuration of points obtained in this way, however, can define different surfaces according to the way the points are connected. For example, the eight points that determine the vertices of a cube can create several spatial configurations, depending on which points are connected to define a surface (figure III-7). From this example it is clear that a single configuration of points, variously connected to form different surfaces, can also generate different perceptual spatial relations—in particular, solid-void and inside-outside relations. These observations represent an intuitive

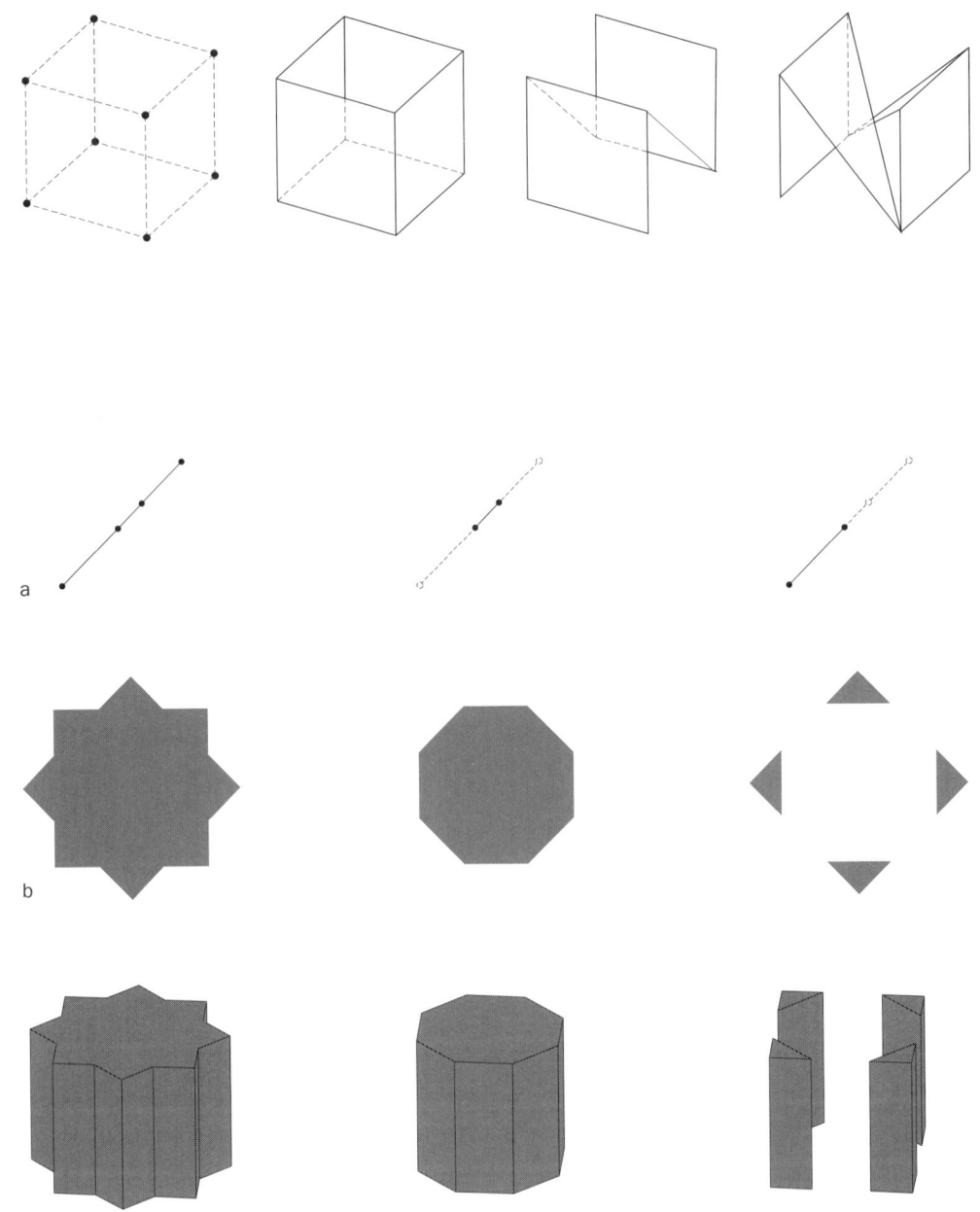

VISUALIZING WITH CAD

view of connectivity, a topological relation that will be discussed later in regard to polyhedra.

BOOLEAN OPERATIONS

Boolean operations (named for George Boole (1815-1864)) are used when two primitives (or parts) are interpenetrating. The Boolean operations are those of **difference**, **union**, and **intersection** and can be used with respect to lines, surfaces, or solids (figure III-8). To be combined according to Boolean operations, primitives must interpenetrate; that is, they must share a portion of a line, surface, or solid, in one, two, or three dimensions respectively. Some CAD systems are provided with solid-modeling features (see Chapter II) that can perform Boolean operations directly from software commands.

AutoCAD offers Boolean operations in a two-dimensional environment as a function of a region modeler from the **model** menu. For Boolean operations in space we have to use AME™, an optional module that can be loaded from AutoCAD. AME™ is provided with solid-modeling capabilities. After loading AME™, we can invoke the Boolean operations to be found in the model menu.

GROUPING OF PRIMITIVES IN PARTS AND SYMBOLS

A set of primitives can be grouped together according to their spatial characteristics or according to their functions to construct constituent parts of a form. For instance, the six surfaces defining a parallelepiped can be grouped together to create a solid component of a larger form. Each part can represent all, part, or a subpart of the whole form. There are different criteria for grouping elements, each motivated by the final form to be created.

Three possible criteria to group primitives
in a symbol:
a. As a set of solids.   b. As an enclosure.
c. As a solid with a hole.

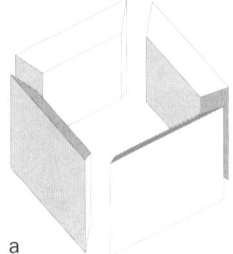  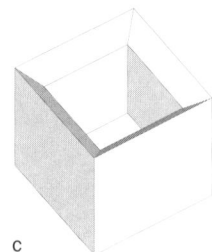

a   b   c

Perception of forms determines one of the most basic criteria by which to group primitives into parts. This criterion will be followed to generate almost all the forms in the sample models. According to this approach, a set of primitives can be classified as (figure III-9):

- the boundary surfaces delimiting a solid;

- the boundary surfaces delimiting a void; or

- a solid with a hole.

Even if this is the most logical way of identifying parts, sometimes it is more appropriate to ignore the solid-void characteristics. For instance, a part may be composed of surfaces that do not enclose a solid but will become boundaries for an enclosure once they are related to other parts. This approach will be clearly demonstrated in the discussion of forms based on recursions.

CAD systems provide a way to identify parts directly in the data structure. Blocks, symbols, and groups are recognizable commands to create a part in some of the most popular CAD systems. Each part (block, symbol, or group) in a CAD model can be named by the designer and manipulated according to the discussed transformations. The name of a part can provide useful information about its semantic content.

The form generation diagram.

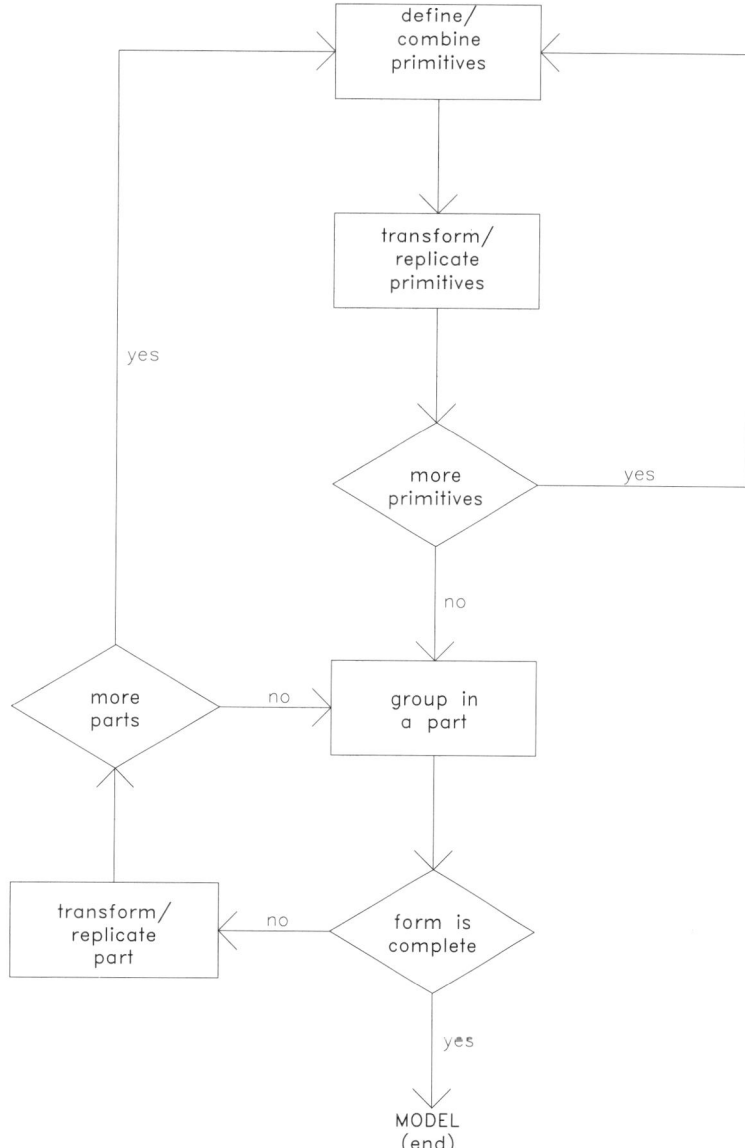

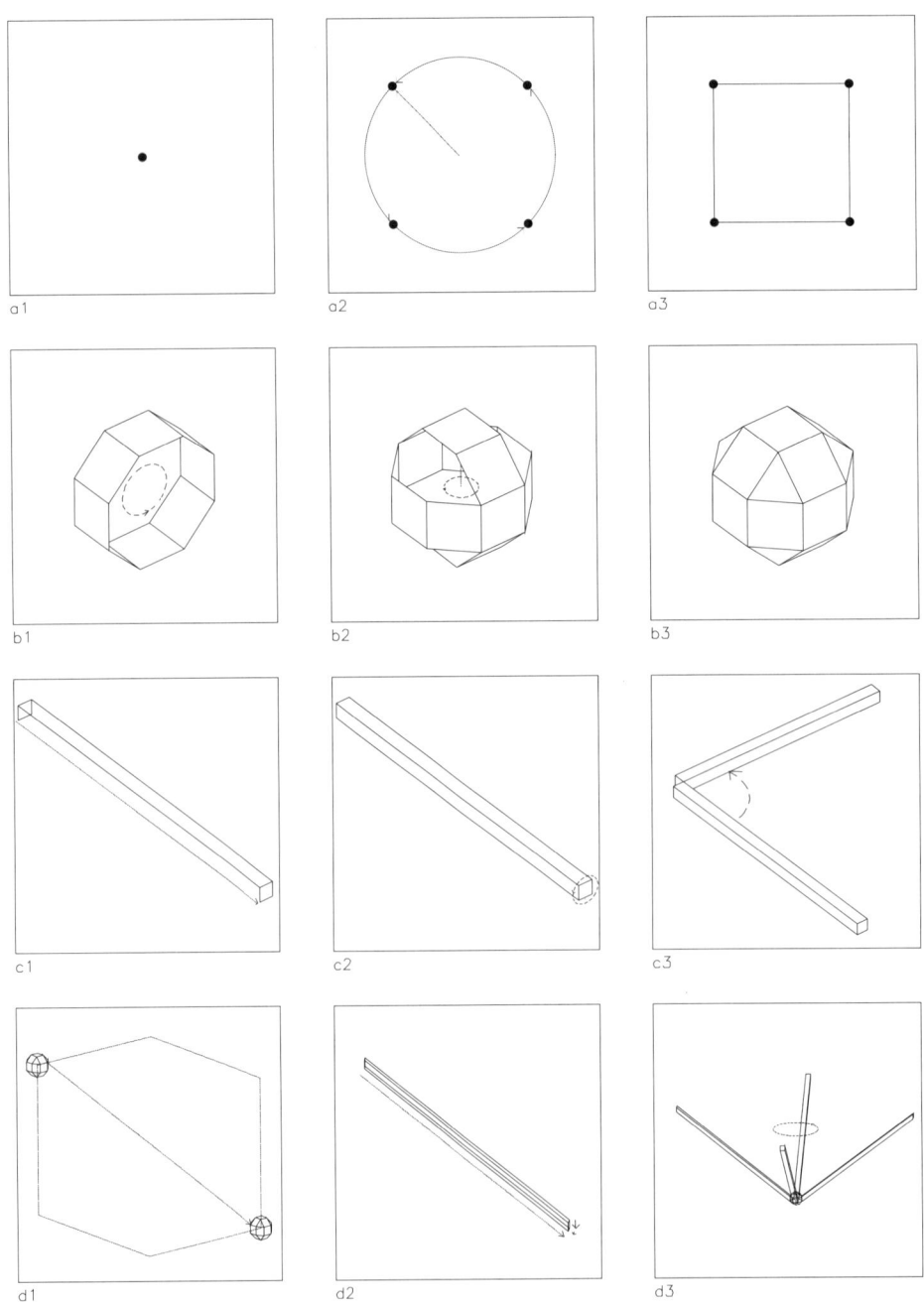

a1

a2

a3

b1

b2

b3

c1

c2

c3

d1

d2

d3

VISUALIZING WITH CAD

Generation of a form as process.

We are ready now to show how complex shapes can be generated from a foundation of geometric primitives in a systematic way using the operations previously discussed. This systematic process, called *form generation and evolution*, determines the syntactic structure. Figure III-10 shows the separate steps in this process, which is described below. Figure III-11 illustrates graphically the generation of a space frame structure according to these steps.

I. Creation of primitives or atoms

- Points can be repeated and transformed to generate boundaries for lines or surfaces (figure III-11a).
- Lines can be repeated and transformed to generate boundaries for surfaces.
- Surfaces.

Geometric transformations of $n-1$-dimensional primitives can be used to generate n-dimensional primitives, which also can be arbitrarily defined. If one or more primitives define a repeatable element, they can be grouped as a **symbol** (figures III-11a3 and III-11b1,2).

II. Replication and transformation of the initial set of primitives

One or more combined geometric transformations (translation, rotation, scaling, and reflection) and Boolean operations can be used to generate and combine primitives (figure III-11b,c,d).

III. Grouping in parts or symbols

After a set of primitives is created, it can be grouped in a part as soon as it satisfies one of the following conditions (partial list):

- The primitives bound an enclosure and thus define a **solid** or a **void** (figures III-11b2,c2,d2).

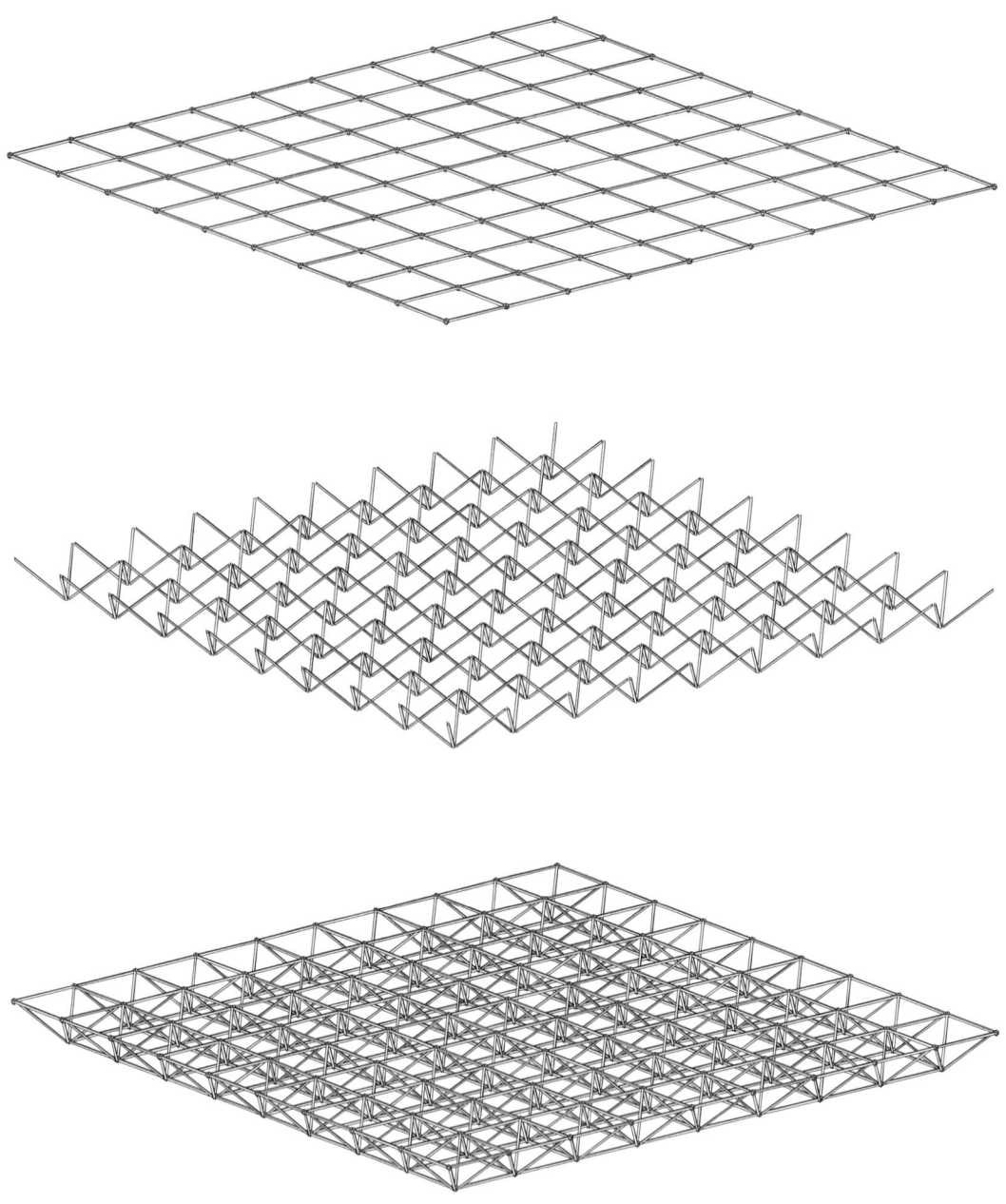

VISUALIZING WITH CAD

- The primitives, though not boundaries for a solid, have a recognizable geometric configuration appropriate for further development of the form.
- The primitives can be associated according to a semantic value, such as a decorative pattern or a structural element (figure III-11b3,c2,d3).

One or more combined geometric transformations (translation, rotation, scaling, and reflection) and Boolean operations are used to generate and combine the parts (figure III-11c3,d3)

Steps I, II, and III are the initial procedures. The designed form can be completed after these steps (or even just after step I) or may grow as part of a more complex form. In the latter case, we have:

IV. Three possible situations

- Replication or transformation of the initial part, using geometric transformations and Boolean operations.
- Creation of a new part (repeat steps I–III).
- Both of the above (figure III-11e).

V. Grouping in parts

Step V may represent the end of our form generation process or just the creation of a more complex part. In the latter case, the next step entails a repetition of step III, with the initial symbol replaced by the last-created symbol (figure III-11f,g). Obviously, a more complex form can be achieved by repeating the described steps in a loop process until the form is completed.

Forms generated according to this process consist of clearly identifiable **parts** and **relations**. Each of the described steps can be generally valid whether the semantic value of the constructed part is geometric or architectural; for the purpose of these steps, the semantic distinction lies in the fact that in a geometric form the parts are depicted by a group of vertices or sides, while in a building the parts consist of such elements as the roof or columns.

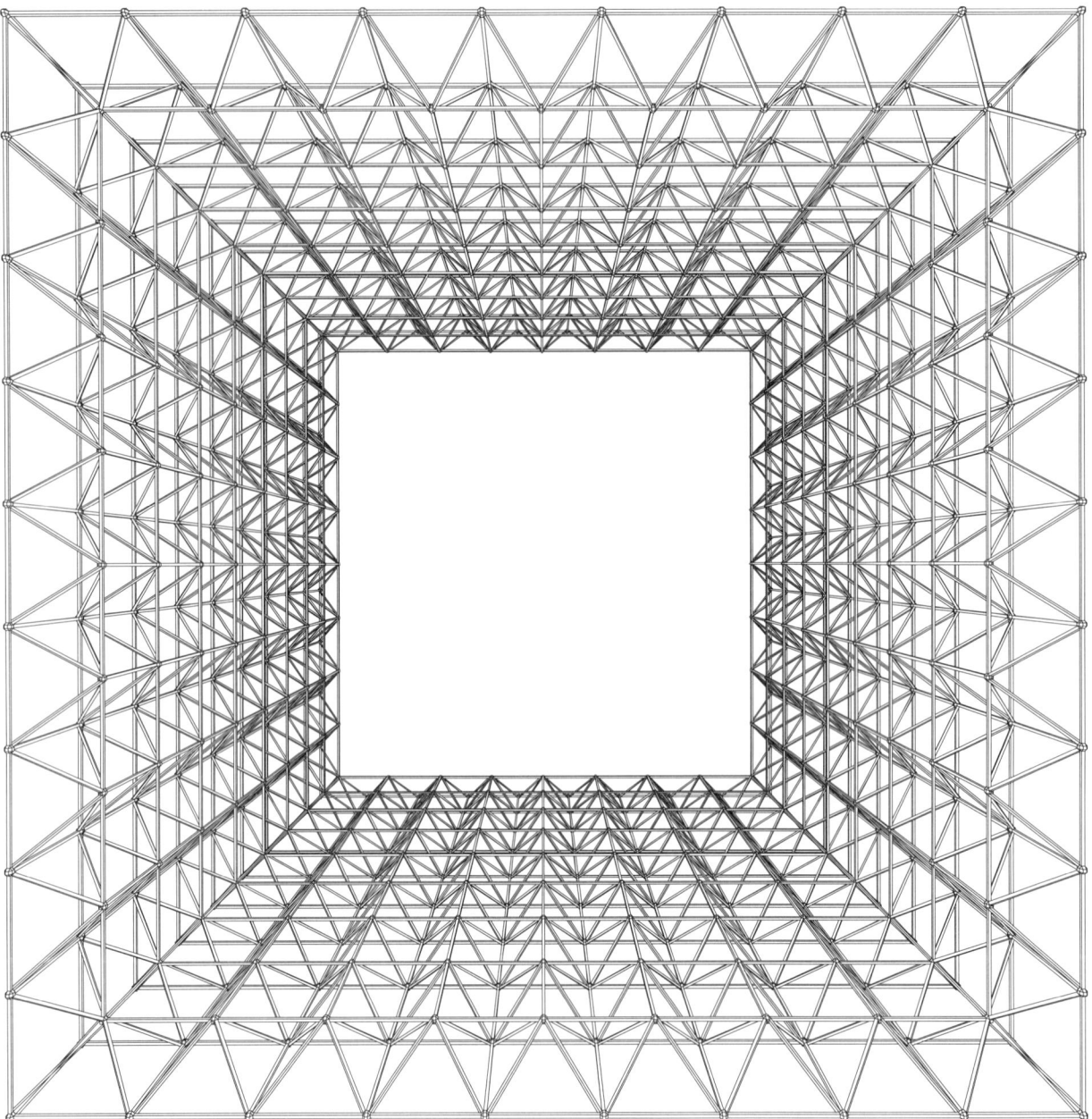

VISUALIZING WITH CAD

The identification of a geometric primitive or part with a symbol is essential for the development of design. Geometric primitives at this stage have two values: The first is that of a geometric entity, the other is that of element with a potential semantic content. In the change from one semantic content to another, that which has been identified with a point may become, for example, a joint in a frame structure (as in the model of figure III-11) or a node in a urban context (see Chapter VI).

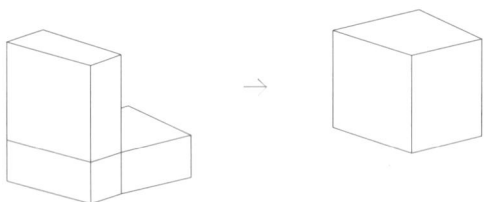

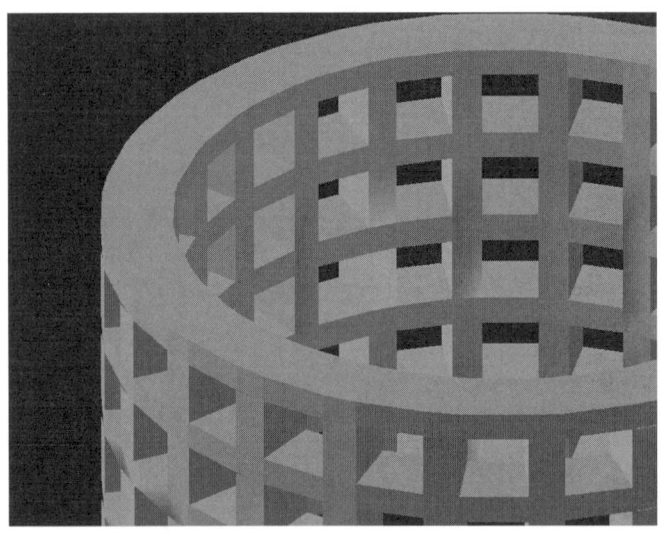

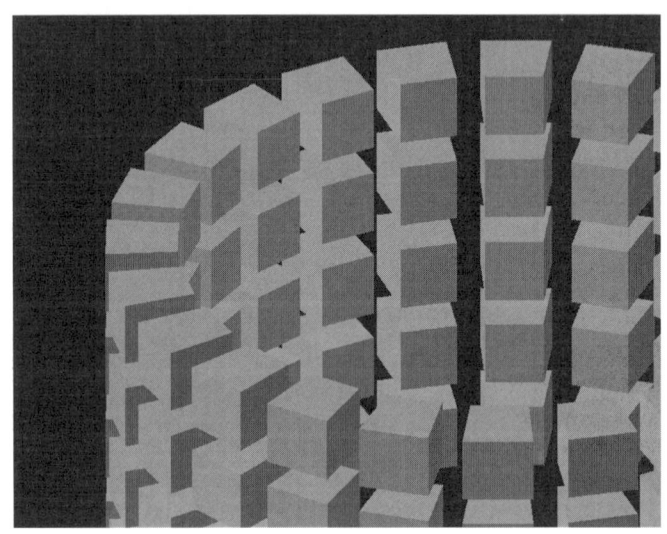

VISUALIZING WITH CAD

Solid-void reversal through symbol replacement.

Substitution, or replacement, is one of the most useful operations in computer-aided design. When the geometric definition of a part in the data structure is replaced with different information, all parts in the model that have the same name as the changed part will be globally changed to the new geometrical configuration. This operation is particularly powerful in designs based on modular repetition since it permits the transformation of the designed form beyond its geometrical definitions and perceptual characteristics. In fact, the replacement of a part that consists of a set of primitives or subparts can even effect the transformation of inside into outside or solid into void (figure III-12).

The AutoCAD implementation of a **symbol** (already discussed in Chapter II) is the following:

> *Command:* **block**
> *Block name (or ?):* **blockname**
> *Insertion point:* **0,0,0**
> *Select objects:* <pick the desired elements >
> *Select objects:* <enter>

The replacement of the content of a symbol is realized with the redefinition of the symbol:

> *Command:* **block**
> *Block name (or ?):* **blockname**
> *The block <blockname> already exists,*
> *Redefine it <N>?* **Yes**
> *Insertion of base point:* <the following prompts are the same as those from the first block definition>

EVOLUTION THROUGH REPLACEMENT

Generation and evolution represent two different stages in the process of creating and defining a form. While the generation process determines the **syntax** of the form, evolution provides its **semantic** transformation. So far we have shown how a higher level of complexity can be obtained through the transformation and replication of a set of primitives or symbols. This process generates a form that is characterized not just by its primitive elements and parts, but primarily by the relations among them.

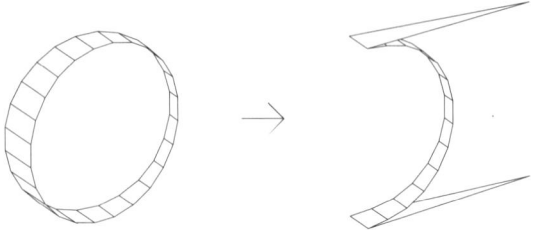

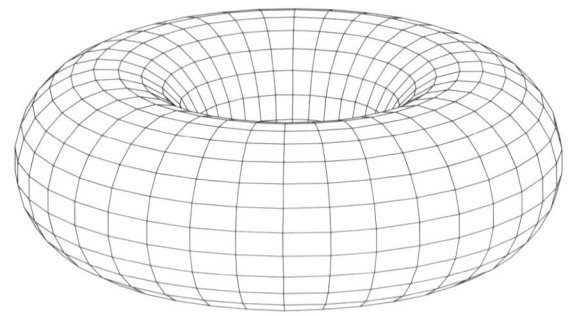
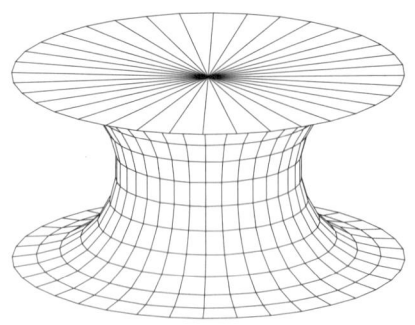

The replacement operation transforms a torus into a shape with different topological properties.

With the replacement operation, forms can evolve in a process typical of computer-aided design, through the substitution of new parts for some or all of the existing ones. In fact, the evolution of a form can produce a new form with spatial attributes (and therefore perceptual characteristics) completely different from the initial one. After its evolution, the identity of a form is given solely by the relations among its parts. Of the parts themselves, only their reference position, or symbol origin, is conserved from the initial form. This allows the possibility of designing with diagrams.

When we use replacement in a vocabulary of geometric forms, we can go beyond the metric, projective, and even topological properties of a form (figure III-13). For instance, a torus can be transformed into a revolution surface, with completely different topological characteristics.

The process of evolution is quite simple. However, the boundary conditions of each part to be replaced are to be considered if the parts touch or create enclosures, and particular attention must be paid to the nested parts, which represent the level of complexity of the form. Several examples will be discussed in this and the following chapters to illustrate the applications and differences between generation and evolution.

FORM GENERATION AND EVOLUTION FINALIZED TO DYNAMIC MODELS

The process of form generation so far described provides a method for geometrically defining a configuration; this description is necessary both for purely geometric shapes and for models in different semantic contexts. Furthermore, the generation and evolution of forms has been structured as a hierarchical composition of parts or symbols. This approach allows the content of individual symbols to be changed without changing the syntactic relations among parts, and for this reason we refer to such models as *dynamic*.

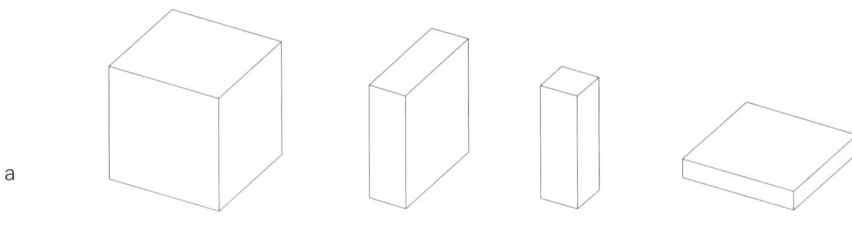

a

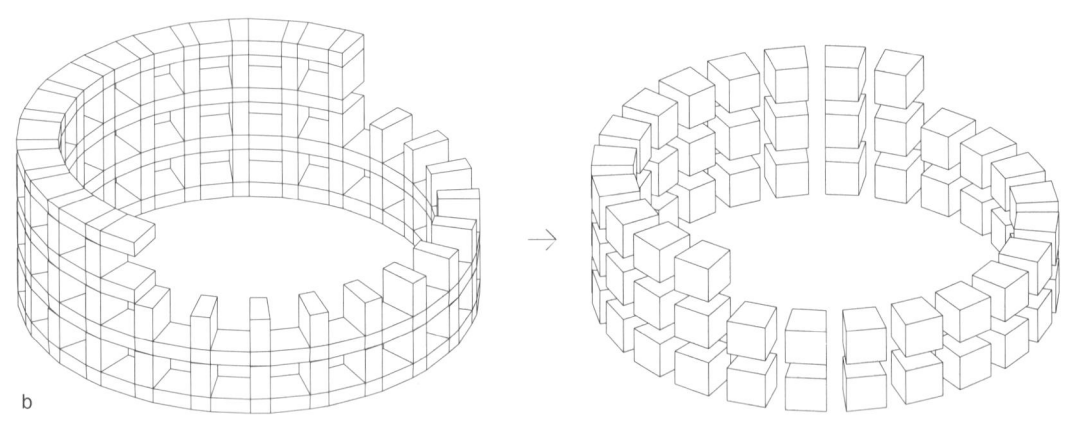

b

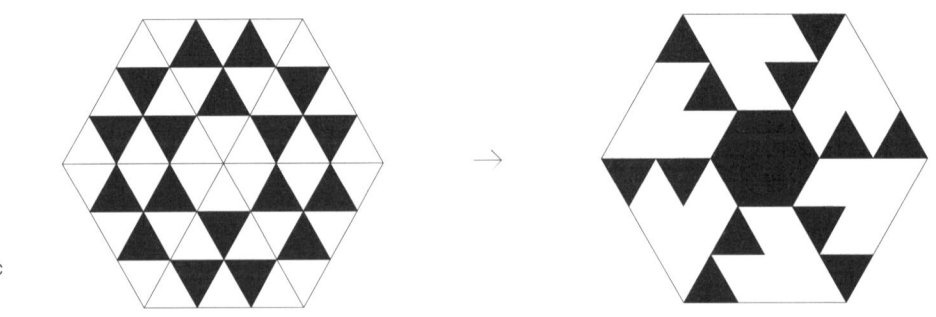

c

There are several ways in which a dynamic model can evolve, according to the different characteristics of each constituent part:

- Each part/symbol, in the whole hierarchical scale, has associated parameters corresponding to its extension—that is, to height, breadth, and depth. The substitution of parameters with different proportionality can generate models with the same topological characteristics but different metric attributes (figure III-14a).

- The geometric, or perceptual, definition of a part/symbol is determined by its solid-void characteristics; thus, if the solid elements of a part/symbol are replaced by voids, a completely different enclosure will result (figure III-14b).

- If a part is comprised of subparts, a variation in form can be obtained by a different arrangement of these subparts (figure III-14c).

HIERARCHY OF FORMS AND SELF-SIMILARITY

As already mentioned, many forms have a hierarchical structure. One or more primitive elements are created, some of which may be grouped in a symbol. The symbol is replicated and transformed. The set of new symbols may be grouped in yet another symbol. This process can be repeated many times. The final hierarchical form can be represented by a tree diagram (figure III-15). An indication of the level of hierarchical organization is given by the number of nested symbols.

A particular type of hierarchical relation is that of self-similarity among sets of elements of different complexity levels; such a relation is called *recursive*. A thorough discussion of self-similar shapes can be found in the fractal geometry developed by Benoit Mandelbrot, mentioned in Chapter I. A CAD model of a self-similar form is based on an initial shape (**generator**) and a **transformation** that determines the self-similarity relation.

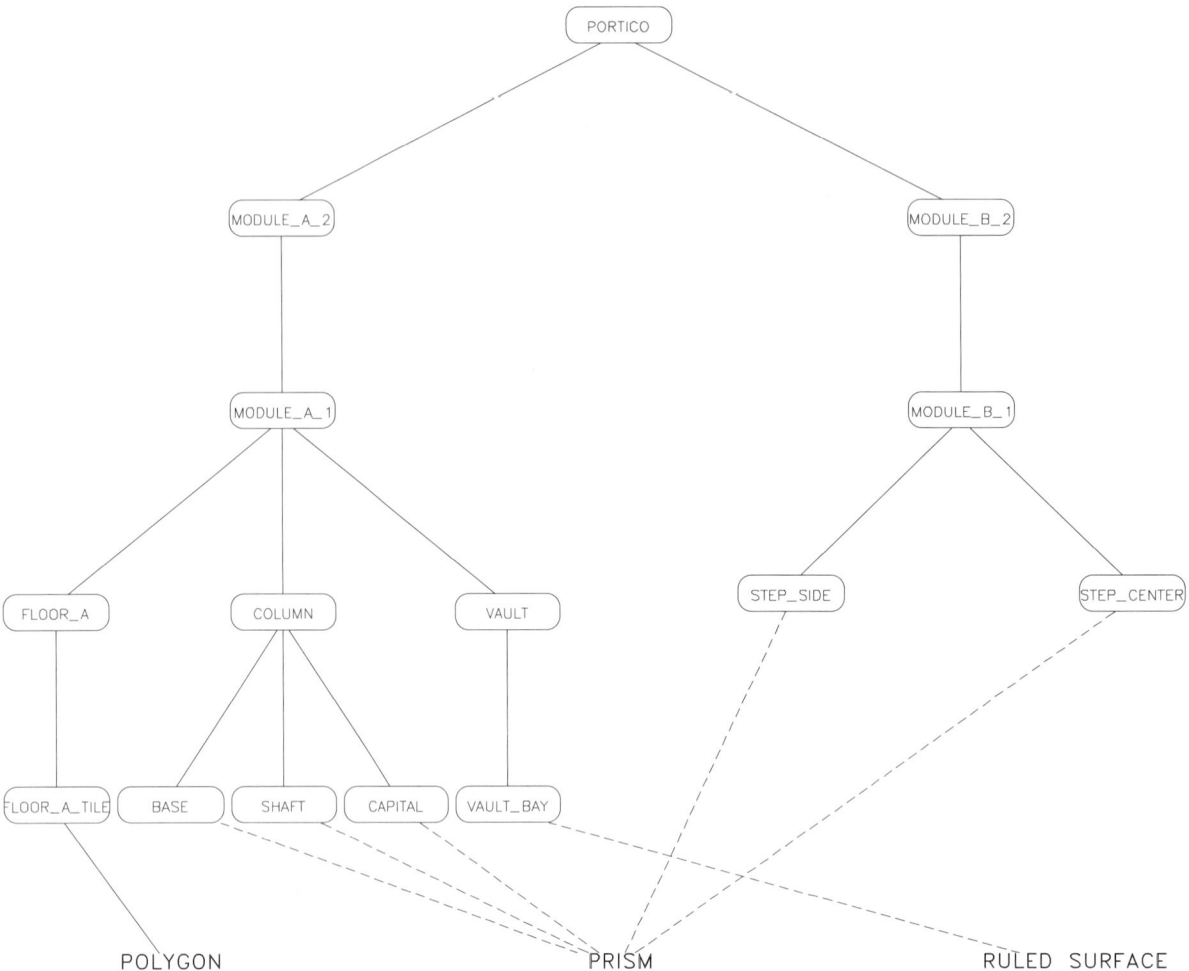

POLYGON   PRISM   RULED SURFACE

VISUALIZING WITH CAD

Hierarchical structure of the
composition shown in figure VI-9.

menger.dwg

Menger Sponge

The Menger Sponge is a recursive shape for which the generator is a cube and, under the transformation rule, at each step in the recursion every cube is replaced by twenty cubes arranged as shown in figure III-16a. Once the original cube is transformed, the transformation rule is applied to each of the twenty cubes that replaced the generator cube. This recursive process can be applied an unlimited number of times. The resulting object has the property that, each time the transformation rule is applied, the ratio of volume to surface area decreases. For the Menger Sponge shown in figure III-16b, the transformation rule has been applied three times, producing a structure composed of 8,000 cubes. The transformation, which could be quite laborious if executed manually, is easily achieved in CAD by substituting for each cube symbol a twenty-cube symbol.

The form may evolve through the replacement of solids by voids. Suppose the generator is still a cube, but the replacement shape is the set of six cubes shown in figure III-17a, corresponding to the void space of figure III-16a. If the twenty-cube symbol of the model of figure III-16 is redefined by this new set of cubes, the previously generated Menger Sponge automatically evolves into the shape of figure III-17b.

Three-Dimensional Web
Inspired by the Sierpinski Gasket

If an equilateral triangle is used as a generator, the resulting recursive figure is called a Sierpinski Gasket (figure III-18a); a three-dimensional version of this form results when the generator is a tetrahedron. The transformation rule is the replacement of the generator by a shape consisting of four tetrahedra, arranged as shown in figure III-18b1. The form shown in figure III-18b2 is generated by applying the transformation rule four times, resulting in a set of 256 tetrahedra. The form may evolve analogously to the Menger Sponge, if voids are replaced with solids, as shown in figure III-19a,b. The polyhedron that corresponds to the void defined by the initial set of tetrahedra is an octahedron.

sierpins.dwg

III-16

Menger Sponge.

a. Transformation rule.

b. Axonometric view.

c. Perspective view from inside.

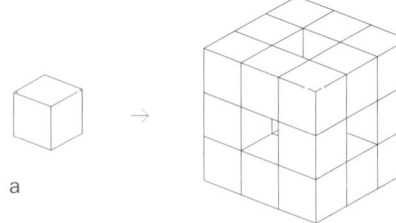

a

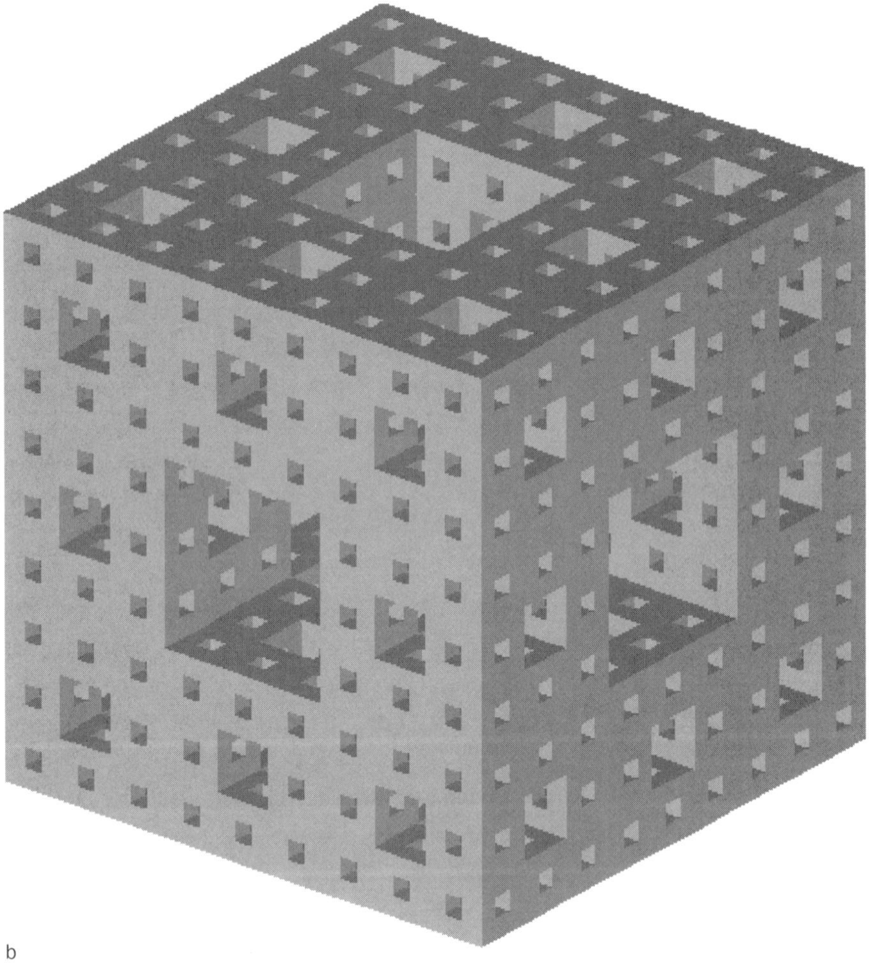

b

C

a. Transformation rule for the evolution of the Menger Sponge obtained exchanging solids with voids.

b. Evolution of the Menger Sponge obtained exchanging solids with voids.

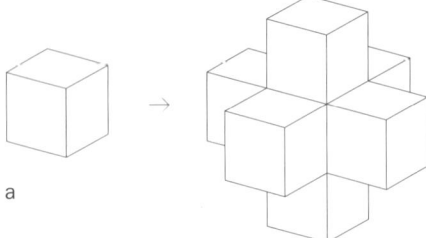

a

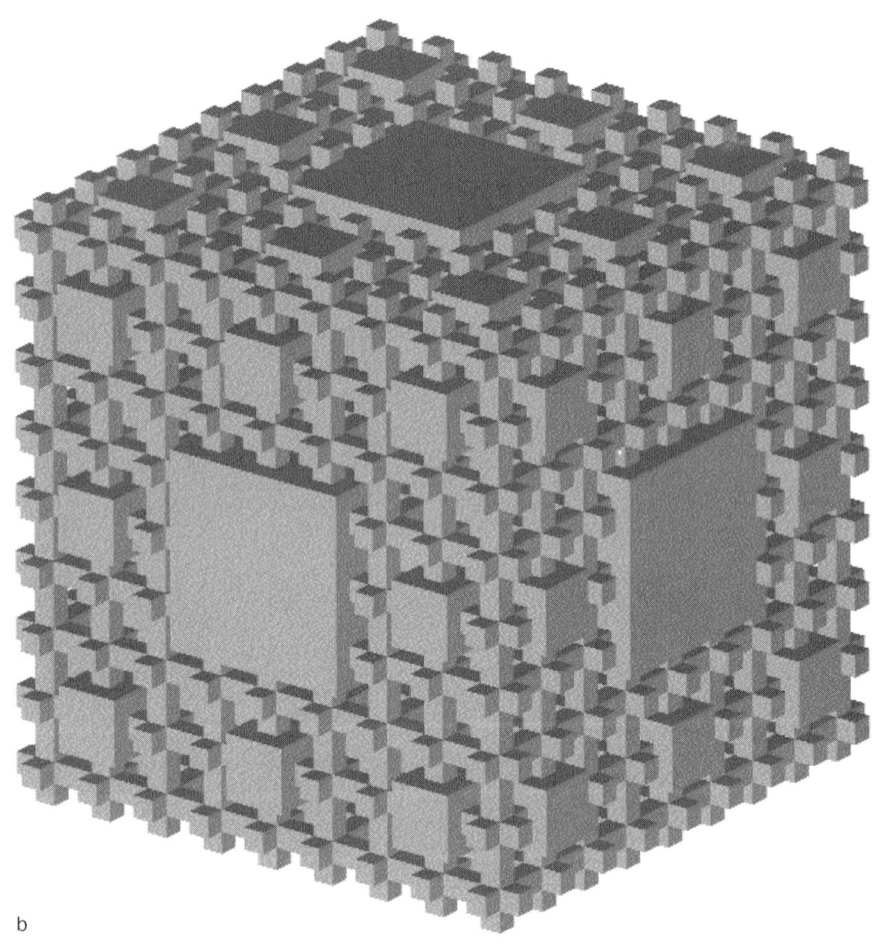

b

III-18A

a1. Transformation rule for the generation of a Sierpinski Gasket.

a2. Sierpinski Gasket.

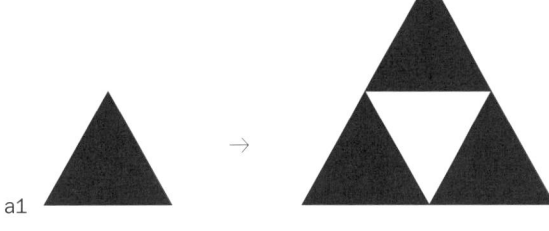

a1

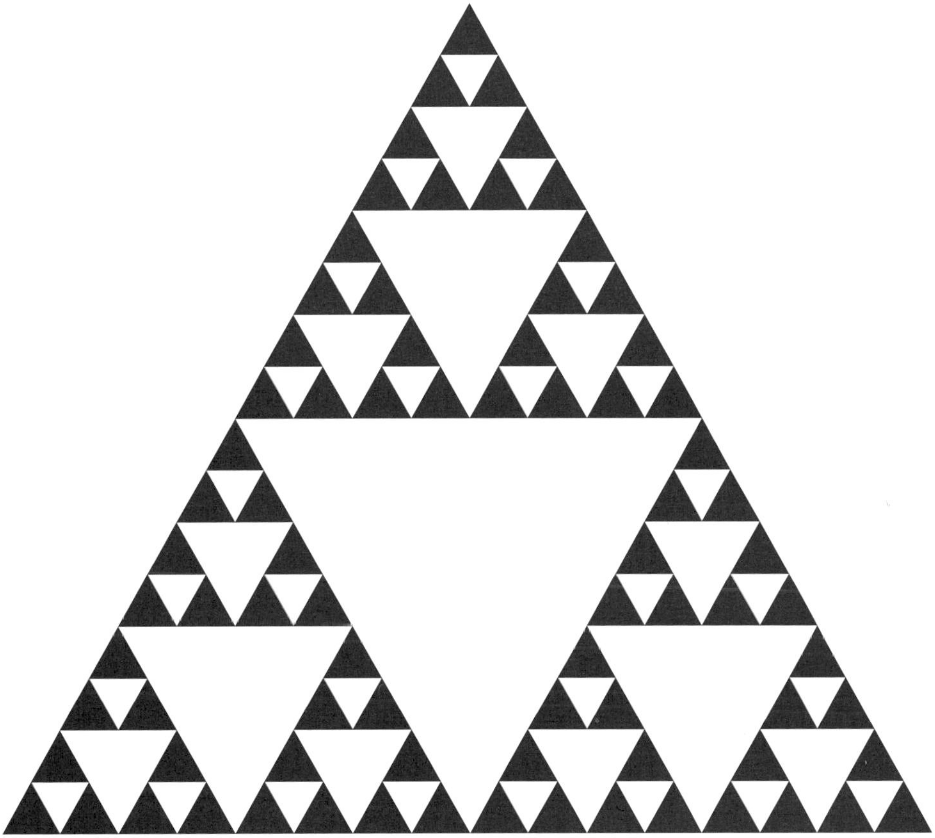

a2

b1. Transformation rule for the generation
   of a three-dimensional web inspired
      by the Sierpinski Gasket.

b2. Three-dimensional web inspired
      by the Sierpinski Gasket.

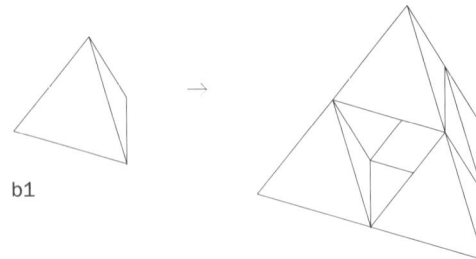

b1

b2

c. A wireframe plan view of the
three-dimensional web inspired by the
Sierpinski Gasket.

c

III-19

Evolution of the three-dimensional web inspired by the Sierpinski Gasket obtained exchanging solids with voids.

a. Transformation rule.

b. Axonometric view.

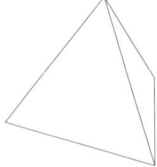  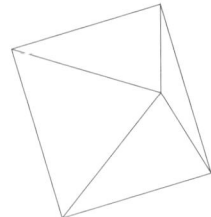

a

b

VISUALIZING WITH CAD

FROM FORM GENERATION TO VISUALIZATION

In the course of this chapter we have developed the spatial definition of a form in terms of geometric relations and primitives. The model created in such a way carries strictly spatial properties given by the coordinates of each point and the type of geometric primitives the points define. Symbols can be associated to primitives to allow an evolution of the form through their replacement. The three-dimensional coordinate description of a form—with or without the associated symbols—provides the most basic spatial definition necessary to visualize the form. In addition, the view from which we see the model (plan, elevation, isometric, perspective) must be defined; this is accomplished by a standard operation given by a software command (see Chapter II). These are the essential procedures to create and visualize an electronic model.

Note that the images of the forms used here as examples—the Menger Sponge and the Sierpinski Gasket—are different views of the same models. Our discussion of these forms has been concerned only with the generation of the model in terms of geometric primitives; but more information can be provided to the visual model. Each piece of this additional information is associated with spatial points in a way similar to coordinate definition. For this reason it is often referred to as *dimension* (not to be confused with the three dimensions of Euclidean space). According to this terminology, physical qualities determined by light, such as color and surface texture, provide these additional dimensions. It is not necessary to include these dimensions in the initial spatial definition, but they can be easily added for a more realistic visualization (see Chapter II). In the next chapters, several complex geometric and architectural models will be investigated and visualized in the accompanying images.

BIBLIOGRAPHY

AutoCAD® Release 12 - Advanced Modeling Extension®
    Release 2.1 Reference

AutoCAD® Release 12 - Command Reference

Autodesk 3D Studio™ Release 2 - Reference Manual

Coxeter, H. S. M., and S. L. Greitzer, *Geometry Revisited*,
    Random House, New York, 1967

Coxeter, H. S. M., *Regular Polytopes*, Dover, New York, 1973

do Carmo, Manfredo P., *Differential Geometry of Curves and Surfaces*,
    Prentice Hall, Englewood Cliffs, New Jersey, 1976

Foley, J. D., and A. Van Dam, *Fundamentals of Interactive Computer
    Graphics*, Addison-Wesley, 1984

Greenberg, Marvin Jay, *Euclidean and Non-Euclidean Geometries*,
    W. H. Freeman, New York, 1980

Mandelbrot, Benoit B., *The Fractal Geometry of Nature*,
    W. H. Freeman, San Francisco, 1977

Modenov, P. S., and A. S. Parkhomenko, *Geometric Transformantions*, Academic Press, New York, 1965

Yaglom, Y. M., *Geometric Transformations*, Random House, New York, 1962

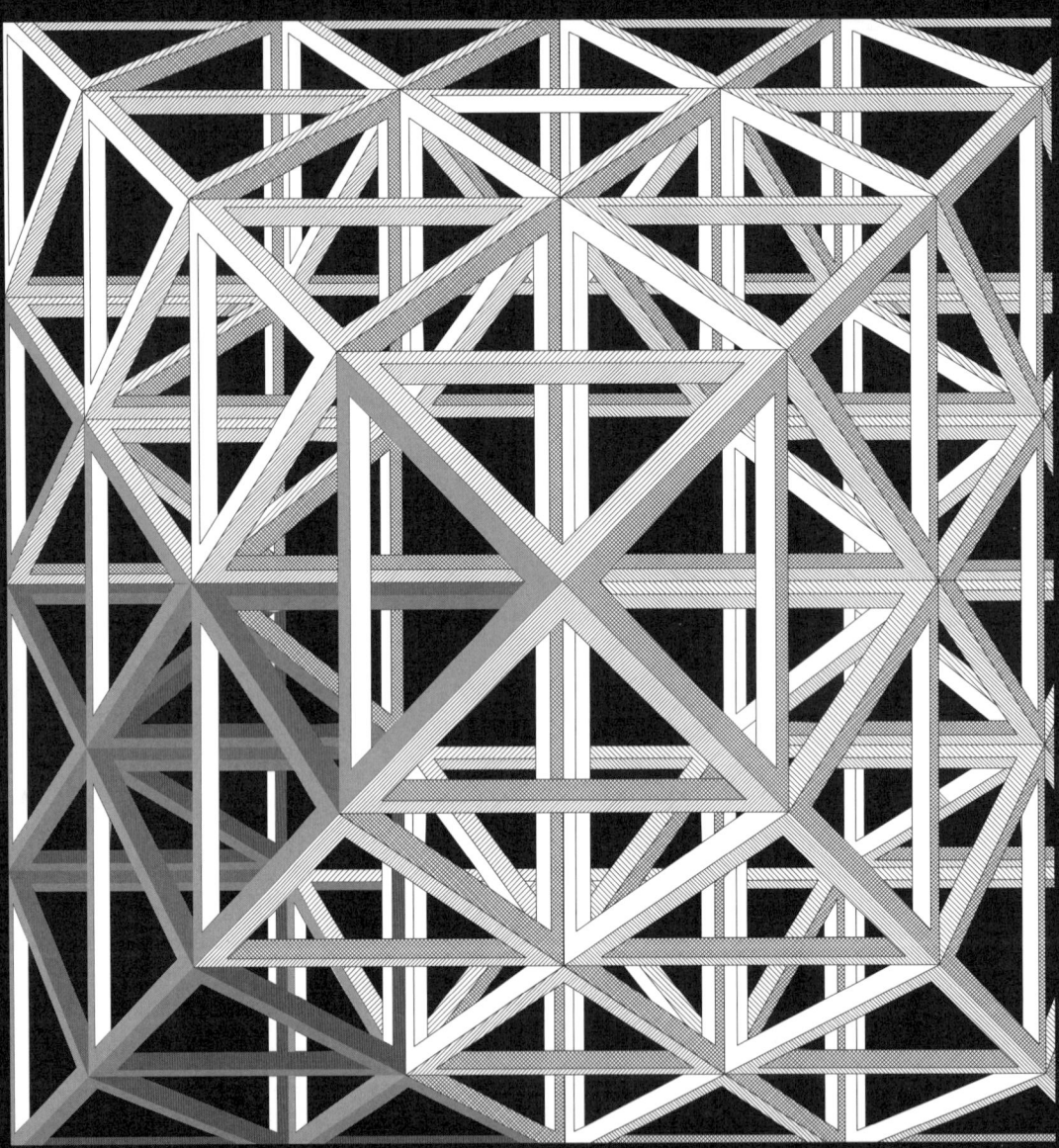

IV

To earth let us give the cubic form . . . that

solid which has taken the form of pyramid

shall be the element and seed of fire;

the second in order of generation

[octahedron] we shall affirm to be air, and

the third [icosahedron] water.

Plato

In this chapter we will provide applications of the syntactic rules established in Chapter III for the generation and evolution of geometric forms. Most of these two-dimensional and three-dimensional forms are either taken directly from classical geometry or derived from geometric shapes, but some of the models have been inspired by more abstract geometries, such as topology, differential geometry, and projective geometry.

Some of the generated forms can also be interpreted at a semantic level other than that of geometry—in particular, that of architecture, since architectural forms are often inspired by geometry. In the transition from geometric shapes to designs of architectural shapes that are eventually built, physical properties such as gravity, mass, materials, and connections, as well as functional requirements, must be considered. Any geometric shape satisfying all these requirements and limitations can be used as a formal model for a building in a second process that considers its structural stability in relation to the materials to be used. In this transition, the form can be given a different semantic content; a revolution surface becomes a column, for example, or a helicoidal shape is interpreted as a spiral staircase (figure I-29). The transition from one semantic description to the next, from geometric content to architectural, will be discussed in the last two chapters. The forms investigated in this chapter comprise a "library" (see Chapter VI and diskette) of dynamic models based on geometry. These shapes can evolve as a CAD model for architectural components, with the inclusion of further definitions for some properties belonging to architecture.

As mentioned in Chapter I, a straightforward visualization of a geometrically defined space is sometimes impossible. The geometric property of continuity, for example, is impossible to represent because of the physical limitations of our devices, so continuous geometric shapes will be approximated by discrete forms that preserve other geometric characteristics. In this way, any complex geometric model can be constructed out of primitives that are grouped and identified by symbols in the computer data structure. This approach will also be valuable when the geometric model becomes architectural and the primitives are replaced by elements with different semantic descriptions.

GRIDS AS REGULAR CONFIGURATIONS OF POINTS

Grids are a perfect example of geometric systems based on discrete elements since they are characterized by elements, such as points, separated by regular relations of distance. The following discussion will deal with connected grids—that is, grids defined by the connections of the points—and grids consisting of discrete points.

The semantic content of grids goes beyond the geometric one. We can construct diagrammatic models of grids the points of which are defined by symbols in such a way that, in architectural models, they can be replaced by columns. Alternatively, if a two-dimensional grid is superimposed on an elevation, the diagrammatic points can be replaced by windows. At an urban scale, the grid points can be replaced by buildings, trees, or whatever elements need to be modularly repeated. Another very common architectural example of the use of connected grids is in the formal and structural design of metal space-frame structures. All these examples will be discussed and illustrated in Chapter VI.

Crystallography is another important application of three-dimensional regular grids. The model of an atom can be schematically constructed as a central point with equidistant points radiating from it which correspond to the valence. The aggregations of atoms in molecules is indeed related to the characteristics of the grid formed by the geometric model of the atom.

A quite interesting use of grids, in geometry as well as in design, is that of tessellation or tiling, in which a set of polygons is arranged in a plane in such a way that they cover the entire plane without overlapping [Grunbaum 1987]. Tiling, as the word suggests, addresses problems involving patterns, such as the problem of covering a floor with tiles. The artist M. C. Escher often used tessellations in his artwork.

Two-Dimensional Grids

The simplest two-dimensional grid is the **square** grid (figure IV-1a), generated by the following operations. A point/symbol is replicated and translated $n-1$, $m-1$ times in the x-axis, at a certain distance a, generating a row of n points/symbols. The row is then repeated and translated the same distance a along the y-axis $n-1$, $m-1$ times, generating m rows of n points. The grid created in this way can be thought of as generated by the translation of a square (hence its name), but as shown in figure IV-1b, it can also be generated by the translation of any parallelogram having grid points for vertices (provided the vertices are the only grid points that lie inside or on the parallelogram).

A grid based on an **equilateral triangle** can be constructed as follows. A point/symbol A is translated once at a distance a along the x-axis, creating point B. The new point is then rotated 60 degrees about the initial point, generating point C. Point A is then translated $n-1$, $m-1$ times in the x-axis, at distance a, generating a row of n points. The row is then repeated and translated $n-1$, $m-1$ times, according to a distance determined by making A and C overlap, generating m rows of n points (figure IV-1c). The distance in the y-axis between two successive rows of points is $a \times \sqrt{3}/2$, which is the height of an equilateral triangle with sides of length a.

gr_polar.dwg

An example of grids that are irregular, but of interest for architectural applications, are those obtained by **rotations**. First we generate a row of n points/symbols as for the first row of a square grid, then the row is repeated and rotated $n-1$, $m-1$ times at a certain angle θ (equal to $360°/m$) about the first point of the row (figure IV-1d).

Three-Dimensional Grids

Three-dimensional grids are generated in a similar way as two-dimensional grids. A **cubic** grid is created from the square grid described earlier repeated and translated along the z-axis at the same

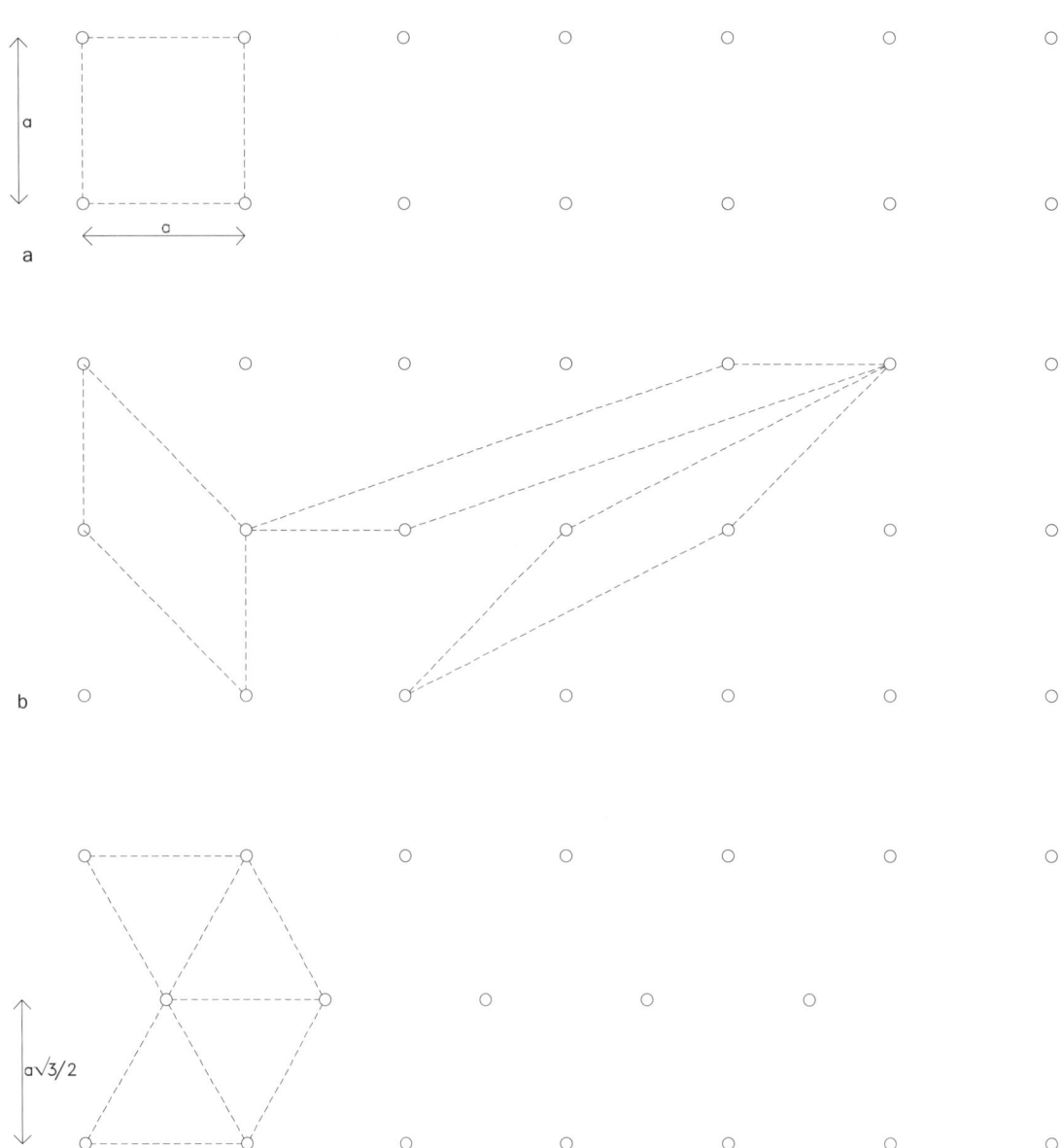

VISUALIZING WITH CAD

gr_eu_ev.dwg

distance *a* used to create the square grid (figure IV-2a). As with the square grid, the cubic grid can also be generated from parallelepipeds having grid points for vertices, with the condition no grid point lies inside the parallelepiped or on its surface (figure IV-2b). From a cubic grid it is also possible to derive several other grids by, for example, connecting the centers of the imaginary squares determined by adjacent points of the grid.

gr_te_ev.dwg

A grid based on **tetrahedra** (see discussion of polyhedra below) can be generated in a similar way, starting from the triangular configuration of points/symbols in figure IV-1c, where the distance between any two consecutive points is *a*. The initial grid is then repeated and translated in a single movement at a distance *a*/2 along the *x*-axis,

IV-1

Two-dimensional grids.

a. Square grid.

b. Generation of square grid by translation of parallelograms.

c. Grid based on equilateral triangles.

d. Grid generated by rotation.

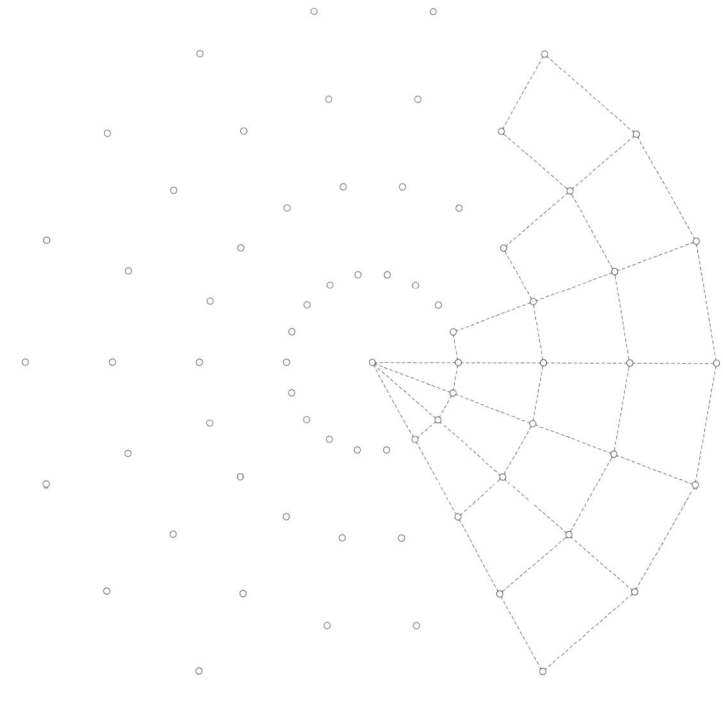

d

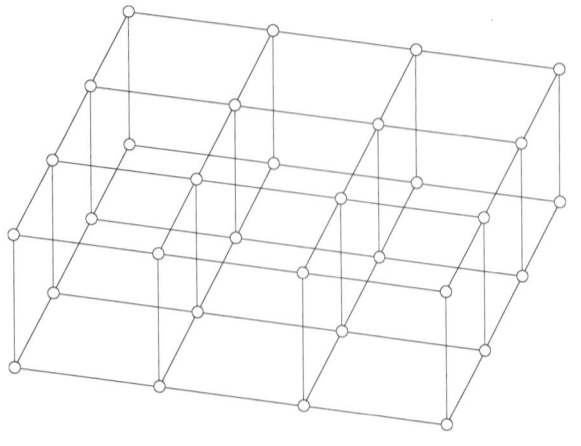

a

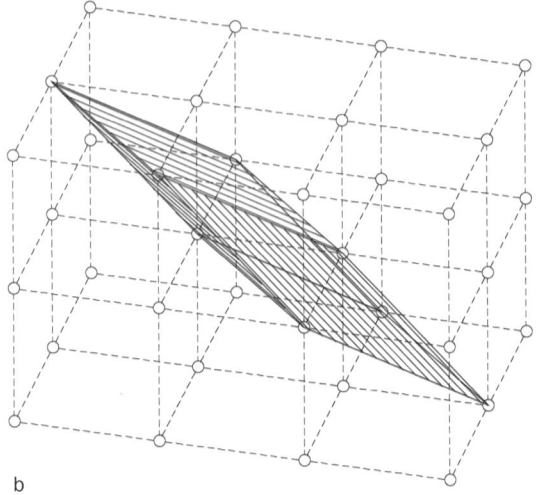

b

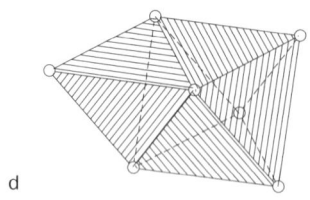

c

d

IV-2

Three-dimensional grids.

a. Cubic grid.

b. Generation of cubic grid by
translation of parallelepipeds.

c. d. Grid based on tetrahedra
and octahedra.

mk_grid.lsp

a x $\sqrt{3}/2$ along the y-axis and a x $\sqrt{3}/4$ along the z-axis. The grid so
defined does not correspond to a division of space into tetrahedra,
however, as might intuitively be assumed by analogy with a grid based
on equilateral triangles. Instead, it divides the space into tetrahedra
and octahedra (figure IV-2d), or into cuboctahedra and octahedra.

Figure IV-3 shows some of the different possible evolutions for a
cubic grid. The grid shown in figure IV-4 is based on tetrahedra; in
this evolution of the original grid, the grid points have been replaced
by tetrahedra and the connecting lines by triangular prisms.

VISUALIZING WITH CAD

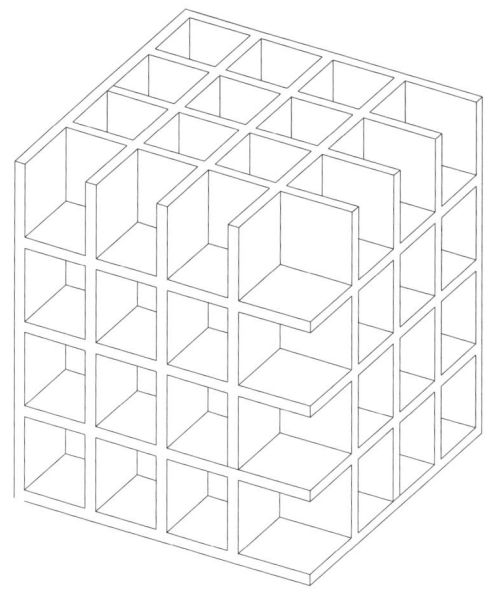

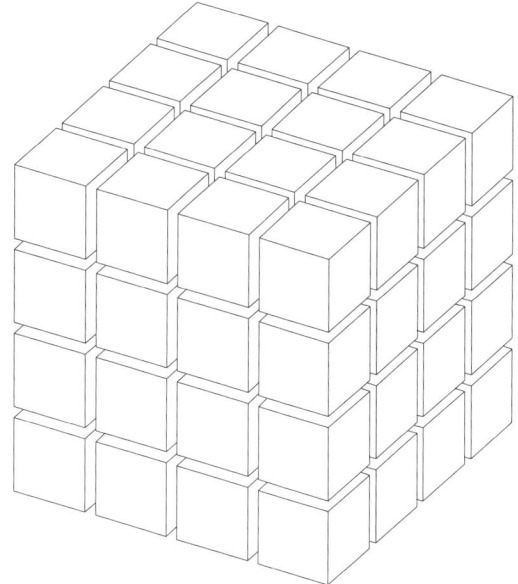

IV-3

Evolutions of a cubic grid.

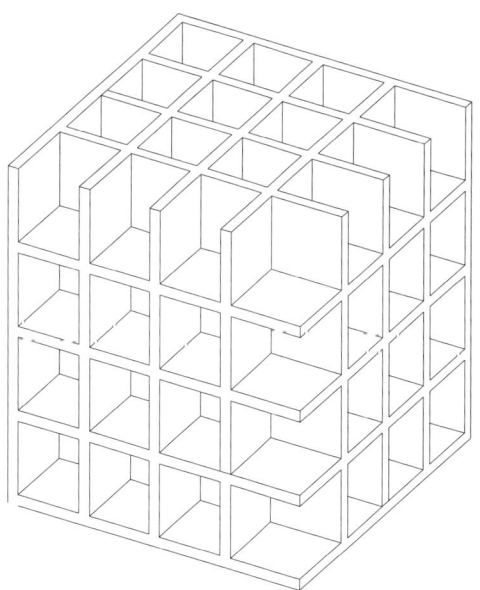

VISUALIZING WITH CAD

Evolution of a grid based on
tetrahedra.

Tessellations.

Tessellations

It is easy to create different types of tessellation using the framework of a regular two-dimensional grid [Grunbaum 1987]. Several points of the grid are connected and become vertices of polygons that, if infinitely repeated, would cover the plane without overlapping. Several examples are shown in figure IV-5. Tessellations are also possible in space, where polyhedra can fill space without overlapping; except in the case of a regular grid filled with cubes, however, the constructions are less immediate.

REGULAR POLYGONS

A polygon is a shape enclosed by a set of segments, or sides, joining consecutive pairs of points, or vertices. Two consecutive sides define an angle. The number of sides is the same as the number of angles and vertices. In a plane polygon, all the vertices belong to the same plane. If the angles (and thus the sides) are all equal, the polygon is called *regular*. Regular polygons, with the exception of the square, are named according to the number of sides: triangle, pentagon, hexagon, and so forth (figure IV-6).

In CAD, the boundaries of a polygon are created by defining an initial point (the center point of the polygon), replicating this point, and rotating the second point about the center point. The rotation angle is given by $360°/n$, where n is the number of sides of the polygon, and the points generated by the rotation transformation are the vertices of the polygon. Thus, to generate a polygon as an enclosed portion of a plane, we have to construct a surface determined by three points (the center of the polygon and any two consecutive points), which will then be replicated and rotated $n-1$ times about the center at an angle $360°/n$.

Regular polygons.

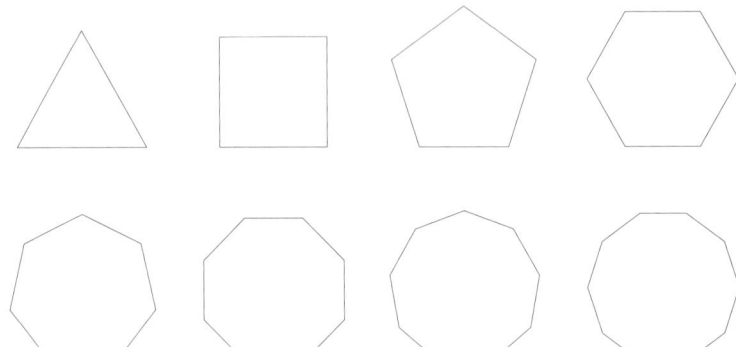

VISUALIZING WITH CAD

A convex polyhedron is an enclosed portion of space, bounded by a set of plane polygons called faces. The term *convex* means that, from the outside of the polyhedron, the angles between faces are all greater than 180°. In a regular polyhedron all the faces are identical regular polygons. Each face shares each side or edge with just one other face and each set of three or more faces meets in a point called a vertex. Certain types of polyhedra are characterized by a numerical relation discovered by the eighteenth century Swiss mathematican Léonard Euler:

$$V - E + F = 2$$

where V, E, and F are the number of vertices, edges, and faces, respectively. All polyhedra for which Euler's formula is valid are called *simply connected*. Euler's formula may seem like an abstract concept, but it can give practical hints about how forms can be evolved by changing their solid-void and inside-out relations. Euler's formula will be further discussed regarding the topographical aspects of form generation.

While there exists an infinite number of regular polygons, only five regular polyhedra—tetrahedron, cube, octahedron, dodecahedron, icosahedron—exist. A detailed explanation of the reasons for this can be found in *Geometry and the Imagination* by David Hilbert and Stefan Cohn-Vossen. Polyhedra have also been widely discussed and illustrated in several treatises [Pacioli 1509, Kepler 1619].

The five regular polyhedra are also called **platonic solids**, because Plato, in his cosmological theory, associated four of them with the basic elements of nature. The tetrahedron was associated with fire, the cube with earth, the octahedron with air and the icosahedron with water. The dodecahedron was considered to be the shape enclosing the whole universe.

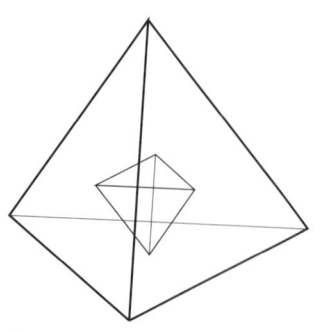 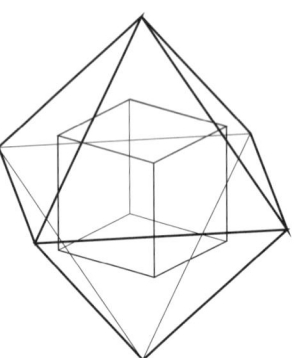 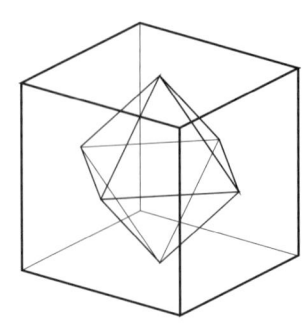

a

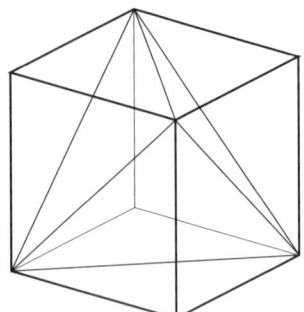 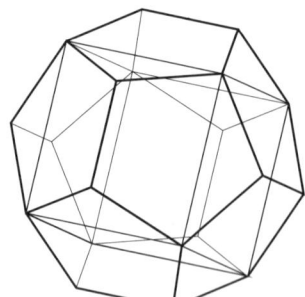

b

VISUALIZING WITH CAD

Relationships among the regular solids.

a. Duality.

b. Cube-tetrahedron and
dodecahedron-cube.

The five solids represent all the possible cases of regular three-dimensional configurations that can be repeated as basic elements of elaborate systems. Therefore the models for the regular solids, once defined, exhaust all the possible schematic representations for regular spatial configurations. The relations that exist in each of the five platonic solids as a result of the regularity of the elements can have several applications in architecture, both theoretical and practical. Examples include the analysis of metal frame modular elements and even the design of domes (since each regular solid can be inscribed in a sphere).

The regular polyhedra have other interesting relationships that can be helpful in the generation of forms. One property is their reciprocal duality; that is, a polyhedron can sometimes be constructed by connecting the centers of the faces of another regular polyhedron. According to this relationship, the cube is dual to the octahedron, the dodecahedron to the icosahedron, and the tetrahedron to itself (figure IV-7a). Also, the diagonals of each face of the cube form the edges of a tetrahedron, and a cube can be inscribed in a dodecahedron (figure IV-7b).

For purposes of form generation, each polyhedron becomes a definer, or **dynamic model**, of geometric relations. CAD models of polyhedra are constructed on the basis of these relations, whereas faces, sides, and vertices are defined as symbols—impermanent elements that can be replaced by others. In a CAD model, each of these elements—faces, sides, vertices (defined as symbols)—has to be identified, not only as a geometric entity characterized by extension and position, but also by a vector: Definition of the direction of the geometric primitives/symbols comprised in the solid is fundamental for its eventual evolutions. The directions of the vectors for each face, side, and vertex are given by the straight line (pointing outward) connecting the center of the polyhedron with a face, side, and vertex. The figures below illustrating the construction of each polyhedron show only the vectors associated with the faces; those characterizing vertices and sides are suppressed to simplify the illustrations.

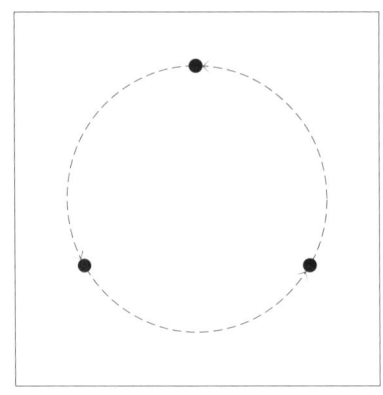
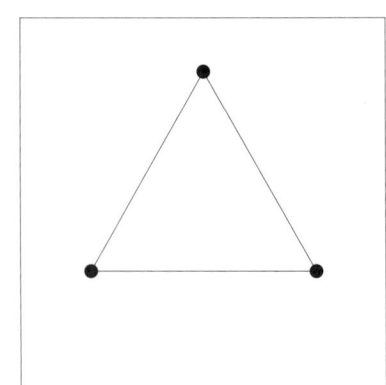
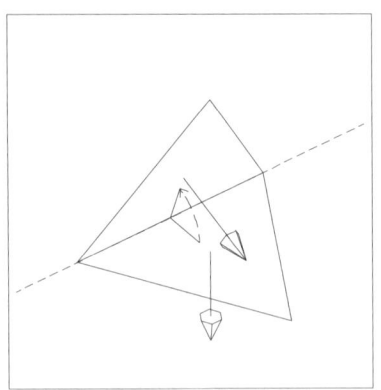
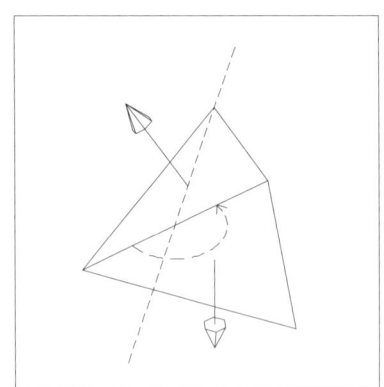
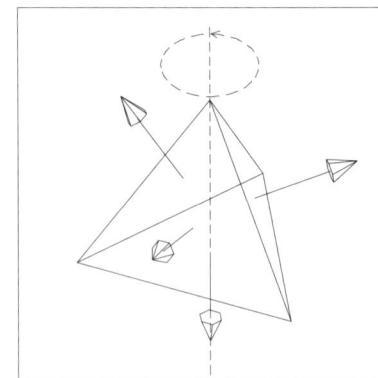
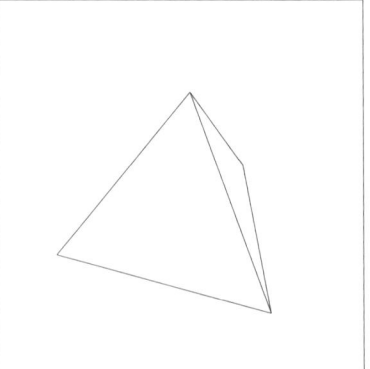

VISUALIZING WITH CAD

Tetrahedron

tetra.dwg

tetra_ev.dwg

The tetrahedron is bounded by four faces that are equilateral triangles, six edges, and four vertices; three faces meet at each vertex (figure IV-8, IV-9a).

A point/symbol on the *y*-axis, different from the origin, is replicated and rotated twice about the origin, each time at 120°. The three points are consecutively connected by a side/symbol (line) and by a face/symbol (AutoCAD "3dface"). These elements are then replicated and rotated at an angle of 70.5333° (70°32') about an axis defined by any sides of the triangle (be careful not to replicate elements in the same position). The new elements so obtained are then replicated and rotated twice about the origin at an angle of 120°.

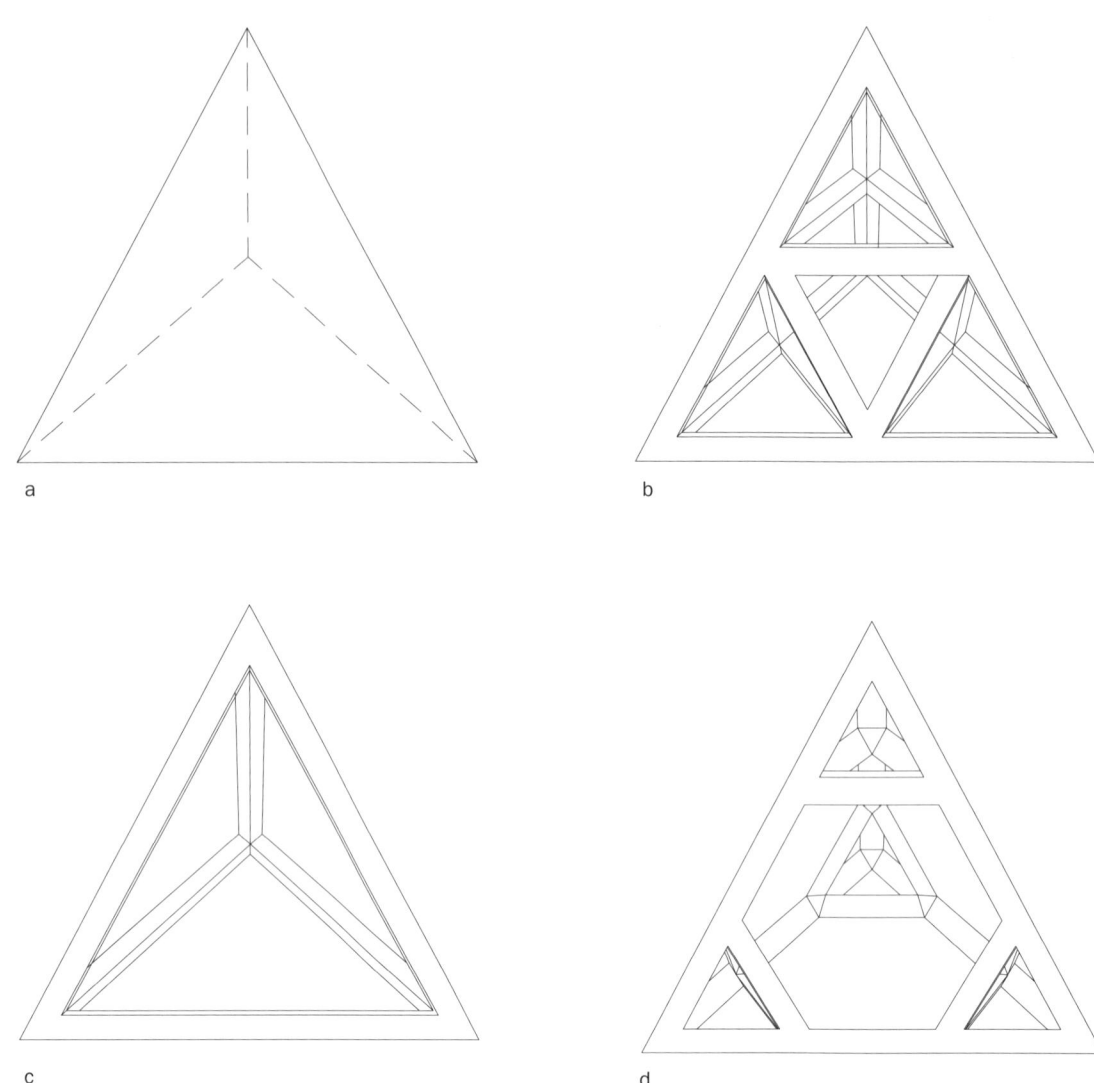

a
b
c
d

IV-9

a–d. A tetrahedron and varied evolutions.

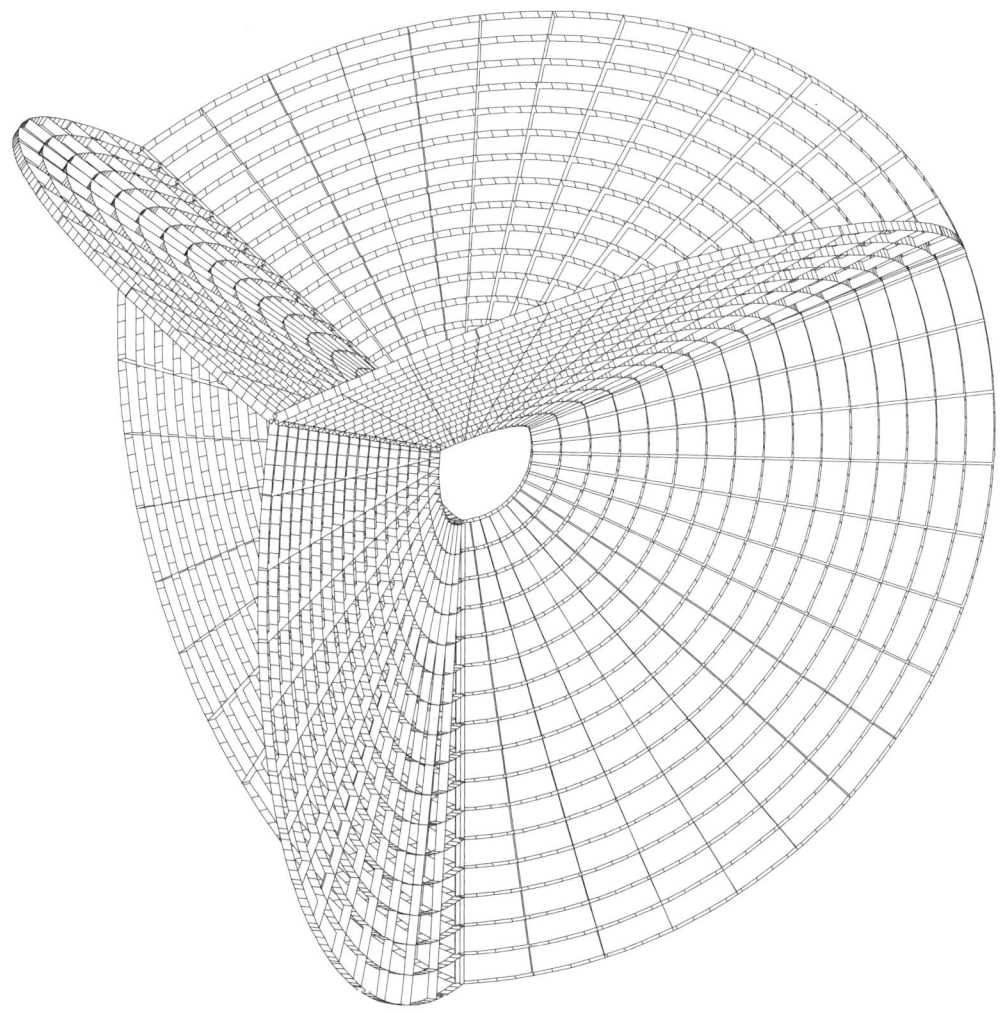

IV-9

e. Further evolution of a tetrahedron.

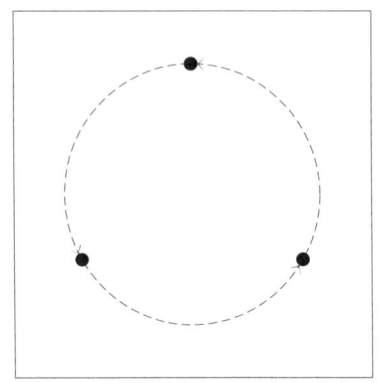
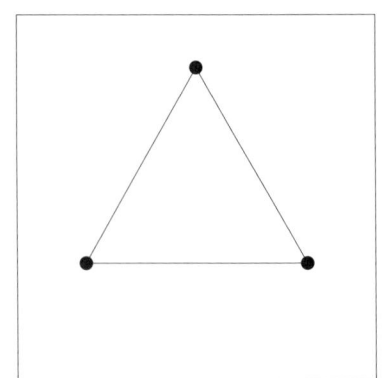
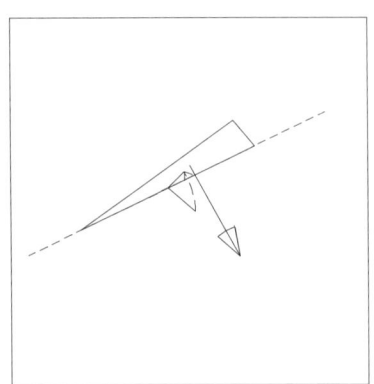
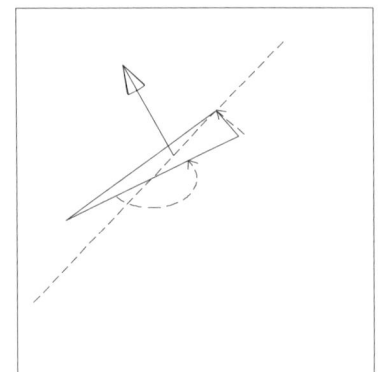
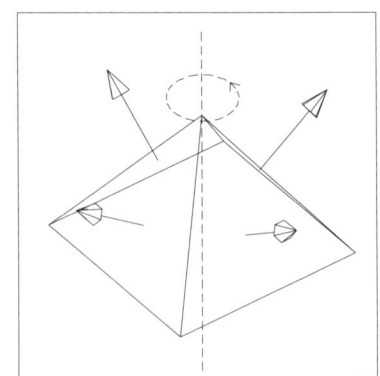
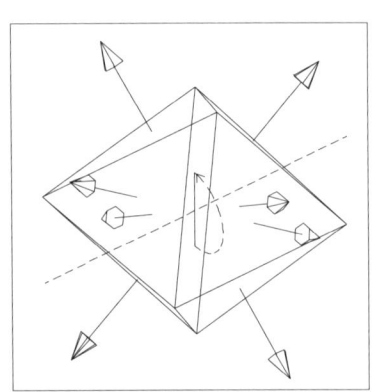
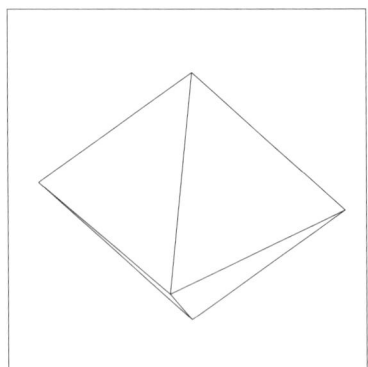

VISUALIZING WITH CAD

Octahedron: sequence of construction.

octa.dwg

octa_ev.dwg

Octahedron

The octahedron has six vertices, twelve edges, and eight faces that are equilateral triangles, meeting in groups of four at each vertex (figure IV-10, IV-11a).

A point/symbol on the y-axis, different from the origin, is replicated and rotated twice about the origin, each time at $120°$. The three points determine three vertices, which are consecutively connected by a side/symbol and by a face/symbol. All the elements are rotated at an angle of $54.7333°$ ($54°44'$) about an axis defined by any side, then translated until the vertex lying above the x-y plane has coordinates $(0,0,z)$ for some z. The translated elements are then replicated and rotated three times about the origin at an angle of $90°$. (Again be careful not to replicate elements in the same position.) All the elements so far obtained, except for the four vertices and the four sides belonging to the x-y plane, are replicated and rotated at an angle of $180°$ about a horizontal line passing through the origin.

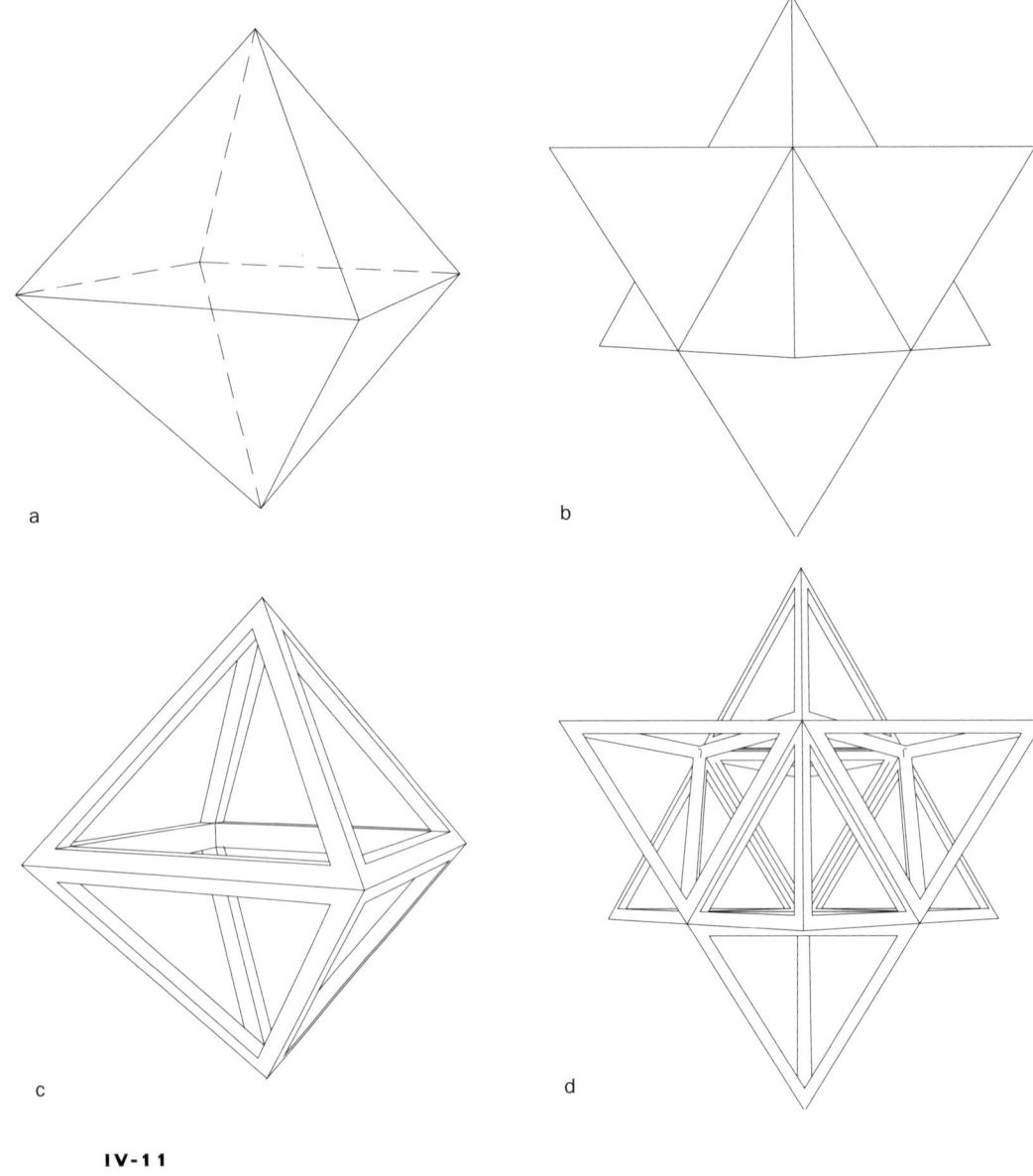

a

b

c

d

IV-11

a–d. An octahedron and varied evolutions.

VISUALIZING WITH CAD

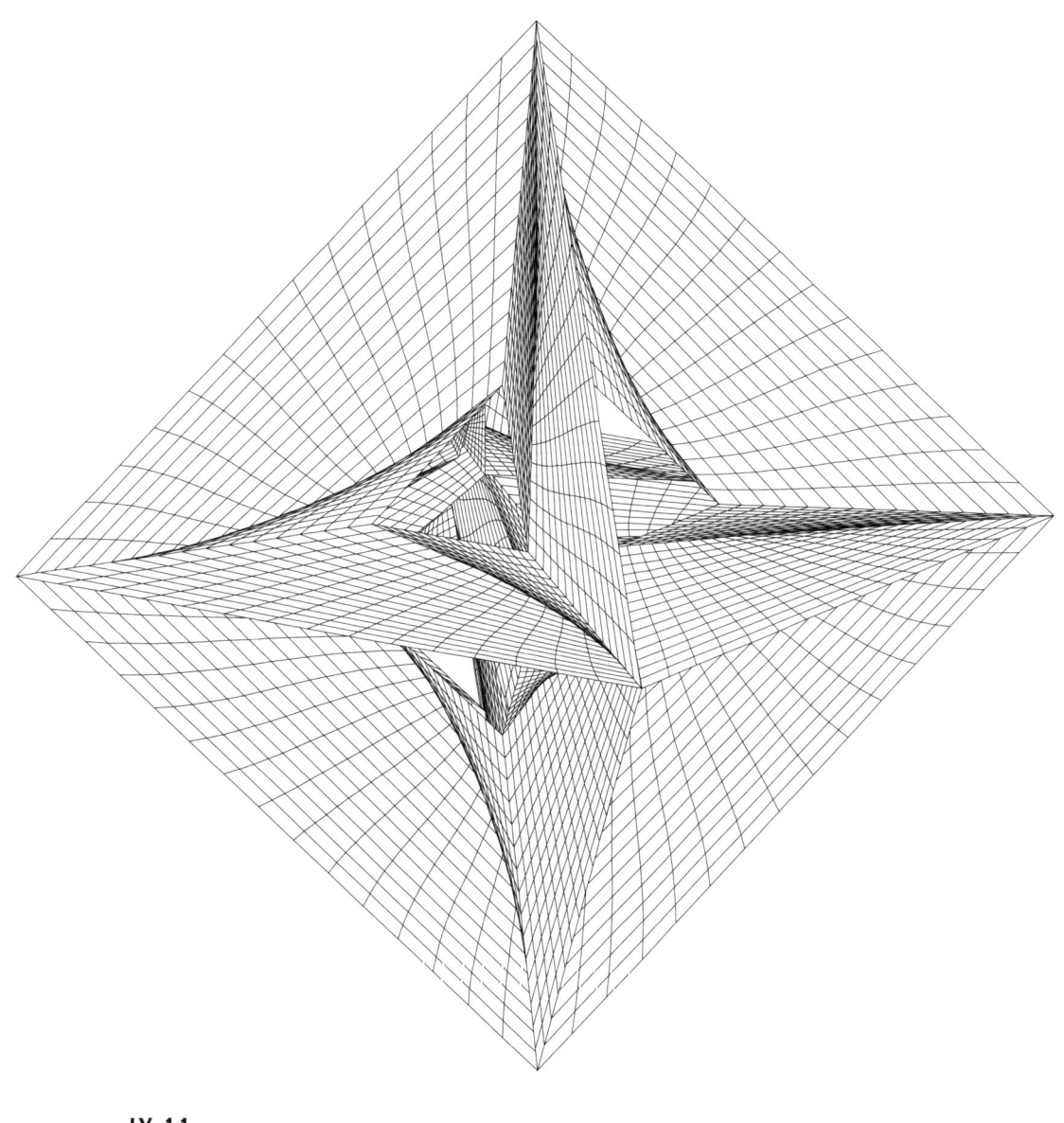

IV-11

e. Further evolution of an octahedron.

 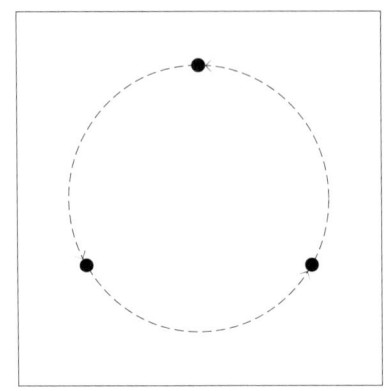 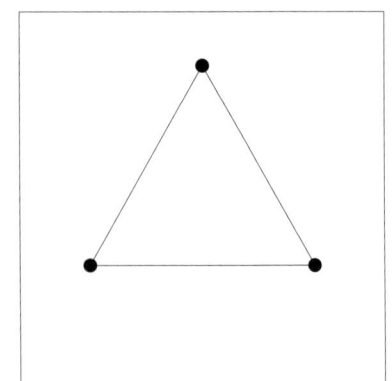

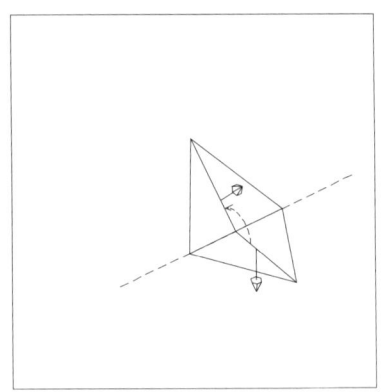 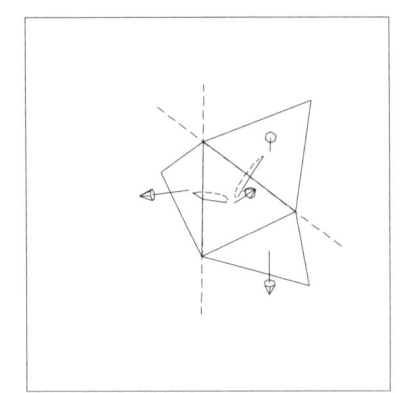 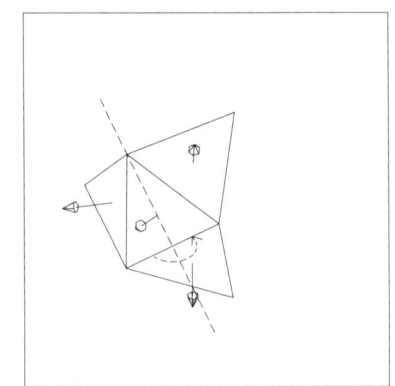

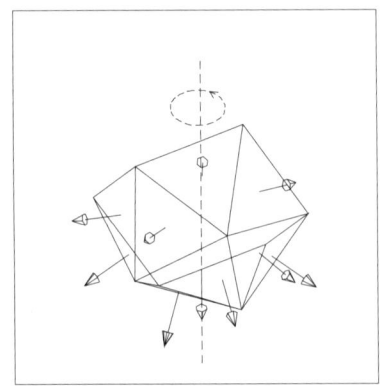 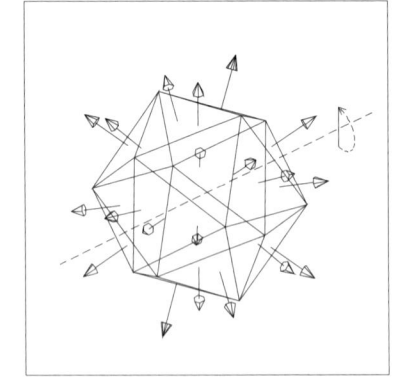 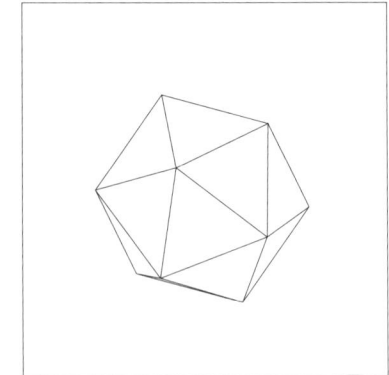

VISUALIZING WITH CAD

Icosahedron: sequence of construction.

icosa.dwg

icosa_ev.dwg

Icosahedron

The icosahedron has twelve vertices, thirty edges, and twenty faces, all equilateral triangles, meeting in groups of five at each vertex (figure IV-12, IV-13a).

A point/symbol on the *y*-axis, different from the origin, is replicated and rotated twice about the origin, each time at 120°. The three points determine the vertices of an equilateral triangle and are consecutively connected by a side/symbol and a face/symbol. All the elements are then rotated at an angle of 138.18333° (138°11') about an axis defined by any of the sides. Then they are twice replicated and rotated at an angle of 138.18333° (138°11') about an axis defined by each of the two sides not belonging to the *x*-*y* plane, with care taken not to replicate elements in the same position. All the elements so far obtained (except those belonging to the *x*-*y* plane) are replicated and rotated twice about the origin, each time at 120°. All the elements are then translated along the *z*-axis until the midpoint of any side that is not the boundary between two adjoining faces has *z* coordinate equal to zero. All the elements, except for the six vertices and the six sides that are not common boundaries between two adjoining faces, are then replicated and rotated at an angle of 180° about the *x*-axis.

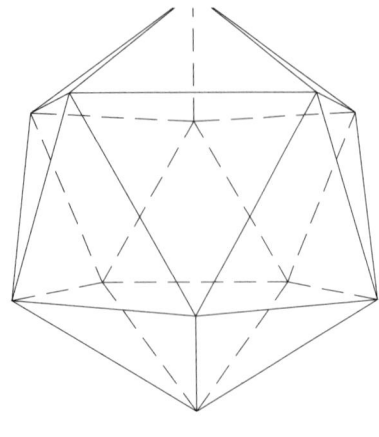

a

c

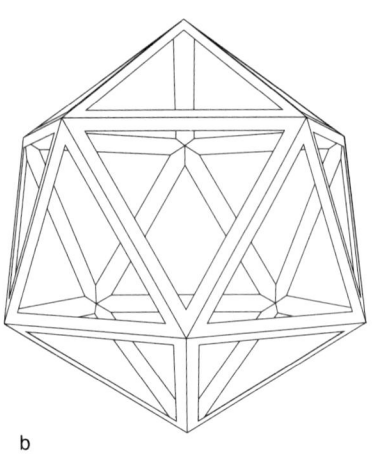

b

d

IV-13

a–d. An icosahedron and
varied evolutions.

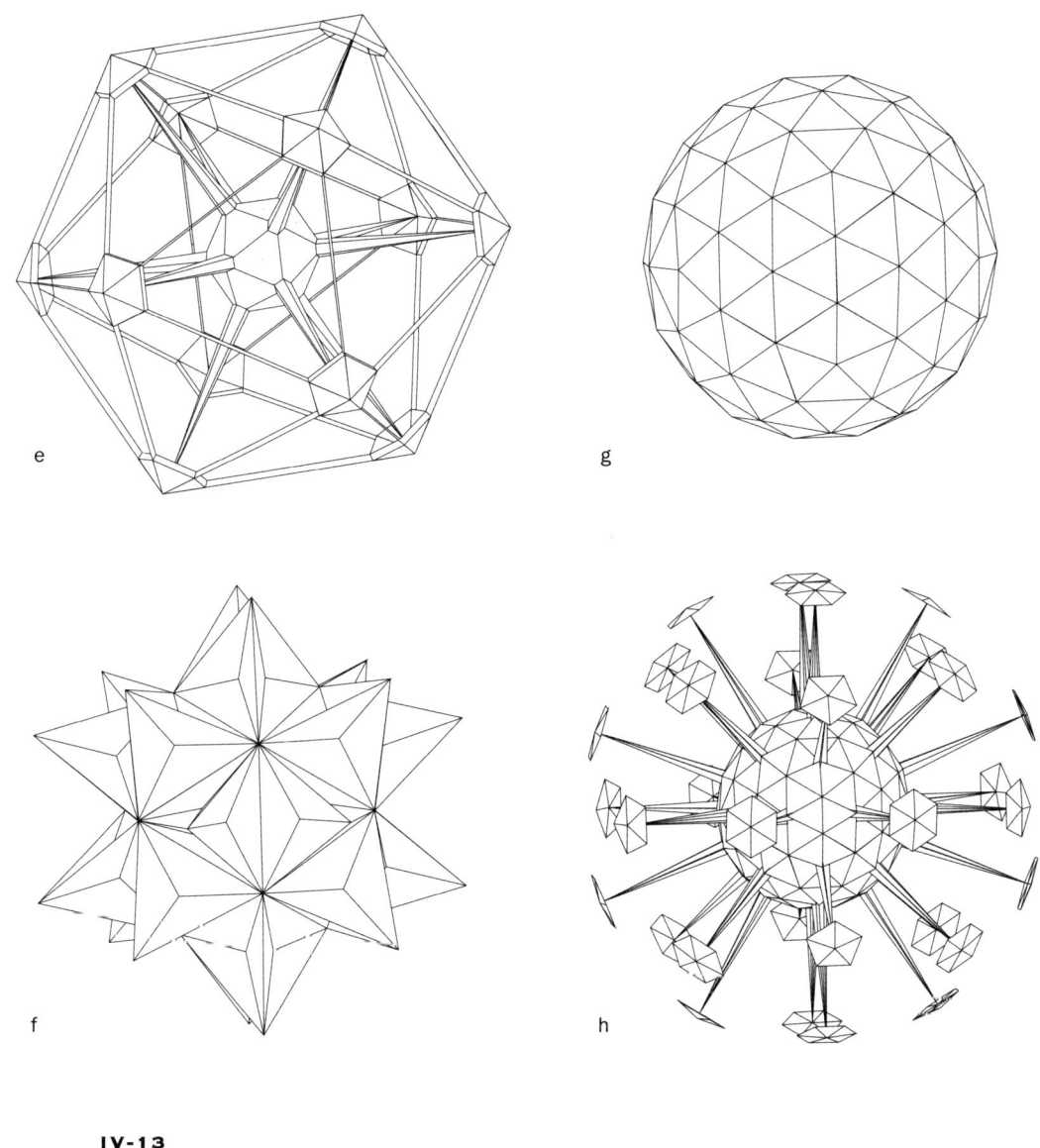

e

g

f

h

IV-13

e–h. Further evolutions of an icosahedron.

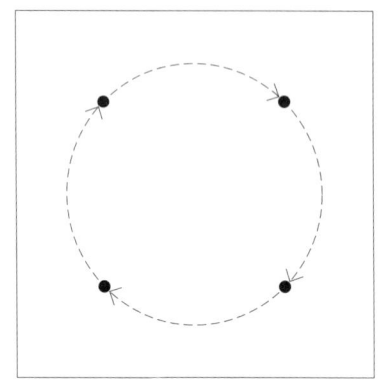
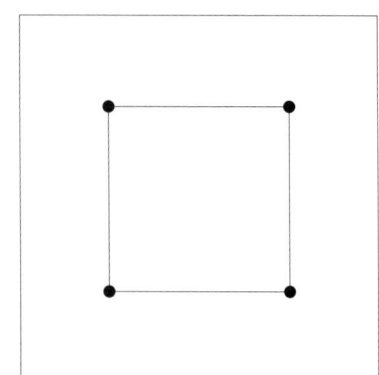
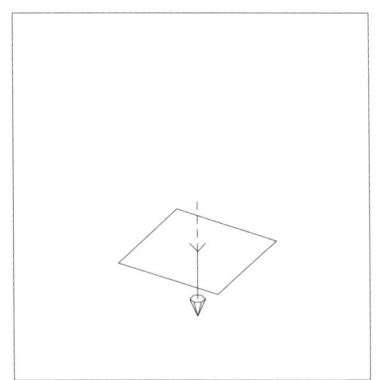
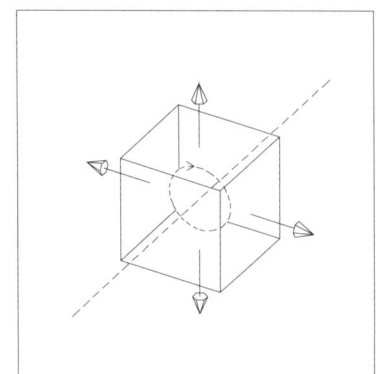
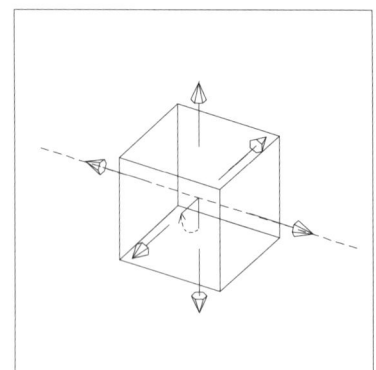
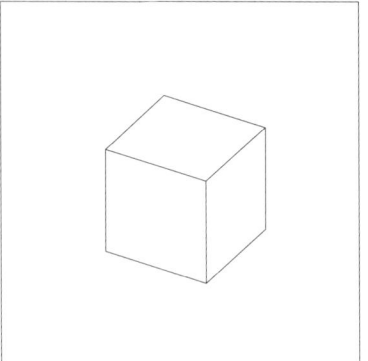

176

176

176

Cube

The cube has eight vertices, twelve edges, and six square faces, meeting in a group of three at each vertex (figure IV-14, IV-15a).

A point/symbol contained in the x-y plane, different from the origin and such that $x=y$, is replicated and rotated three times about the origin, each time by 90°. The four points represent the vertices of a square and are consecutively connected by side/symbols and then by a face/symbol. All these elements are then moved to a different horizontal plane, at distance from the x-y plane equal to half of the length of the side. Then this face is replicated and rotated three times, each time of an angle of 90° about the x-axis, so that no elements are replicted in the same position. The elements contained in horizontal planes are then replicated and rotated about the y-axis at an angle of 90°.

Cube: sequence of construction.

cube.dwg

cube_ev.dwg

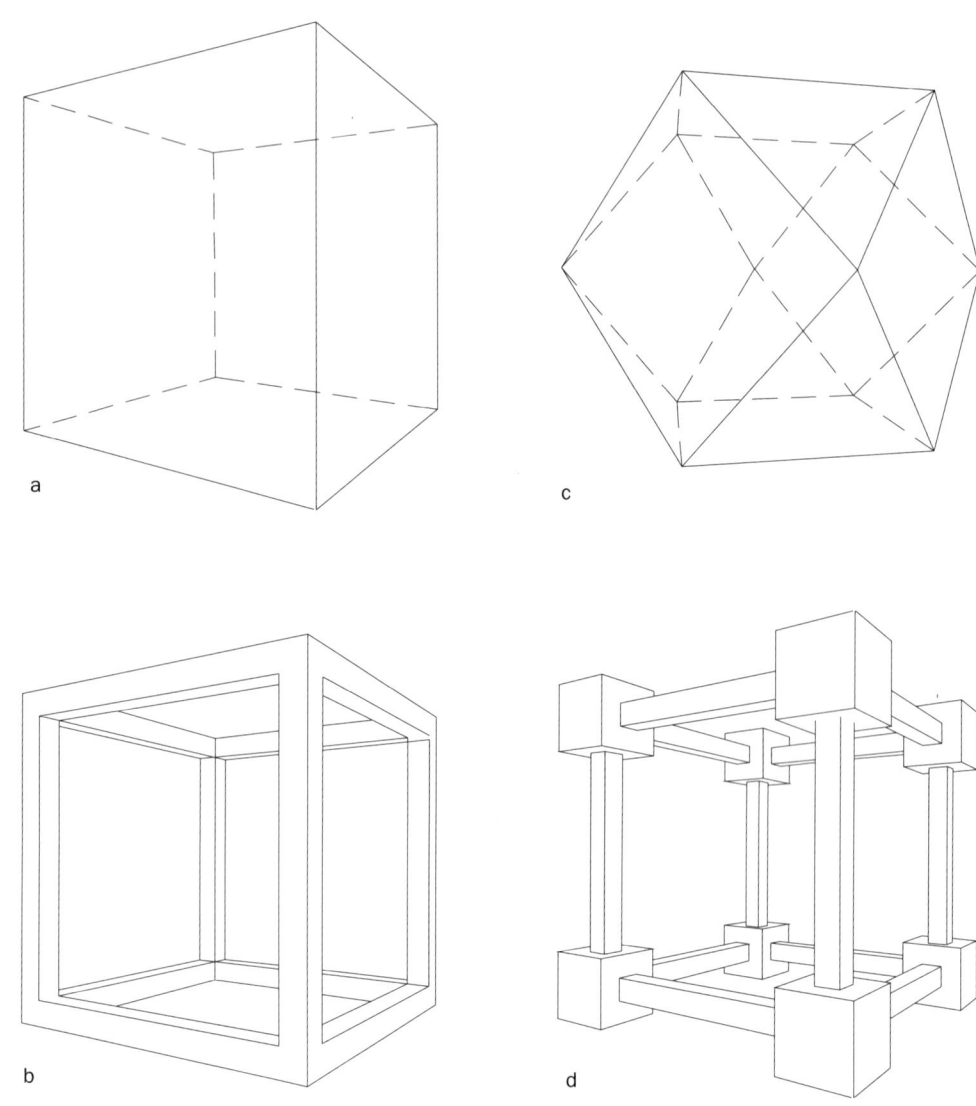

a

c

b

d

IV-15

a–d. A cube and varied evolutions.

VISUALIZING WITH CAD

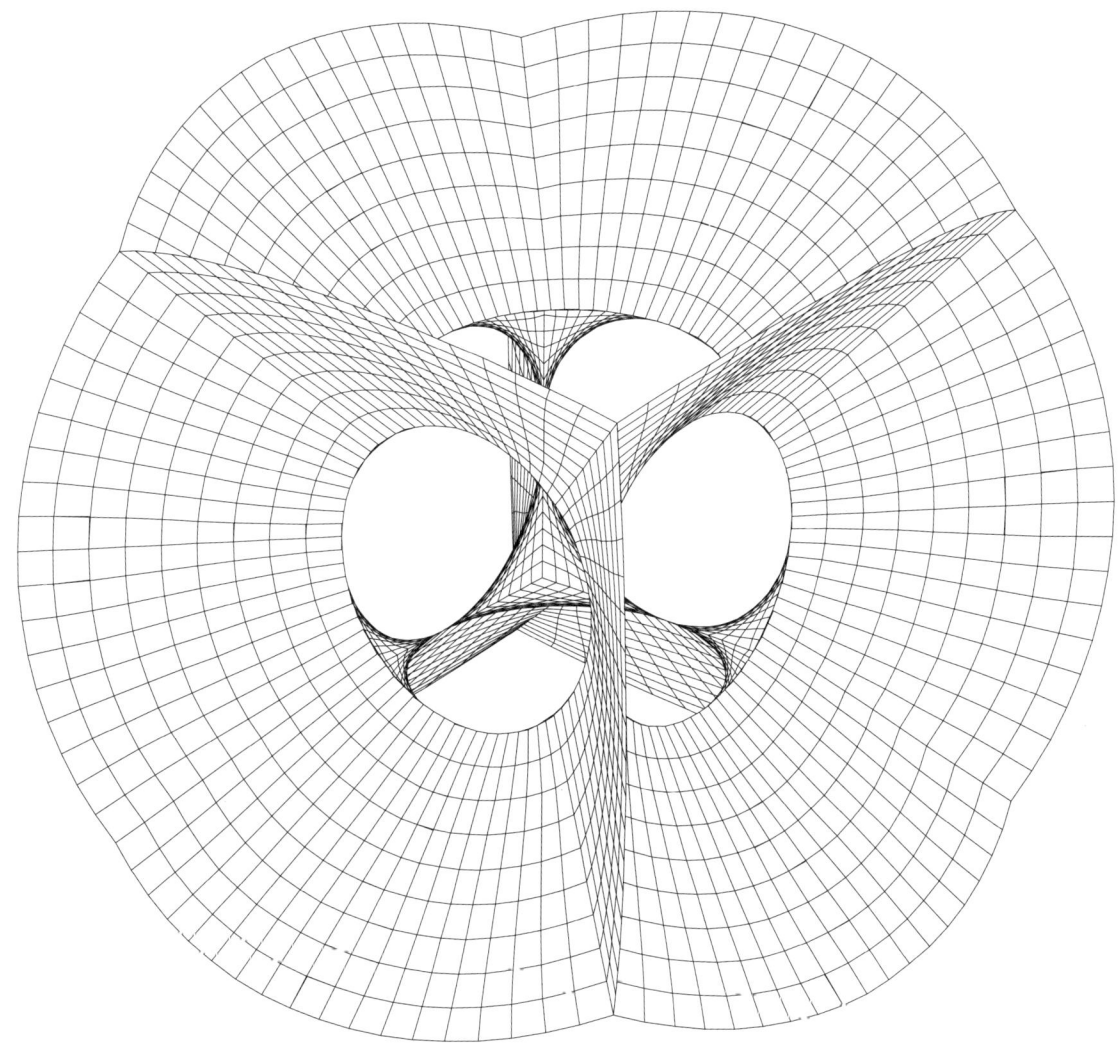

IV-15

e. Further evolution of a cube

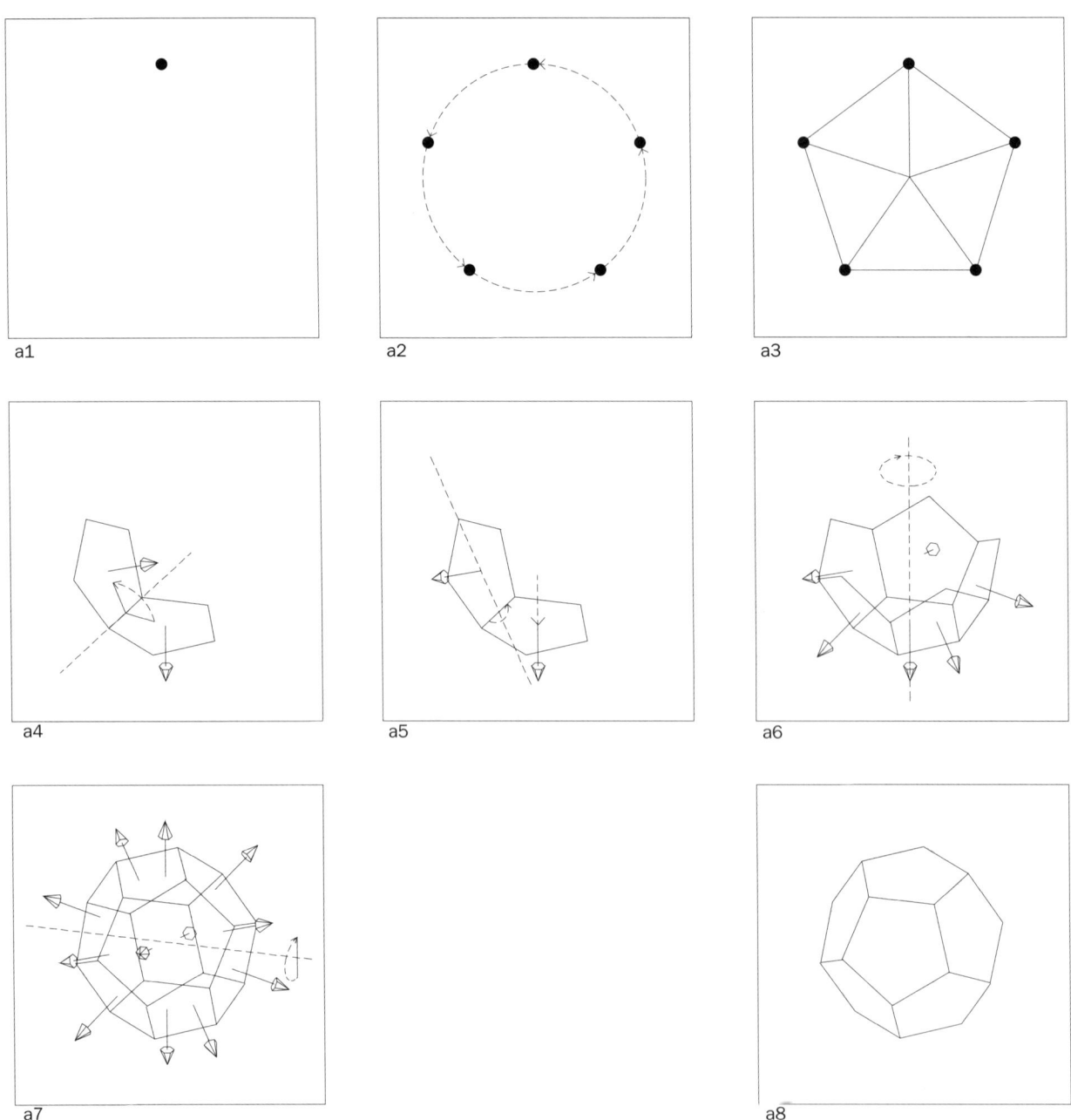

a1

a2

a3

a4

a5

a6

a7

a8

VISUALIZING WITH CAD

Dodecahedron

Dodecahedron: sequence of construction.

dode.dwg

dode_ev.dwg

The dodecahedron has twenty vertices, thirty edges, and twelve pentagonal faces, that meet in groups of three at each vertex (figure IV-16, IV-17a).

A point/symbol on the *y*-axis, other than the origin, is replicated and rotated four times about the origin, each time at 72°. The five points determine the vertices of a pentagon and are consecutively connected by a side/symbol and then by a face/symbol (made of five triangular faces with the common vertices at the origin). The elements so far obtained are then replicated and rotated at an angle of 116.5666° (116°34') about an axis given by any of the sides; then they are four times replicated and rotated at an angle of 72° about the origin. All the elements are then translated along the *z*-axis a distance such that the midpoint of any side that is not a boundary between two adjoining faces has *z* coordinate equal to zero. All the elements, except for the five vertices and the ten sides that are not boundaries between any two adjoining faces, are then replicated and rotated of an angle of 180° about the *x*-axis.

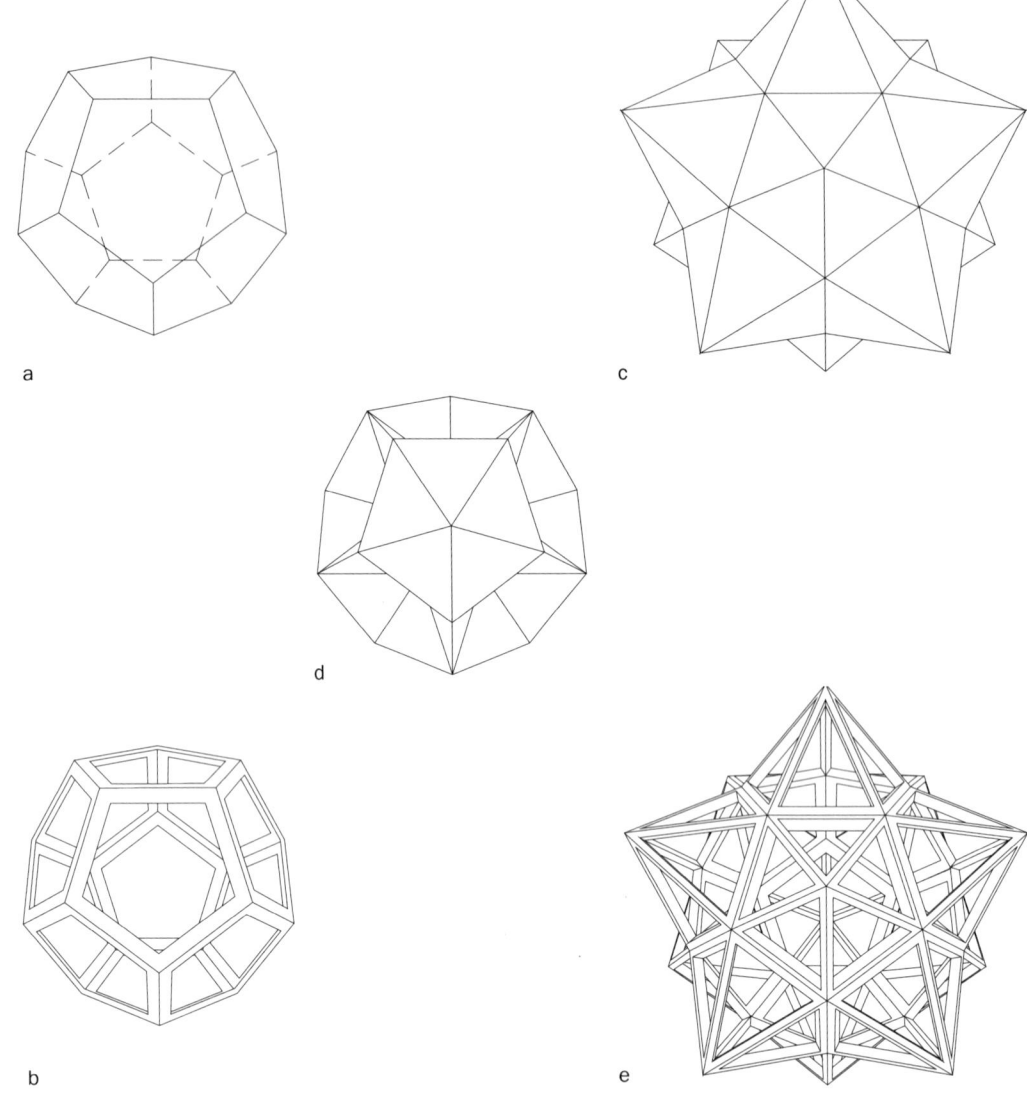

a

c

d

b

e

IV-17

a–e. A dodecahedron and
varied evolutions.

VISUALIZING WITH CAD

IV-17

f. Further evolution of a dodecahedron.

The geometry of the dynamic models of polyhedra described above can evolve in several ways. For example, if each face/symbol is replaced by a hole [Pacioli 1509] and the sides replaced by three-dimensional solid shapes, we obtain a form which is no longer a convex polyhedron, but which still conserves some of its geometric characteristics, such as the position of the vertices and sides. The solid-void relations become inverted and the new form does not define an enclosed space. Examples of such forms are shown in part b of the various polyhedra illustrations.

tetrahol.dwg
octahol.dwg
icosahol.dwg
cubehol.dwg
dodehol.dwg

A cuboctahedron (figure IV-15c) can be constructed as an evolution of a cube if each vertex/symbol is replaced by a new symbol composed of an equilateral triangle with vertices defined by the midpoints of each of the three cube sides joined in the original vertex. Each face is replaced by a new one, a square with vertices at the midpoints of the sides of the original face. The original cube sides are eliminated.

Stellar Polyhedra

An interesting type of evolution is that of stellar polyhedra, which are constructed by substituting for each face/symbol a pyramid, the base of which is shaped like the original face. If the base of the pyramid is an equilateral triangle, as in the tetrahedron, octahedron, and icosahedron, the pyramid can be a tetrahedron. Note that stellar polyhedra are neither convex nor regular. The stellar polyhedra of the illustrations have been generated from an octahedron (figure IV-11c), icosahedron (figure IV-13c), and dodecahedron (figure IV-17c). A different type of stellar polyhedron has also been obtained from the dodecahedron (figure IV-17d), where the pyramids have replaced the faces of the polyhedron in an inside rather than outside orientation. As with the regular polyhedra, each face of a stellar polyhedron can be replaced by a hole, and the sides can evolve into solid shapes, as is shown in figures IV-11d, IV-13d, and IV-17e.

tetrastl.dwg
octastl.dwg
icosastl.dwg
cubestl.dwg
dodestl.dwg

DISCRETE APPROXIMATION OF CONTINUOUS FORMS

The forms so far investigated have been defined in terms of discrete elements. In contrast, the forms generated in the following sections are inspired by continuous surfaces such as those studied in differential geometry. In CAD models, continuous forms are often approximated; a continuous surface can be defined by a set of contiguous triangular or quadrangular planar surfaces. The greater the number of discrete elements used to represent a continuous form, the better is the approximation of continuity, both conceptually and visually. A clear example is given by the model of a sphere shown in figure IV-18a: The approximation of a smooth surface improves with the number of faces of the polyhedron. An analogous two-dimensional example is the approximation of a circle by a polygon (figure IV-18b). In CAD, a continuous form can be approximated by a model constructed of modular elements (symbols) positioned

IV-18

a. Discrete approximation of a sphere.

b. Discrete approximation of a circle by a polygon.

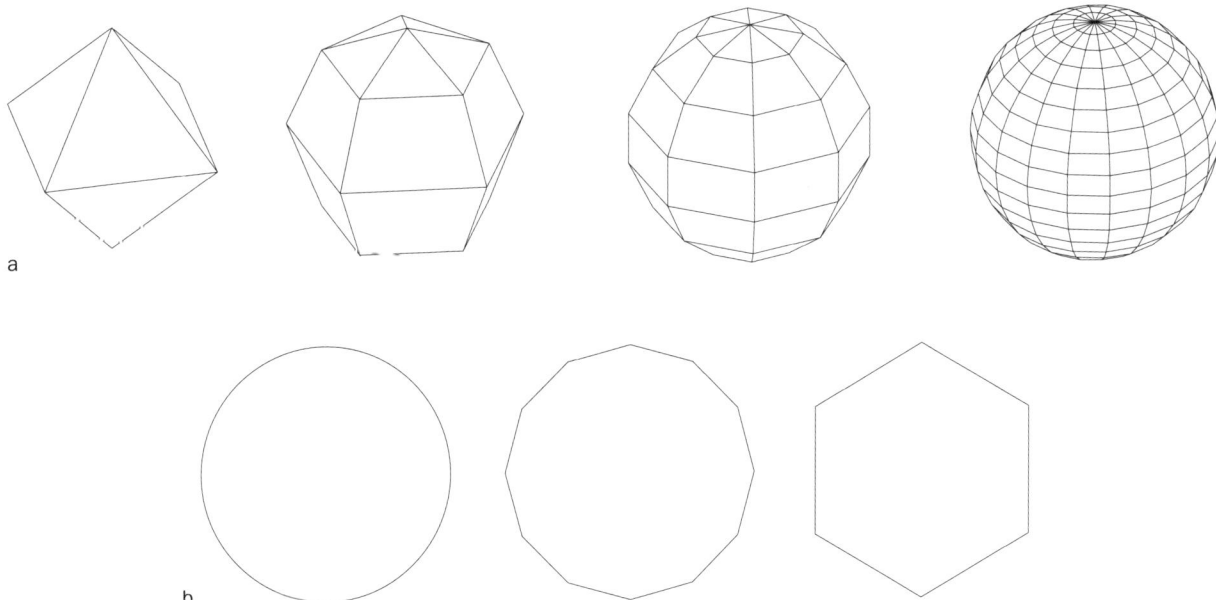

a

b

and replicated by means of the fundamental orthogonal and similarity transformations. This approach is used to construct the models shown in the rest of this chapter, and by using a sufficiently large number of discrete elements to approxiate continuous forms, we can generate CAD models of any form. In some of the discrete models of "continuous" forms (figure IV-19), the number of elements is intentionally kept to a minimum, to make the process clear.

A similar technique is often used in mathematics, where continuous functions, such as those found in integral and differential cal-

Discrete approximation of
a continuous form.

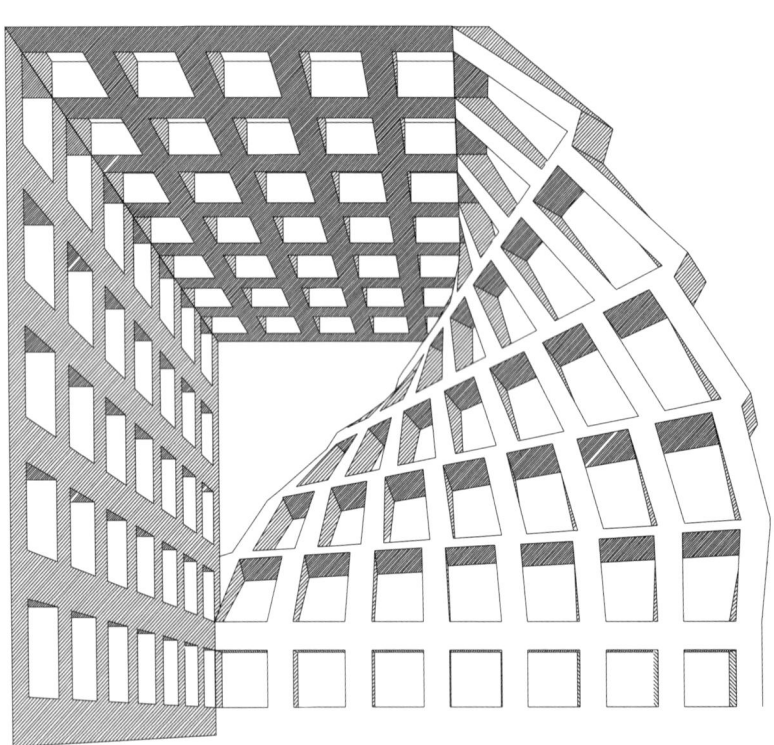

VISUALIZING WITH CAD

culus, can be approximated by finite sets of discrete elements. This type of approximation allows the use of numerical methods in place of integrals and differential equations, which are often cumbersome or even impossible to solve. Discrete numerical methods are often used for the mathematical solution of physics problems.

In the evolution of forms, the substitution of complex shapes for primitive elements also gives an intuitive idea of how continuous forms can be composed from separate elements. At a different semantic level, the physical construction of a designed form provides another example: A tensile structure, the geometrical model of which is given by a minimal surface, can be constructed as a cable net. The cable is shaped as a straight line segment connecting the intersection points of the net, which are points belonging to the minimal surface of the geometric model.

HELICOID

As mentioned in Chapter III in regard to combined transformations, a helix is the curve traced by a point moving continuously in space according to a twist. An architectural example of a helix is the rail of a spiral staircase. A two-dimensional analogue is a helicoid, a surface generated by a straight line moving in such a way that it passes through a fixed helix (figure IV-20a).

A generic helicoid can be generated with a **line/symbol**, replicated and rotated at a certain angle about a vertical axis. After each rotation the line is translated at a displacement z along the same vertical axis of rotation. The line in the first symbol is then replaced by a surface defined by itself and the adjacent line/symbol. The form so generated is a discrete approximation of a continuous helicoidal surface. This dynamic model of a helicoid may further evolve into several others in which the the surface in the symbol is replaced by a configuration of three-dimensional elements (figure IV-20a).

helix.dwg

helix_ev.dwg

a. Helicoid.

b. Evolution of a helicoid.

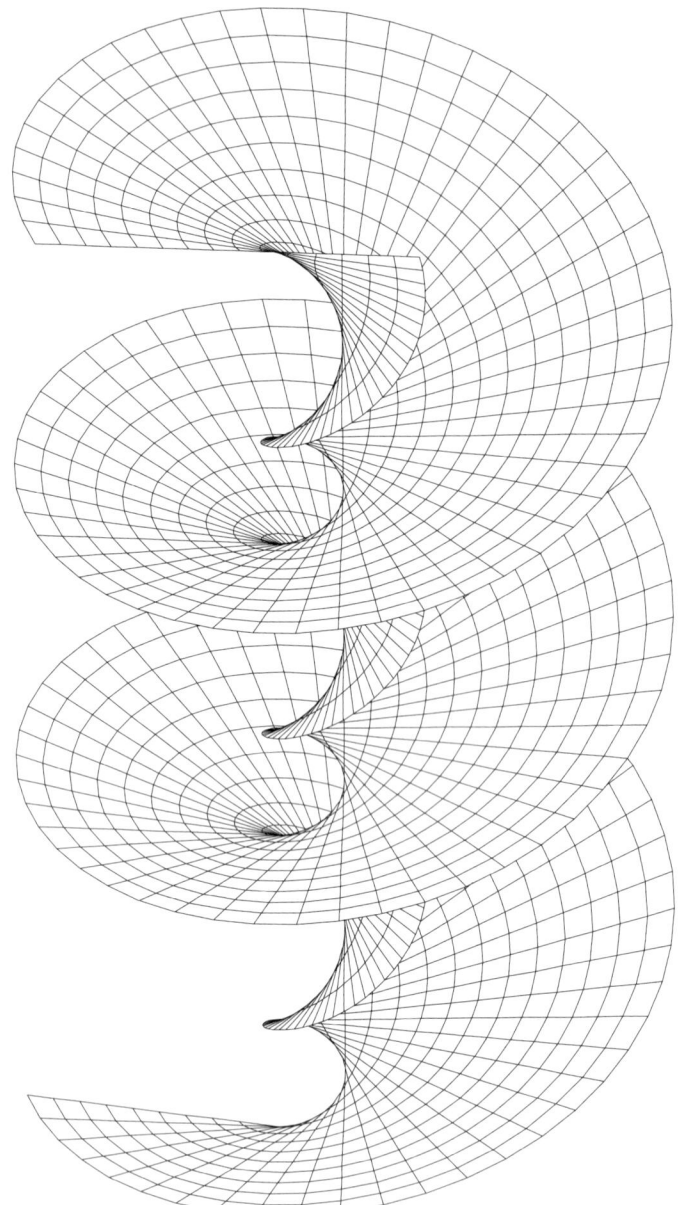

a

b

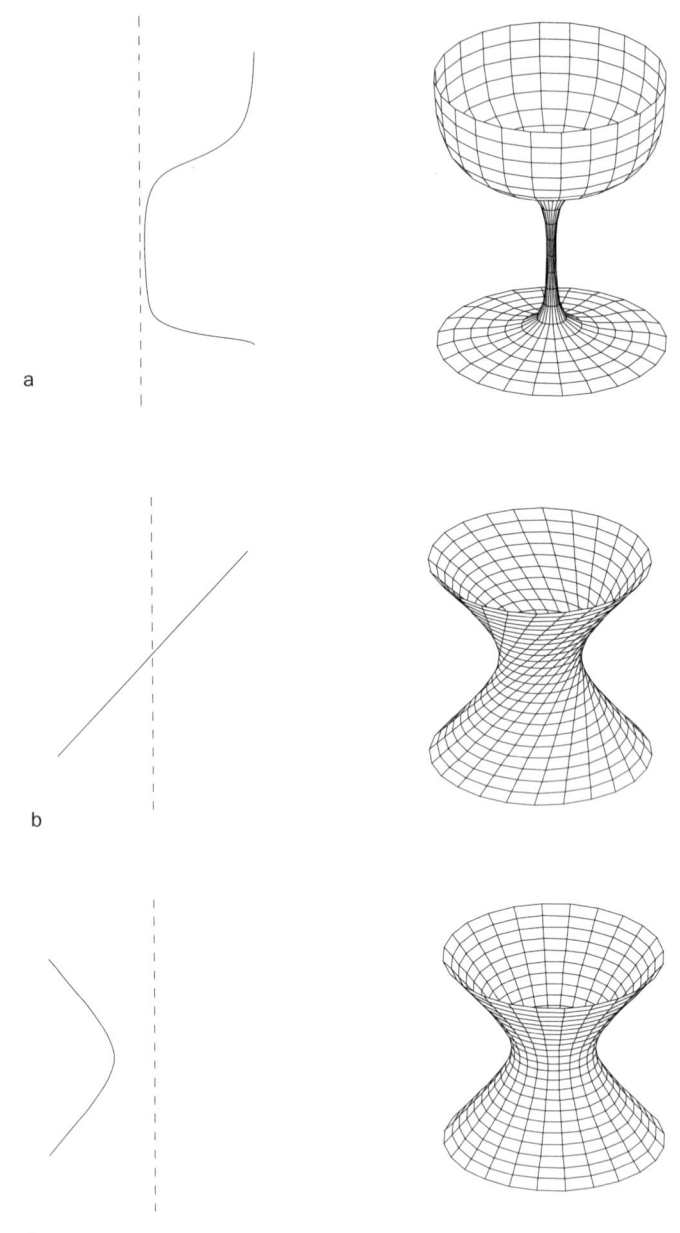

a

b

c

VISUALIZING WITH CAD

SURFACES OF REVOLUTION

Revolution surfaces

a. Surface obtained by the revolution of a two-dimensional curve around an axis contained in the same plane.

b. Surface obtained by the revolution of a two-dimensional curve around an axis contained in a different plane.

c. Surface obtained by the revolution of a hyperbola around an axis lying in the same plane; this surface is equilvalent to that shown in b.

Surfaces of revolution are generated by the rotation of a two-dimensional curve (lines are the simplest example) around a generic axis contained in the plane of the curve (figure IV-21a). The two-dimensional curve is a symbol that can be generated by transformations of other simpler element/symbols. Everyday life affords many examples of revolution surfaces, such as glasses, bottles, vases, dishes, pencils, and so on. In the universe of architectural forms, examples of surfaces of revolution are given by anything with a circular or circle-derived plan, such as columns, towers, and amphitheaters.

If a straight line is rotated about an axis lying in a plane different from that of the line itself (figure IV-21b), the resulting form is the same as would be generated by rotating a hyperbola about an axis lying in the hyperbola plane (figure IV-21c). Once again, the straight line that is rotated must be identified by a symbol to allow later evolution.

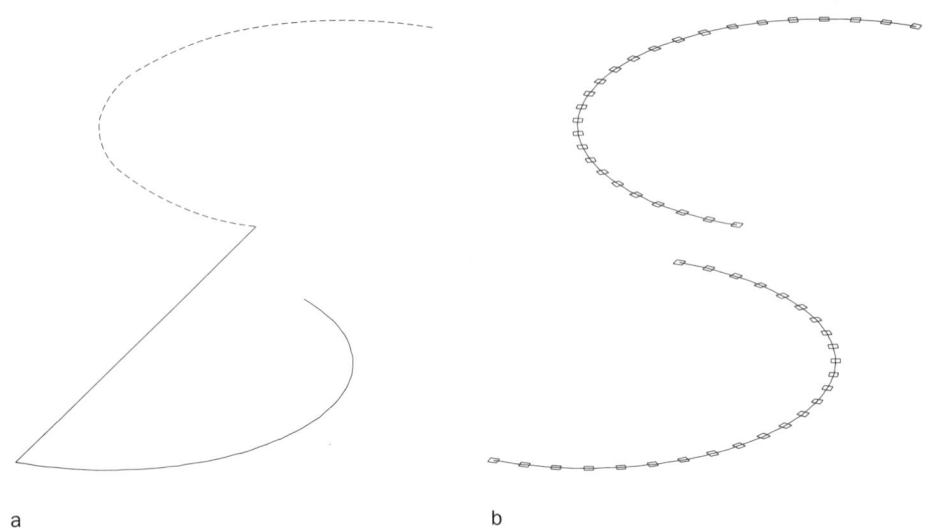

a

b

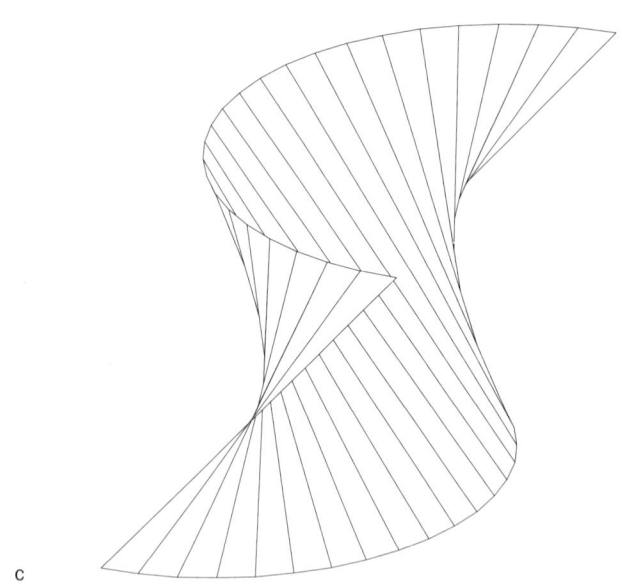

c

VISUALIZING WITH CAD

Ruled surfaces.

Ruled surfaces are generated by a straight line moving along a path (figure IV-22a). The discrete model of a ruled surface can also be described as follows. Two generic paths (curved or straight) in three-dimensional space are each divided at equal intervals by points/symbols (figure IV-22b). A pair of consecutive point/symbols from one path and the corresponding pair of point/symbols from the second path determine a surface, and the set of all these constituent surfaces creates a ruled surface (figure IV-22c). The sides of these constituent surfaces that are not part of the original paths represent the moving straight line that generates the ruled surface according to the definition above. Again, definition of the boundary points as symbols will allow the dynamic model of this form to evolve.

Ruled surfaces are widely used in architecture and engineering for the design of large span roofs, in which regard they will be discussed again in Chapter V.

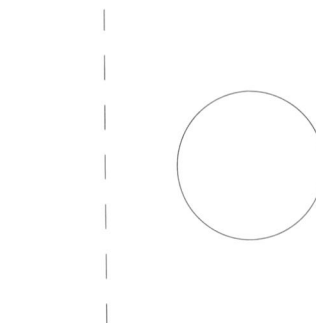

a

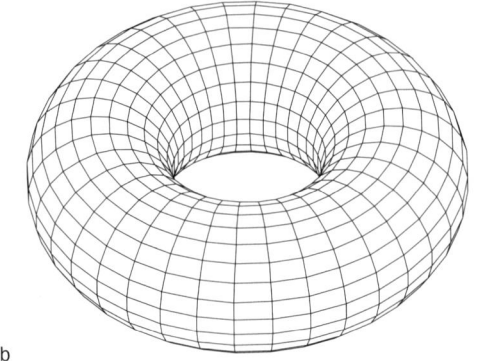

b

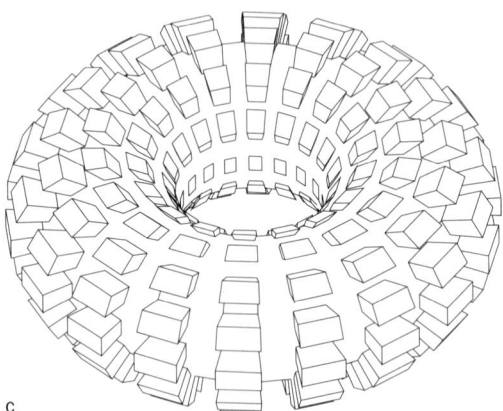

c

VISUALIZING WITH CAD

Torus.

torus_ev.dwg

The torus is a surface shaped like a donut (figure IV-23b). Geometrically the torus is defined as a surface of revolution generated by the rotation of a circle around an axis that lies in the plane of the circle but does not intersect it (figure IV-23a). The torus is therefore generated by a double rotation transformation: the rotation that produces the circle and the rotation that generates the torus itself. For the toris to evolve as a dynamic model (figure IV-23c), the circle must be a **nested symbol**, composed of a **circular array** of symbols.

The torus is treated here separately from the other revolution surfaces because of its topological characteristics. Recall that topology is the branch of geometry concerned with the continuous transformation of one form into another. The original form may be distorted in any way except by "tearing" or "gluing"; the only relation that must be preserved during the transformation is the continuous connectivity of the original form. In this logic, a cube can be transformed into a sphere (figure IV-24a), but not into a torus, because that would require the cube to be "torn" to create a hole, which is not allowed. The prismatic block (a prism with a hole) in figure IV-24b, however, can be deformed into a torus. A criterion that indicates which forms are equivalent under this type of deformation is given by the so-called *connectivity number h*, which is related to Euler's formula. As previously mentioned, the relation of the numbers of vertices, edges, and faces given by Euler's formula (V - E + F = 2) characterizes the class of forms called simply connected convex polyhedra [Hilbert 1952, Cundy 1961]. By contrast, the heptahedron in figure IV-24c has seven faces, twelve edges, and six vertices, which produce in Euler's equation the value V - E + F = 1, which indicates that the heptahedron belongs to a different topological class from the cube and sphere. Euler's formula can be generalized, however, in the form

$$V - E + F = 3 - h$$

where, again, *h* is the connectivity number. Although the geometric meaning of *h* is beyond the scope of this discussion, it has a very imme-

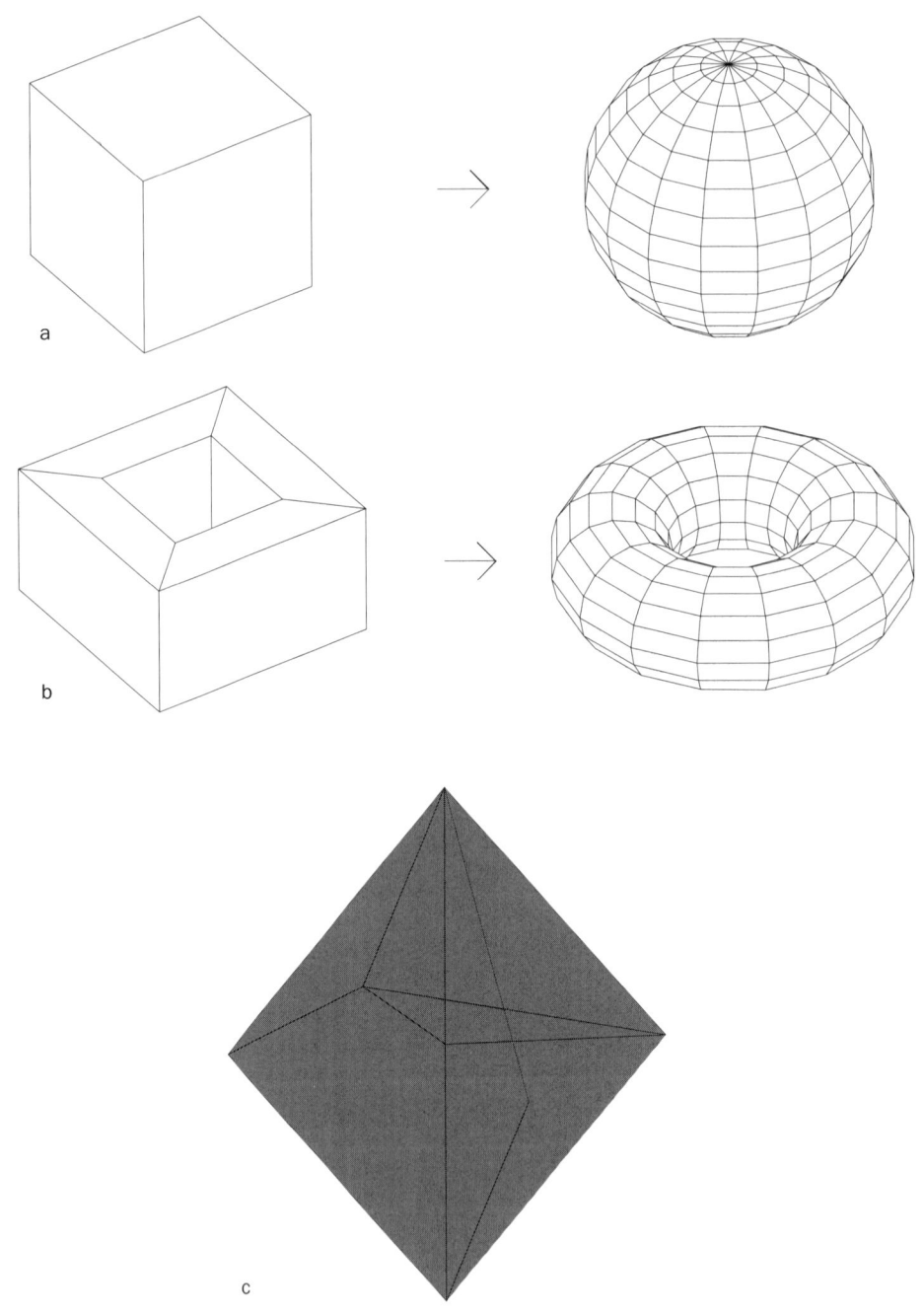

a

b

c

VISUALIZING WITH CAD

diate and intuitive meaning: In fact, $h = 2p + 1$, where p is the number of holes in the form. Thus, Euler's formula provides a criterion for classifying topologically equivalent forms—those that can be deformed into one another—on the basis of the number of holes they have.

MOEBIUS STRIP

All the shapes we examined so far can be considered two-sided surfaces, with boundaries defined by curves or line segments. In two-sided surfaces it is impossible to trace a path from one side of the surface to the other without crossing the boundary or piercing the surface. In **one-sided surfaces** such as the Moebius strip (figure IV-25a), however, it is possible to do so, with the surprising result that such shapes do not have an inside and outside. The Moebius strip has another interesting characteristic: Its boundary is a closed curve, which can be deformed into a circle. Thus, an arrow, perpendicular to one edge, will, if it is moved to a point half way along the boundary, be pointing in exactly the opposite direction of its original position, which shows that this shape presents no distinction between up and down.

moebius.dwg

The Moebius strip's construction can be approximated by a line rotated by 360° (and simultaneously replicated) about a generic axis belonging to the same plane, in $n–1$ movements of constant angular increment equal to $360°/n$. After each incremental movement the new line is also rotated (but not replicated) by an increment equal to half of the previous angle about the perpendicular axis that passes through its midpoint and is coplanar with the first rotation axis. The line can be made a symbol and evolved into several different shapes (figure IV-25b c).

helix_ev.dwg

ITERATIONS

In mathematics an iterative function is a function $f(x)$ in which the value of the variable x is itself determined by the function f, once it has been assigned an initial value. Once the initial value of x has been defined, the iteration can be repeated endlessly. An example is pro-

a. Moebius strip.

b. c. Shapes evolved from a Moebius strip.

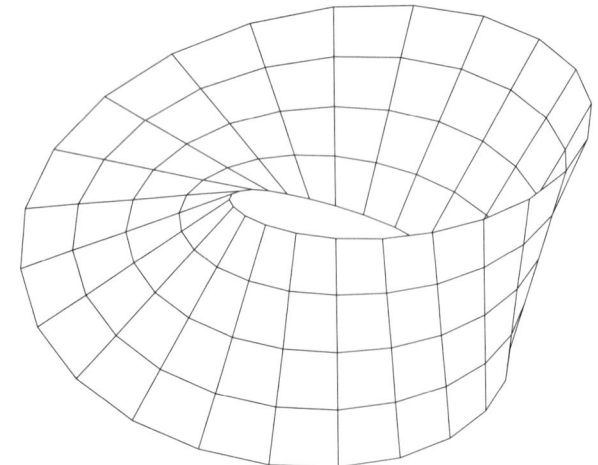

a

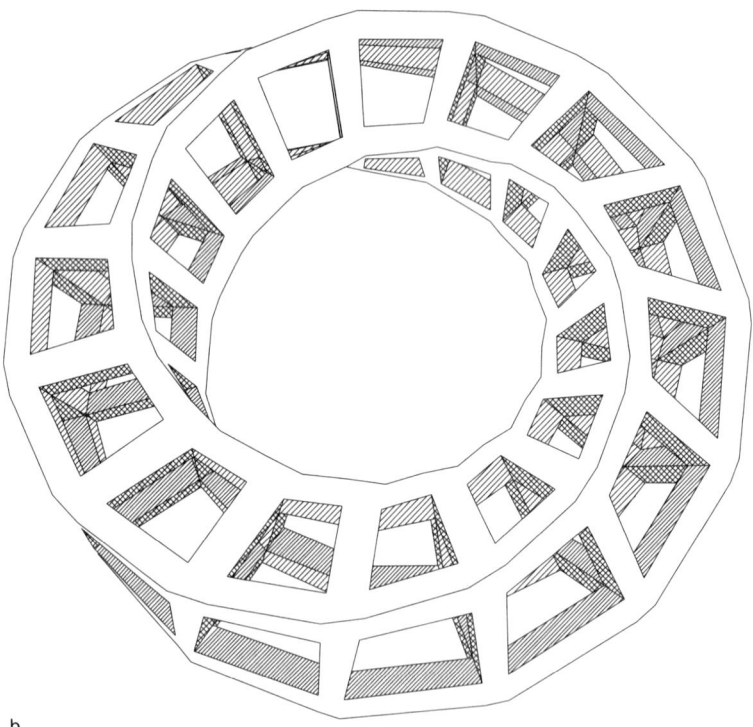

b

VISUALIZING WITH CAD

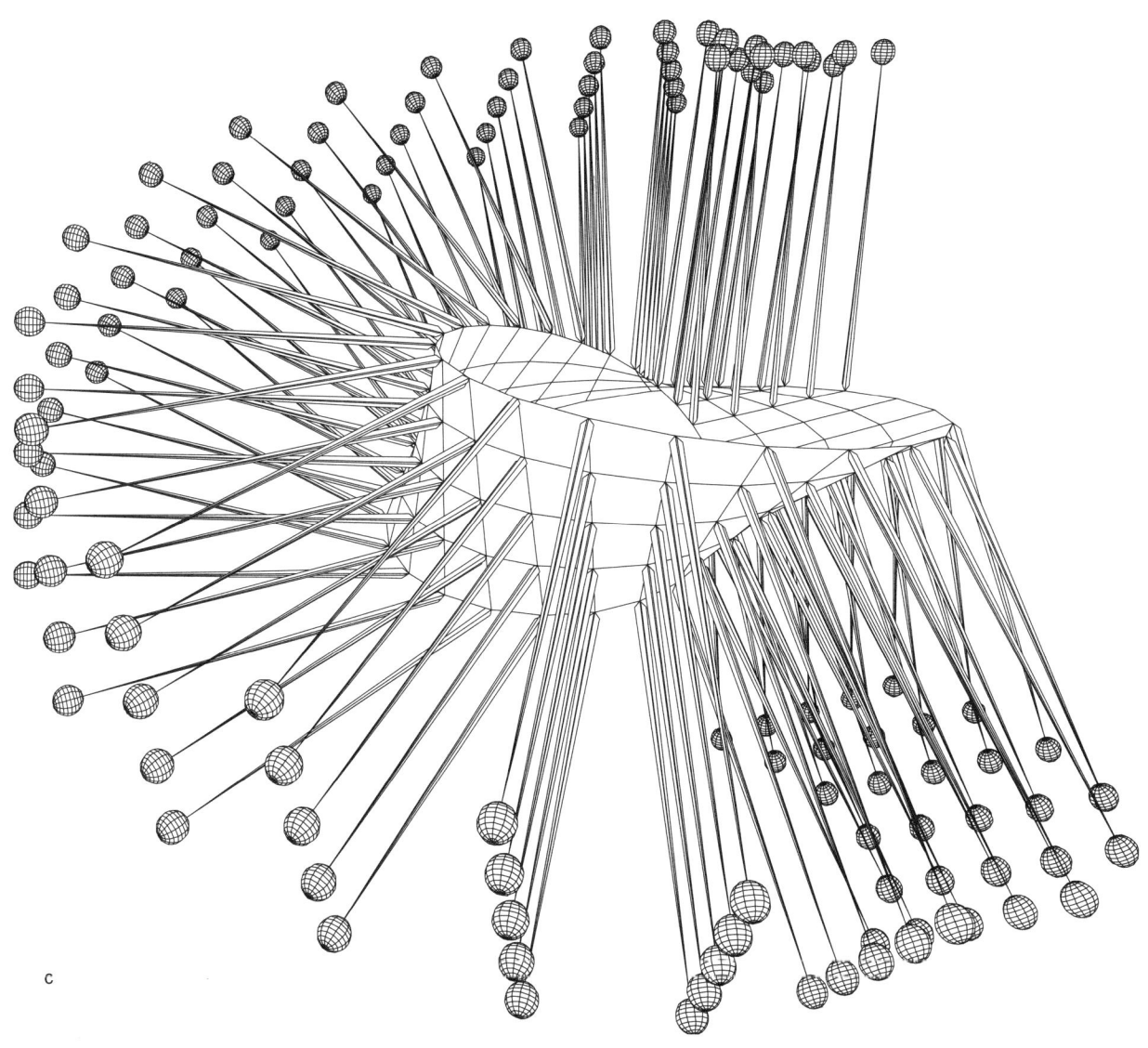

c

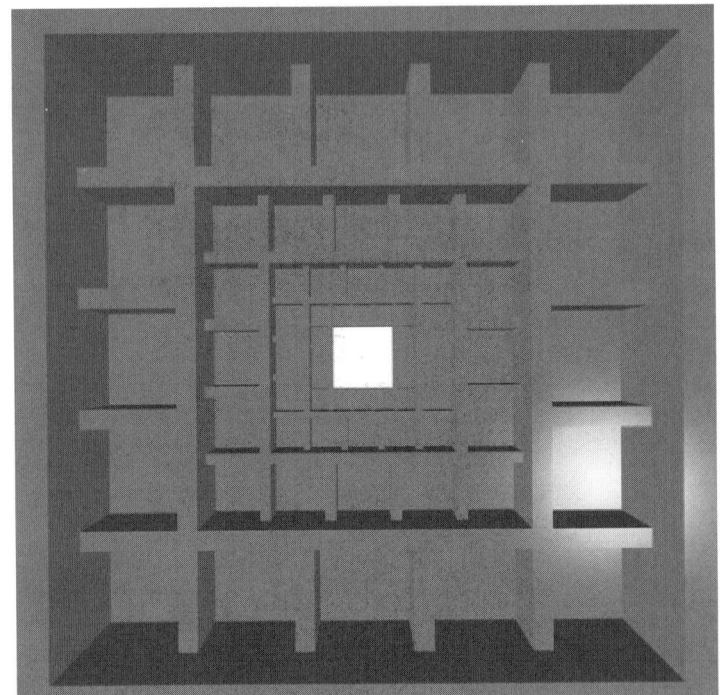

a

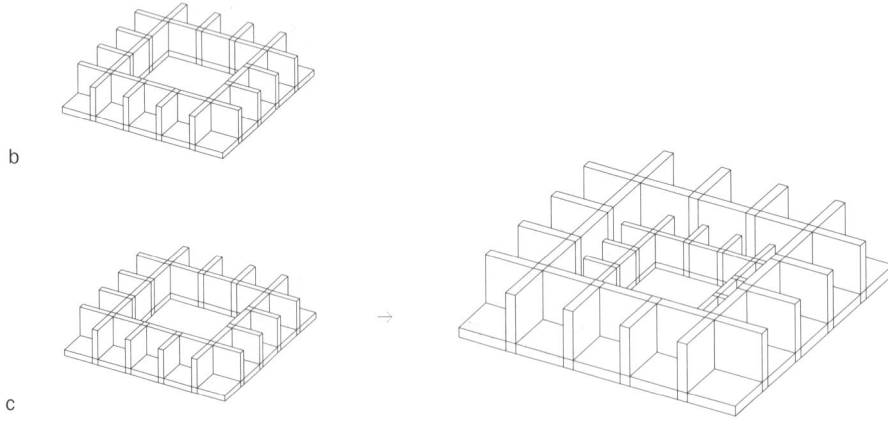

b

c

$\rightarrow$

VISUALIZING WITH CAD

Shape generated by iterations.

b. c. Initial shape and iteration process
for the shape shown in 22-a.

iterat.dwg

vided by the Fibonacci sequence (0,1,1,2,3,5,8,13,21, . . .), for which the initial variable values are defined to be $x_0 = 0$ and $x_1 = 1$, and the function is defined by $f(x) = (x_{i-1} + x_{i-2})$ (for $i = 2,3, . . .$).

In computer-aided design, an iteration can be efficiently constructed by making use of geometric transformations and symbols. A graphic example is provided by the form shown in figure III-26a, where the initial **shape/symbol** was given by a set of parallelepipeds (figure III-26b). The iterative function is defined so that the initial shape/symbol is enlarged by a scaling factor that makes the inner perimeter of the new shape coincide with the outer perimeter of the initial shape; the process is repeated in subsequent iterations (figure IV-26c). The model of the form generated in this way can easily evolve with the replacement of the shape contained in the nested symbol.

The spiral shape of figure IV-27 offers another example of iteration. In this case the iterative function is given by the combined rotation and scaling of the initial shape/symbol. The use of symbols is particularly powerful for the evolution of this shape. The scaling factor and rotation angle for each repeated symbol are such that the outer boundaries of the iterated shape coincide with the inner boundaries of the previous shape (figure IV 27a). Figure IV-27b shows the new symbols used in the transformation operations by which the initial form evolves into the shapes of figures IV-27c through IV-27g.

FURTHER DEVELOPMENTS

In the course of this chapter we have finally progressed to applications of the theory discussed in Chapter III. The electronic models thus far generated provide a framework for exploring several geometric forms that may be cumbersome to visualize without the use of CAD. We have also seen how these initial models can evolve into new forms that can be topologically completely different, even though the same geometric relations among primitives are maintained. The reader is invited to explore new forms derived from the provided models.

These geometric models can be considered fairly comprehensive of the principal shapes explored in visual geometry, and, as shown in

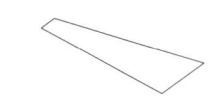  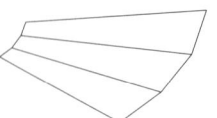 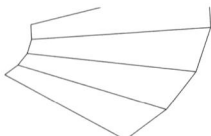

a

IV-27

a. Initial shape and iteration process.

b. Transformation rules for the evolution into the shapes of figures IV-27c-g.

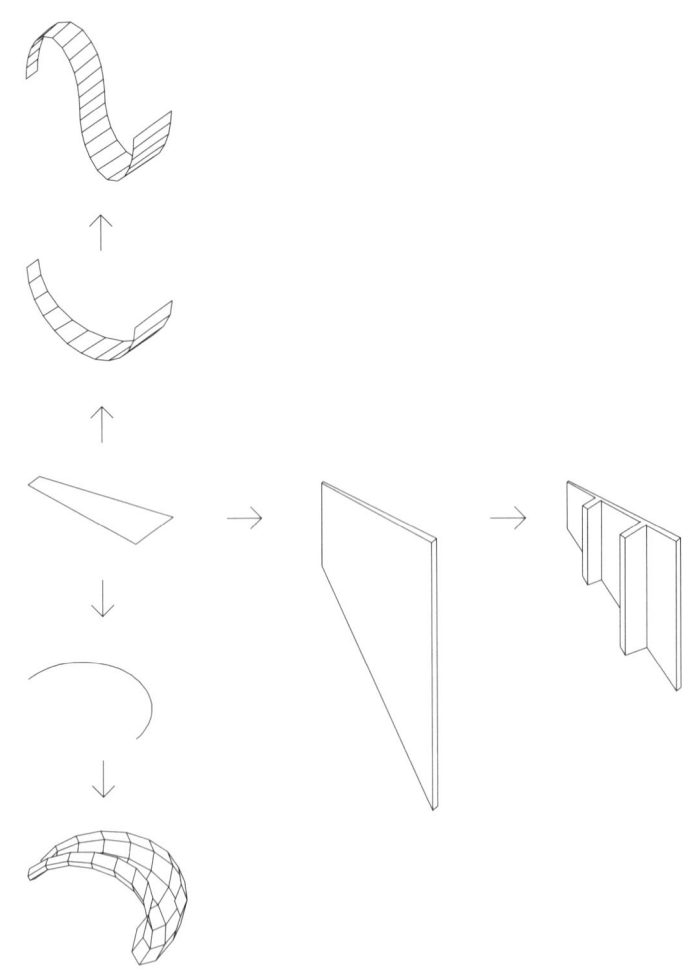

b

VISUALIZING WITH CAD

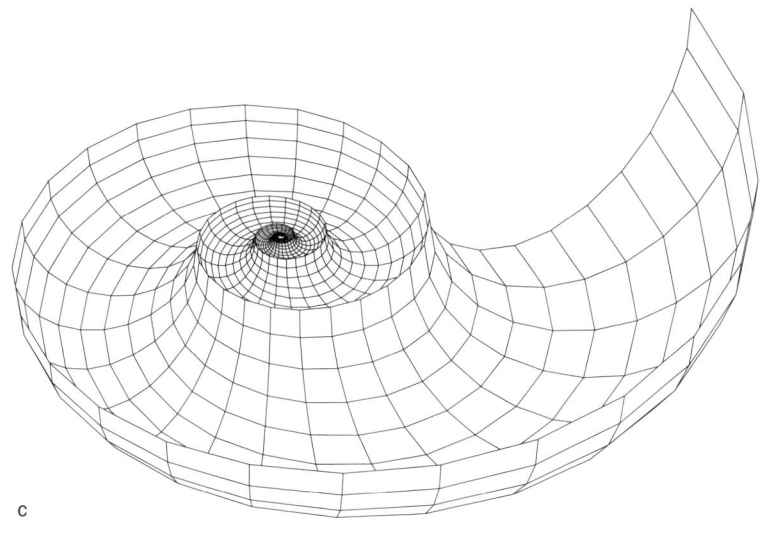

c

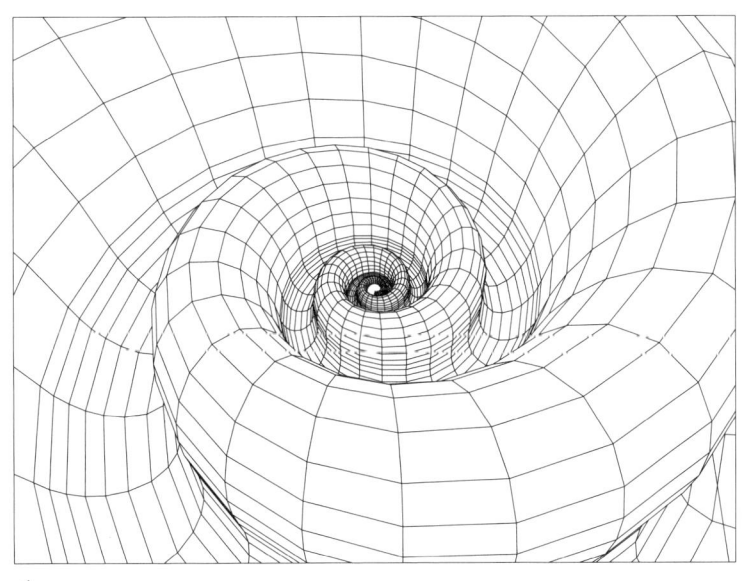

d

e

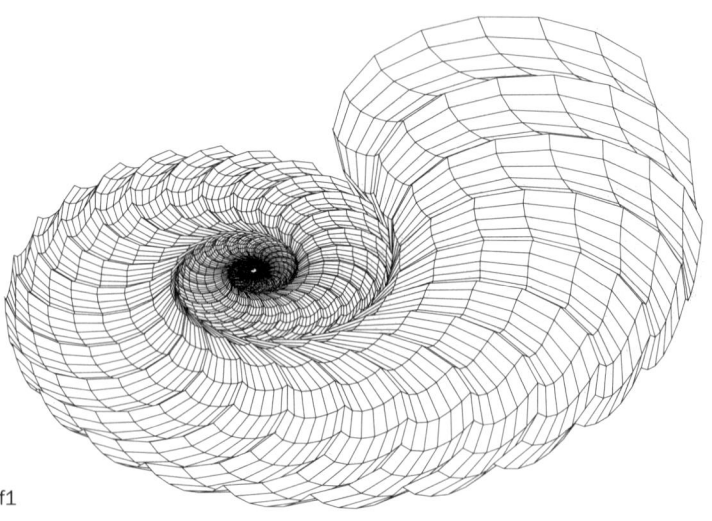

f1

VISUALIZING WITH CAD

f2

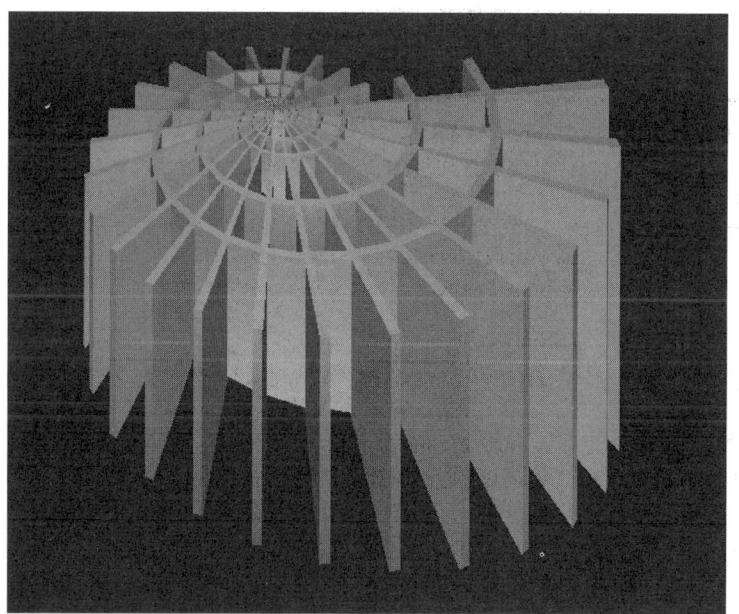

g

Chapter VI, they can also be evolved into architectural compositions. Several other geometric configurations, however, such as saddle surfaces, Klein bottles, heptahedra, minimal surfaces, and flexagons can also be generated as straightforward applications of the theory discussed in Chapter III. In addition, the bibliography offers further reading for exploration of new geometric models.

BIBLIOGRAPHY

Coxeter, H. S. M., and S. L. Greitzer, *Geometry Revisited*, Random House, New York, 1967

Cundy, H. Martin and A. P. Rollett, *Mathematical Models*, Oxford University Press, Oxford, 1961

Escher, M. C., *The Graphic Work of M. C. Escher*, Meredith Press, New York, 1960

Fuller, Buckminster, *Synergetics*, Macmillian, New York, 1975

Ghyka, Matila, *The Geometry of Art and Life*, Dover, New York, 1977

Grunbaum, Branko and Shephard, G. C., *Tilings and Pattern*, W. H. Freeman, New York, 1987

Haeckel, Ernst, *Art Forms in Nature*, Dover, New York, 1974

Hilbert, D. and S. Cohn-Vossen, *Geometry and Imagination*, Chelsea, New York, 1952

Kepes, Gyorgy (ed.), *Structure in Art and Science*, Braziller, New York, 1965

Kepler, Johan, *Harmonices Mundi*, Lincii Austraie, 1619

Mandelbrot, Benoit B., *The Fractal Geometry of Nature*,
W. H. Freeman, San Francisco, 1977

Pacioli, Luca, *De Divina Proportione*, Venitiis: A. Paganius Paganinus
characteribus eligantissimis accuratissime imprimebat, 1509

Plato, *Timaeus*, Penguin Books, Baltimore, 1965

V

ARCHITECTURE depends on Order,

Arrangement, Eurythmy, Symmetry,

Propriety and Economy

Vitruvius

The project of this chapter is to construct models of basic architectural elements that can be considered isolated components, or parts, of a more complex architectural composition. The elements considered are still geometrically defined, but they represent, in a different semantic description, architectural primitives. Once again, the term *primitive* should not be confused with the graphics primitives already encountered in the discussion on CAD. The components developed in this chapter do not represent a comprehensive architectural vocabulary, but rather some of the most recognizable forms, and their construction illustrates how models of other architectural elements can be created.

A work of architecture cannot be created merely as a sum of basic elements. Design in architecture is a synthetic operation, since an architectural artifact is a whole and cannot be created through the random assembly of primitives as though they were children's blocks. To suggest otherwise would be to suppose, in another linguistic analogy, that communication requires only the random combination of words from a vocabulary. Nevertheless, even though the design process is synthetic, an architectural composition can be analytically defined through the use of these primitives, and, for the designer, these ready-made elements can provide a vocabulary. The next chapter will focus on how these elements can be articulated through different types of relations to generate more complex compositions and create the synthesis that characterizes a work of architecture. In this context, we will also introduce the concept of style, which is determined not just by the relations of the basic elements in a composition, but by the characteristics of each individual element as well.

PARAMETERS

Many of the elements discussed below are provided as AutoCAD electronic models and some of them also have an associated **macro** (see Chapter II). It is important to note that the dimensions used in this AutoCAD implementation are not realistic since the models

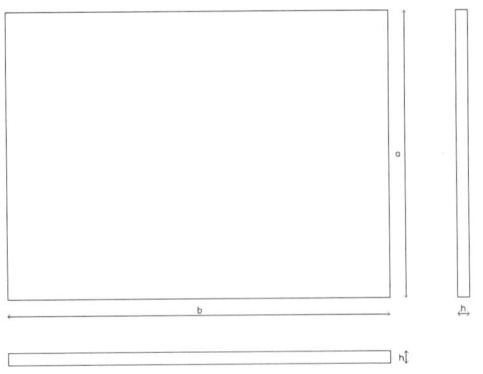

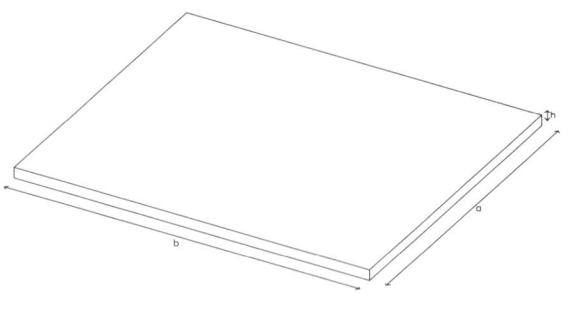

a

b

c

suggested are idealizations. For example, for purposes of simplicity, integer numbers have been used. It is very common in CAD to use the scaling operation to adjust existing elements to new ones with desired dimensions, and it will be easy to adapt these ideal components in the same way. Furthermore, these models do not have a fixed proportionality but can be adjusted to the design requirements. The scaling transformation can be applied globally or in just one direction, which allows proportions to be changed parametrically.

Often, architectural elements are composed of simpler parts. In this case, the whole element and its parts are both treated as symbols; that is, the whole element will be represented by a symbol consisting of nested symbols. The use of parameters will affect both the whole element and its parts.

FLOORS

The floor is an essential element of an architectural enclosure. By virtue of gravity, the floor is the only element with which we are forced to have contact. In natural space the floor is represented by the ground [Norberg-Schulz 1980]. Geometrically, a floor can be defined as a horizontally extended solid—that is, a solid in which width and length predominate over height (thickness)—and it can be easily generated as a set of surfaces with boundaries given by the lines of intersection with the walls (figure V-1a). The parameters are the **thickness** h, the **length** a, and the **width** b.

Floors can become more complex design elements if the floor surface is textured, as in a wood or marble floor. Interesting geometrical problems arise if the floor is covered with a pattern generated as two-dimensional tiling (see Chapter IV). Checkerboard and hexagonal patterns are among the most common patterns used for floors (figure V-1b,c). The most efficient way to generate a CAD model of a tiled floor is to treat each tile as a symbol and then repeat it according to a translatory movement in two directions.

a

b

VISUALIZING WITH CAD

COLUMNS

Columns are the most basic vertical element associated with a structural function, and constitute the supporting elements in the frame of a skeleton system. As opposed to floors and ceilings, columns are not boundaries of an architectural enclosure; their role is solely to support the ceiling or roof. Geometrically, a column is defined as a solid bounded by surfaces. For modeling purposes, columns can be grouped according to two classifications: *simple* columns and *classical* columns, which are made of three parts.

Rectangular and Circular Columns

Simple columns can be further classified in two types according to whether their sections are rectangular or circular. The circular section is approximated by a polygon since, as discussed in Chapter IV, we are restricted to discrete entities. Thus, in the circular column type we include all those columns that appear in section as a regular polygon.

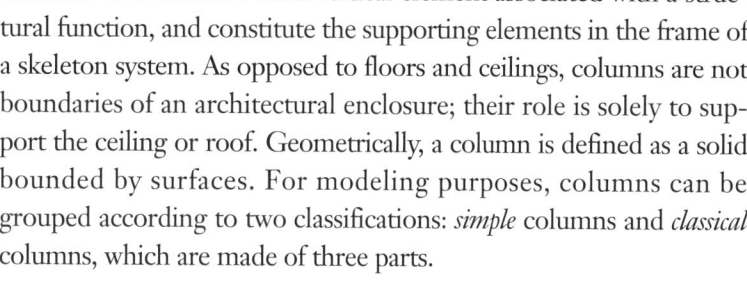

pol_colu.dwg

A column with rectangular section (figure V-2a) can be defined by three parameters: its height h, which is the prevalent dimension, and the two dimensions a and b associated with the section. The geometrical shape used is the parallelepiped, defined by six faces enclosing a volume.

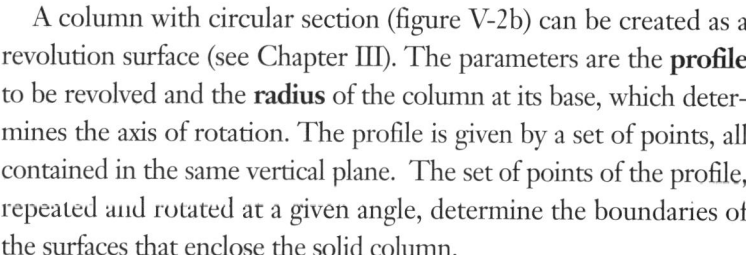

dor_colu.dwg

A column with circular section (figure V-2b) can be created as a revolution surface (see Chapter III). The parameters are the **profile** to be revolved and the **radius** of the column at its base, which determines the axis of rotation. The profile is given by a set of points, all contained in the same vertical plane. The set of points of the profile, repeated and rotated at a given angle, determine the boundaries of the surfaces that enclose the solid column.

The Classical Column and Orders

The classical column consists of three parts: **base**, **shaft**, and **capital**. The forms of these parts are associated with the classical orders. According to Vitruvius, the three parts of columns should be proportioned and shaped according to the three orders of Greek architecture: Doric, Ionic, and Corinthian. In his treatise, Leon Battista Alberti offers clear illustrations in elevation and plan for the shapes and proportions of base, shaft, and capital in each of the three orders (with the exception of the Doric column, in which there is no base). The composite or Roman style is also included in Alberti's discussion. The architectural styles of columns are not limited to these orders, however. Almost every historical architectural period— Cretan, Egyptian, Tuscan, Byzantine, Romanesque, Moorish, Gothic, Renaissance, Baroque, and so on—has its own characteristic column.

Our modeling of these columnar parts according to the Greek orders will be based on definitions borrowed from Vitruvius, who provided very precise proportions and detailed descriptions. The column shown in figure V-3 is a schematic model of an Ionic column. The reader can easily implement models for the other styles by following the methods used here.

The Base

The base of the column represents the transition from the floor to the main structural element, the shaft.

Geometrically a base can be defined as a revolution surface (figure V-3a); therefore, the parameters are the profile and the number of elements needed to accomplish a complete revolution. The profile is the element that differentiates the different orders. Sometimes a parallelepiped shape called a *plinth*, with the parameters width, length, and height, is added to the base.

The Shaft

The shaft is the essential structural element of a column (figure V-3b). It is a vertical element, in which the height dimension prevails over the base dimension. In almost every style, the profile of the shaft presents a convex curvature, called *entasis*. The purpose of this curvature is not structural, but aesthetic; in fact, if the profile of the shaft were straight, it would produce the optical illusion of concavity.

The geometric model for a shaft is a revolution surface with parameters given by the profile and the number of repeated parts in a complete revolution. The surface of the shaft is sometimes treated with vertical concave grooves, called *fluting*. If our model of a shaft incorporates flutes, they will be represented as the basic symbol/part. Geometrically a flute is defined by curved surfaces that in horizontal section appear as arcs. Vitruvius suggested twenty-four flutes for the Ionic order and twenty for the Doric.

The Capital

The capital is the upper conclusion of a column, representing the connection between the supporting structural element and the supported elements above, which can be a ceiling, roof, or just an arch or beam.

The Ionic capital (figure V-3c) consists of the **abacus**, and below it, two symmetric **volutes**. The volutes are shaped as spirals, extended laterally. Vitruvius gives a very detailed explanation of how to generate the spirals, including the right proportions. Recall from the previous discussion on geometrical shapes that spirals can be generated by a combined transformation consisting of rotation and scaling.

Column base.

a1. Plan.　a2. Elevation.

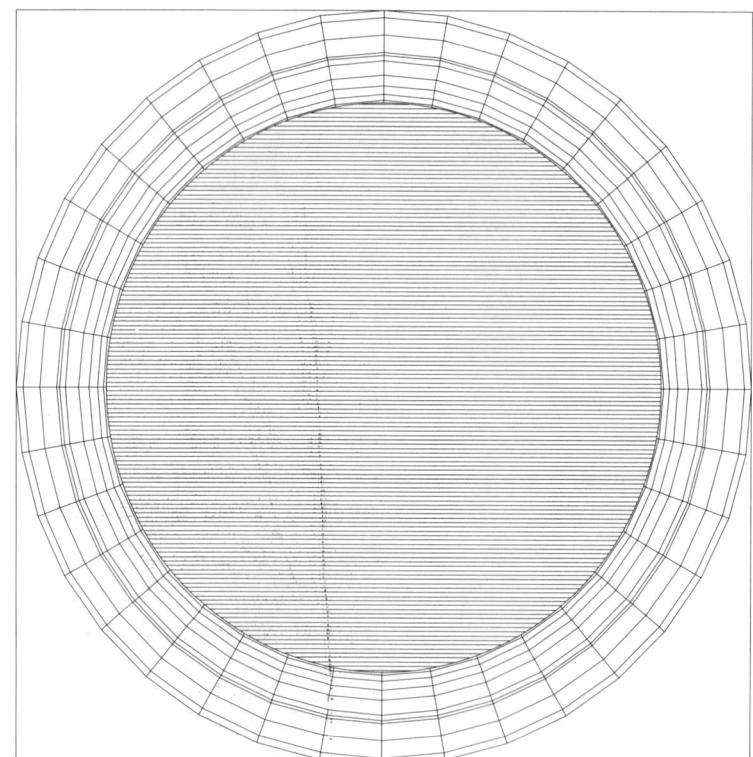

a1

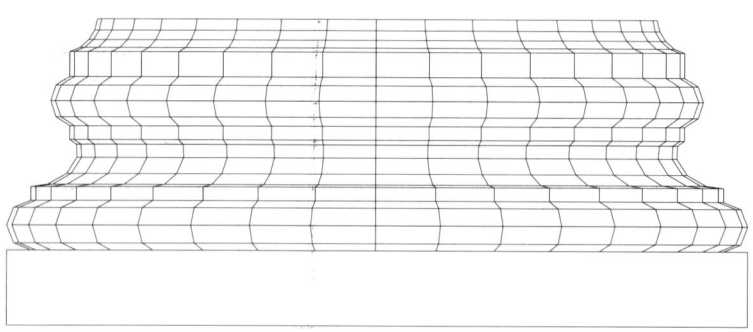

a2

VISUALIZING WITH CAD

Column shaft.

b1. Elevation.   b2. Plan.

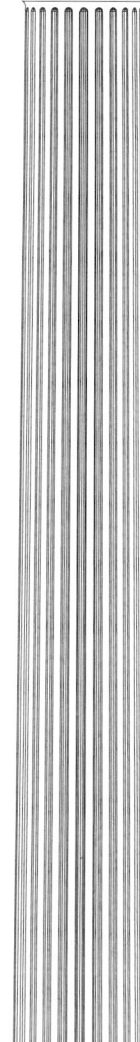

b1

b2

Column capital.

c1. Elevation.   c2. Plan.

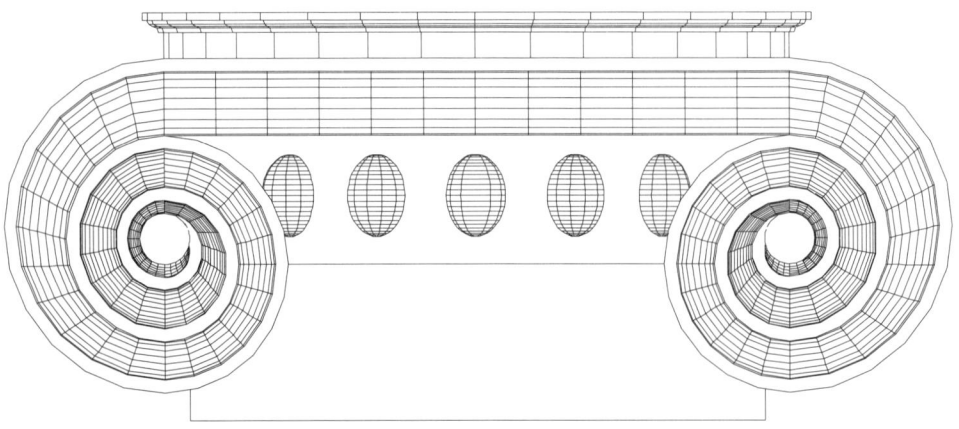

c1

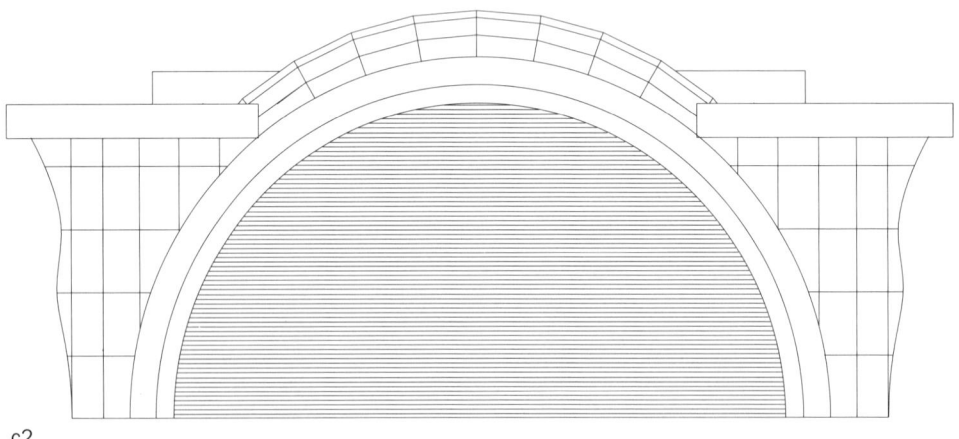

c2

Ionic column.

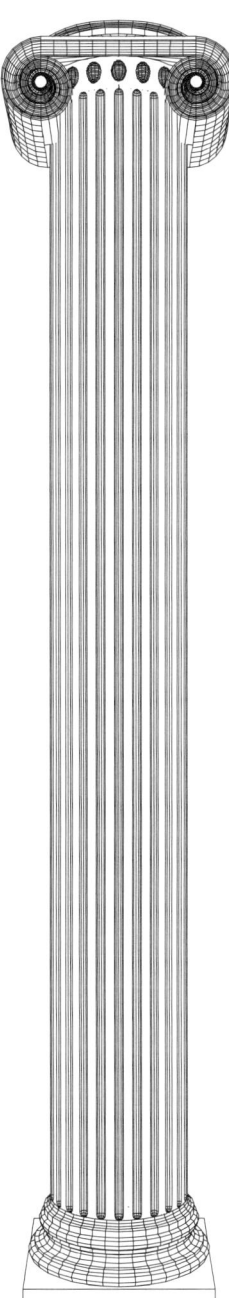

a

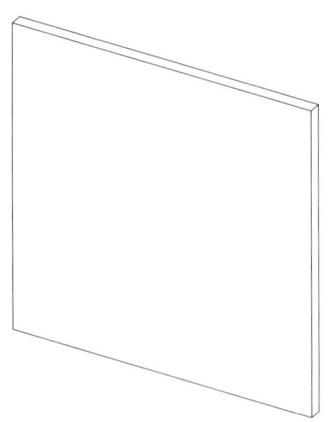

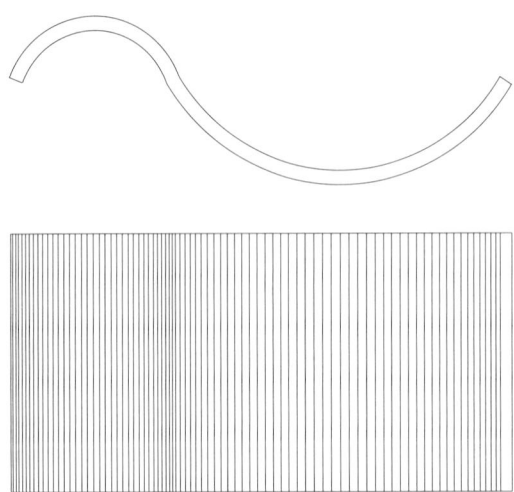

b

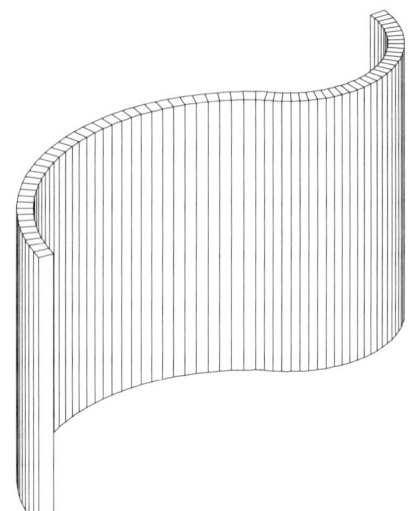

WALLS

Walls are vertical architectural elements that create boundaries; together with ceilings and floors, they enclose a space, separating the inside from the outside. Walls may or may not be part of the structure.

Geometrically, walls have extension mainly in two dimensions, width and height, while the third dimension, depth, is proportionally only a fraction of the other two. Walls can be flat or curved. A flat wall (figure V-4a) can be approximated, in a diagrammatic form, by a plane in three-dimensional space. The simplest case is a solid wall that can be represented geometrically by a parallelepiped defined by six bounding surfaces. The variable parameters are **height**, **width**, and **depth**. A curved wall (figure V-4b) can be generated as a vertical extension of the curvilinear segments that represent the plan section of the wall.

a

b

VISUALIZING WITH CAD

a. Solid-void proportions in walls.

b. Walls obtained from
Boolean operations.

mk_wall.lsp

Walls are usually not completely solid, however; instead they are often articulated through the repetition of opening elements, such as windows and doors, the proportions of which determine the rhythm of a solid-void composition (figure V-5a,c). This type of wall results from the **array** repetition of the simple wall described above. A CAD model of a wall can be constructed by means of three different parts (identified by symbols): **horizontal** and **vertical** parts, representing solid elements, that can be modeled as parallelepipeds, and **opening definers**. This last element, which may appear superfluous, is necessary to define the openings that will be replaced with windows and doors. Given a square grid in elevation, an infinite number of different compositions can be obtained just by altering the relation of solid-void, which can easily be done by changing the parameters (breadth, height, and depth) in the part/symbols components of the wall. In the model of the wall shown in figure V-5a, the relation between opening and solid is 1:2 and follows a square grid in elevation. The reader can experiment by changing the block definition of the basic wall elements to generate several design alternatives based on different solid-void proportions.

The CAD model illustrated in figure V-5a can also be generated by means of Boolean operations. The operators are given by a parallelepiped A (representing the solid wall) of specified height, width, and depth, and a set B of parallelepipeds having the dimensions of the openings. The operation is the subtraction $A-B$ shown in figure V-5b.

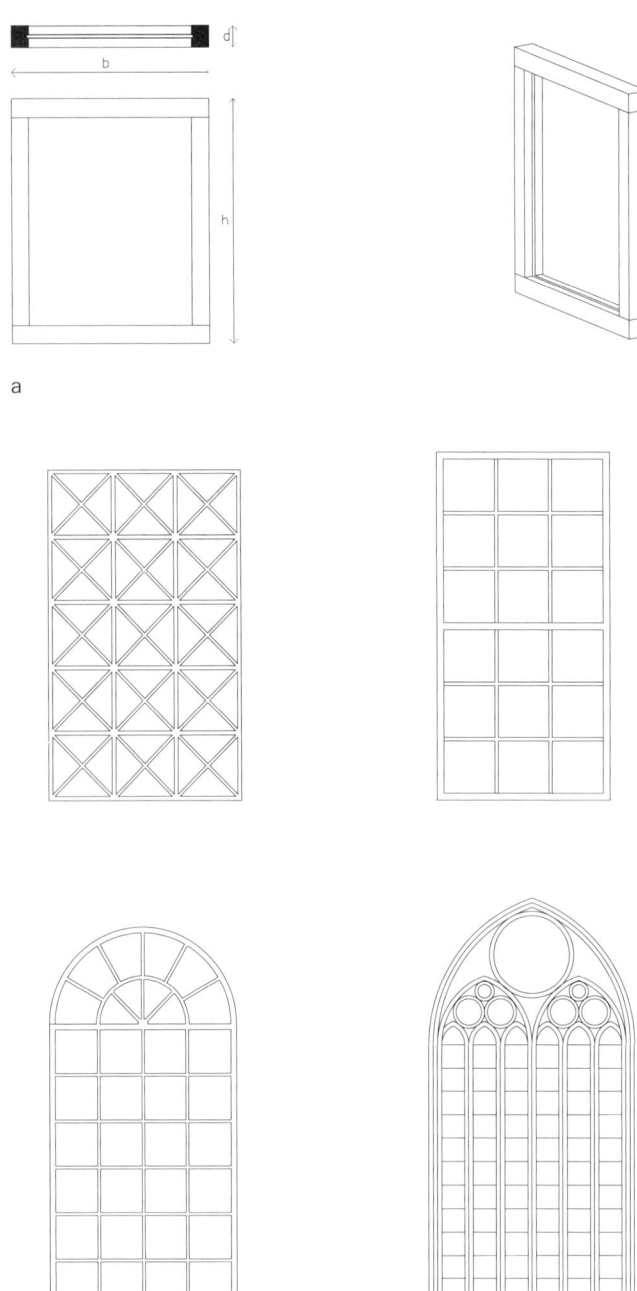

a

b

VISUALIZING WITH CAD

Windows.

WINDOWS AND DOORS

In a wall, regarded as a system of solids and voids, the voids are holes or openings that connect the inside with the outside, otherwise separated by the wall boundaries. The windows and doors that fill the openings in the wall, define diaphragm-like boundaries between inside and outside. Doors fill floor-level openings; they are the elements that, while separating an inside space from the outside, allow it to be entered. Similarly, windows allow the passage of air and light. Stylistically, windows and doors contribute to the character of both the outside and inside elevations.

The geometric models of windows and doors can be treated in the same way. In both, the **frame** is distinguished from the **pane** (figure V-6a). The pane can be represented by a parallelepiped, the depth of which is a fraction of its height and width. The model of a frame can be generated from a vertical orthogonal connected grid (see Chapter IV). The connected elements are parallelepipeds of different heights. Figure V-6b shows elevations of several models of windows.

The model of a window or door can be integrated with that of a wall. In fact, it can be considered part of the symbol denoted as *opener definer* in the wall model. The opener definer can assume different geometric forms, including curvilinear shapes.

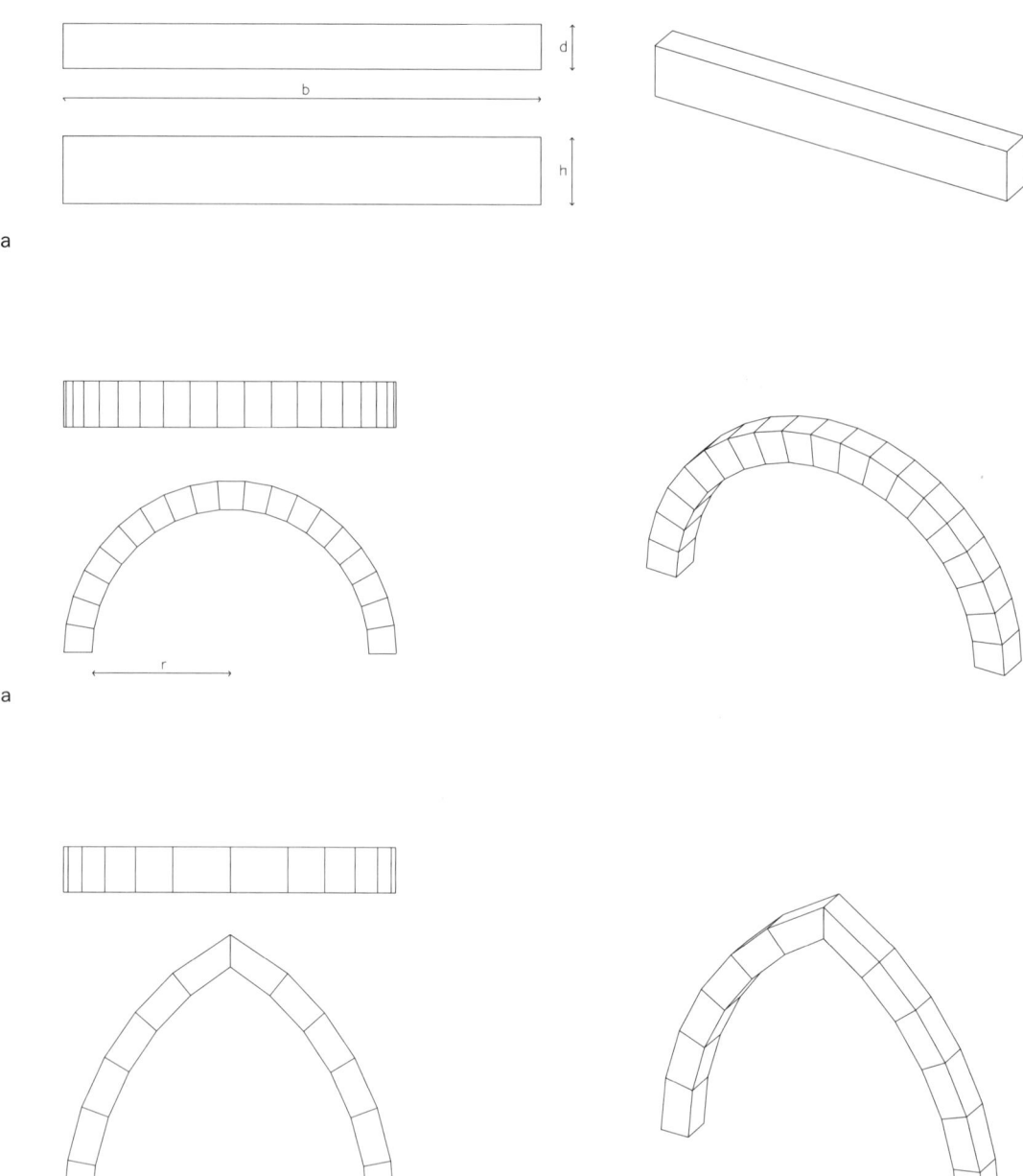

a

a

a

VISUALIZING WITH CAD

BEAMS AND ARCHES

Beams and arches are architectural elements that, in their structural functions, both support and are supported by parts of a skeleton system. A straight beam (figure V-7a) can be geometrically represented by a parallelepiped with a vertical section that is much smaller than the horizontal. The parameters are, once again, **width**, **depth**, and **height**. Arches are characterized by their curved shape and by their structural ability to carry their own weight and other external loads by internal compression. The principal geometric shape associated with arches is the arc, or segment of a circle. In fact, all the different types of arches (drop, elliptical, equilateral, depressed, basket, tudor, horseshoe, pointed, stilted, and so on) can be constructed as **sum** and **intersection** of **segments of circles**. The arch will be generated here in terms of the segments that appear in the arch section in the x-z plane. Each such circle segment can be approximated by the rotation and repetition of a simple polyhedral element, treated as a symbol in the CAD model. (In a similar way, a built arch is made of voussoirs, the wedge-shaped stones that form the elemental structural unit.) The geometric parameters of the arch are the **radius r**, the **total angle θ** of the segment, the **number of voussoirs**, the dimension **a** of the **intrados** (the inner concave surface) and the dimension **d** of the **extrados** (the outer curved face).

The simplest arch is the round, or Roman arch characterized by one 180° arc of a circle. The model shown in figure V-7b is made of nineteen voussoirs. For a more complex arch, such as an equilateral arch (figure V-7c), it is helpful to generate the model of the voussoir first, defining it as a symbol. A repetition and rotation of voussoirs defines a segment of the arch. Two or more segments of the arch will create the final model.

Beam and arches.

a. Straight beam.

b. Round arch.

c. Equilateral arch.

rou_arch.dwg

equ_arch.dwg

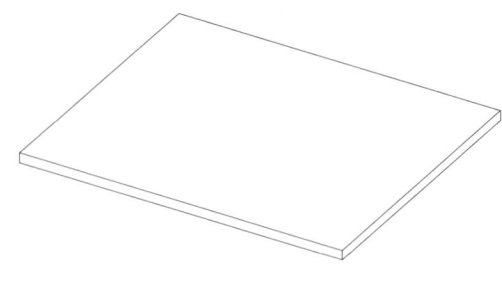

a

b

VISUALIZING WITH CAD

SIMPLE ROOFS

a. Flat roof.   b. Gable roof.

A roof is the top of a building; its function is to enclose the volume already defined by walls and floor and shelter the inner space from the outside. A flat roof (figure V-8a) is geometrically defined as a parallelepiped, the height of which is a fraction of its length and width. Flat roofs are characteristic of modern architecture. Gable roofs (figure V-8b) are characterized by their sloping sides, shaped to get rid of rain and snow efficiently. Their geometric models are given by portions of sloping planes. The roof model parameters are the **angle of slope** (if any), **width**, **length**, and **height**. The intersection of two or more gables can generate design motifs.

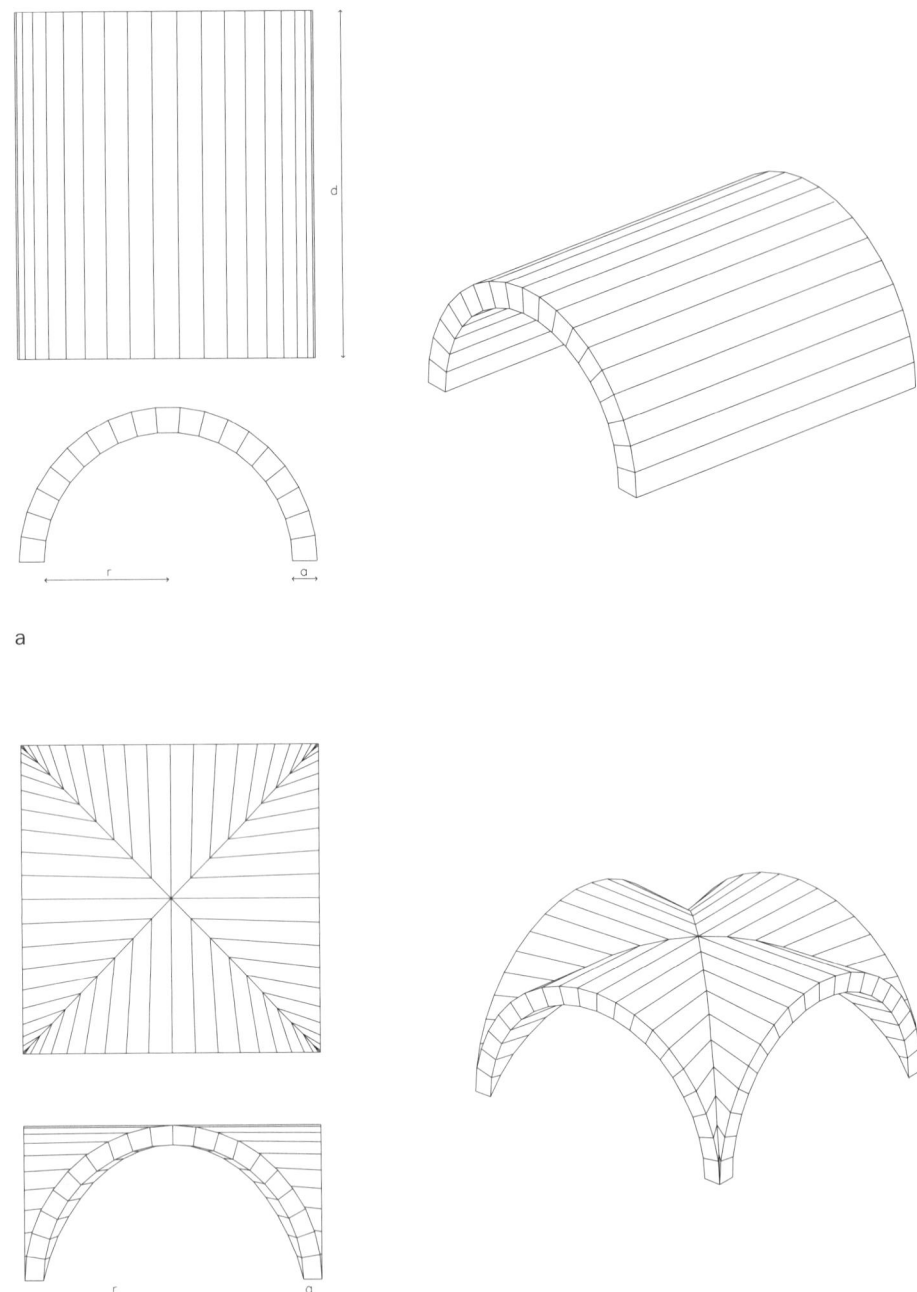

a

b

VISUALIZING WITH CAD

Vaults.

a. Barrel vault.   b. Groin vault.

Vaults are curved ceilings or roofs and can be classified geometrically according to their curvature. In this section we deal with singly curved vaults; doubly curved vaults will be treated in the discussion of domes below. There are several types of vaults, defined according to the arch apparent in the *x-z* section.

Barrel Vault

bar_vaul.dwg

A barrel vault, the simplest form, consists of a round arch laterally extended along the *y*-axis (figure V-9a). The creation of CAD models for vaults is quite trivial once the arch has been constructed. The only parameter that changes is the dimension **d** along the *y*-axis.

Groin Vault

gro_vaul.dwg

More complex vaults are generated by the repetition or intersection of two or more simple vaults. One well-known example is the groin vault (figure V-9b), which is formed by the orthogonal intersection of two barrel vaults of identical shape. The simplest geometric part in a groin vault is a modular element, which is repeated eight times according to different geometric transformations to generate the complete vault. This basic element is generated as a ruled surface (see Chapter IV). The model of the complete vault is obtained through the **mirror** repetition of the module, which generates a symmetric couple that is repeated and rotated four times at an angle of $90°$.

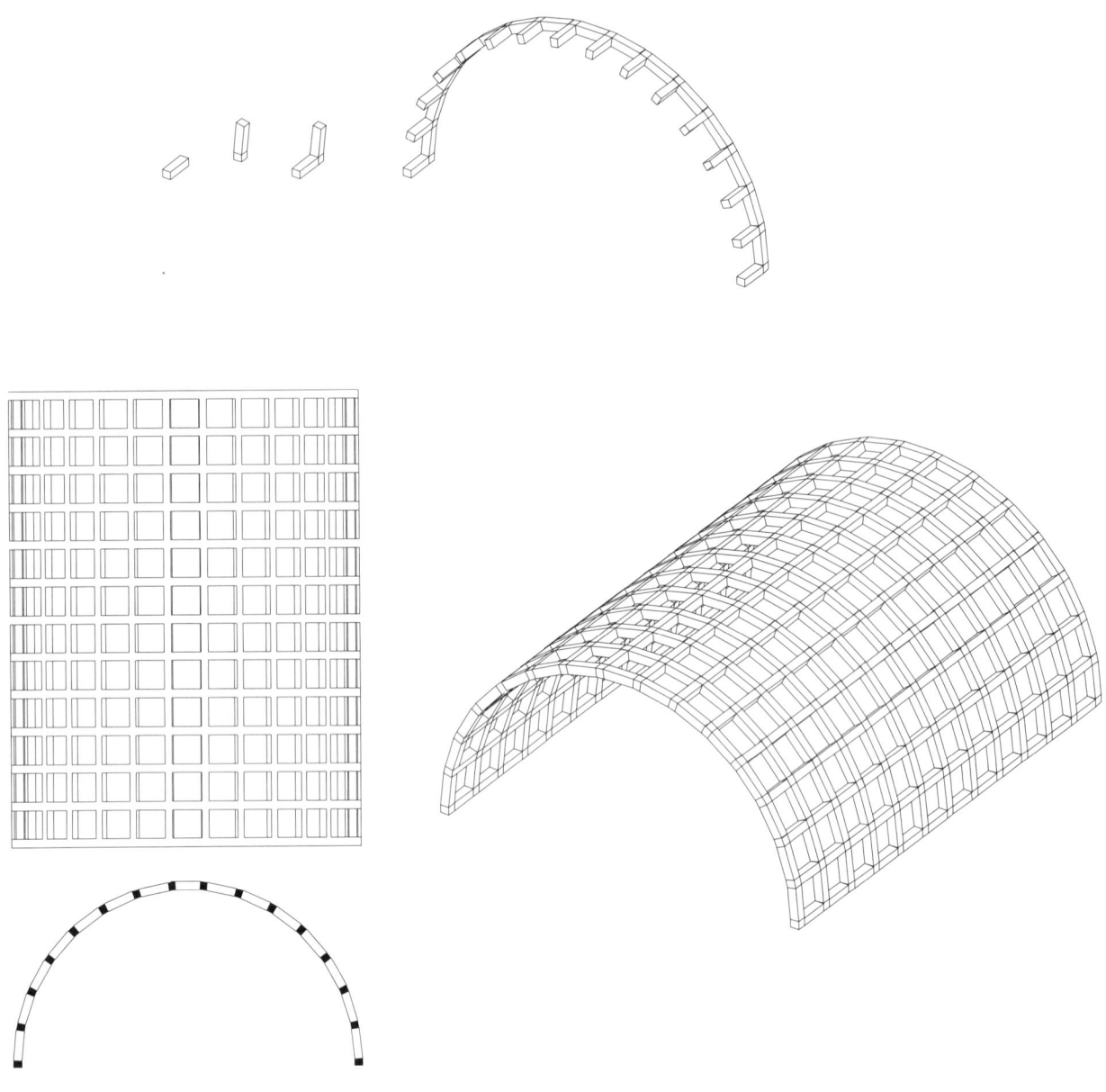

VISUALIZING WITH CAD

Coffered Vault

Vaults.

c. Coffered vault.

rib_vaul.dwg

A coffered vault (figure V-9c) is a vault decorated with sunken polygonal panels, which can be constructed as polyhedral elements that define a module. The polyhedral elements are constructed as solids enclosed by surfaces. The points bounding the surfaces are determined by the repetition and rotation about the *y*-axis of an initial set of points in the *x-y* plane. The rotation angle α is given by $180°/n$, where n is the number of modules in the *x-z* section. The repetition (*n*-1 times) and rotation (at the angle α) of the defined module will generate another characteristic component, called the *bay*. The bay can then be repeated according to a translation movement.

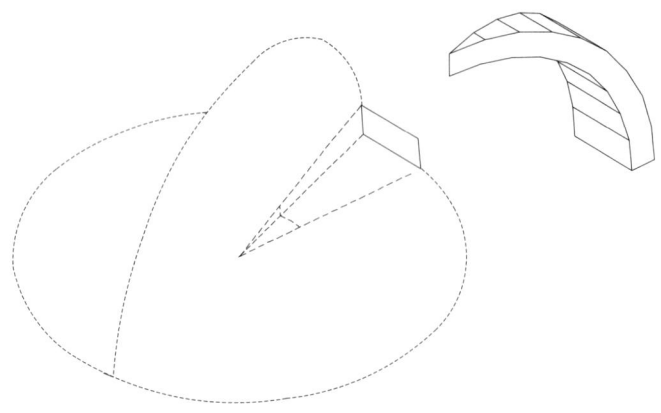

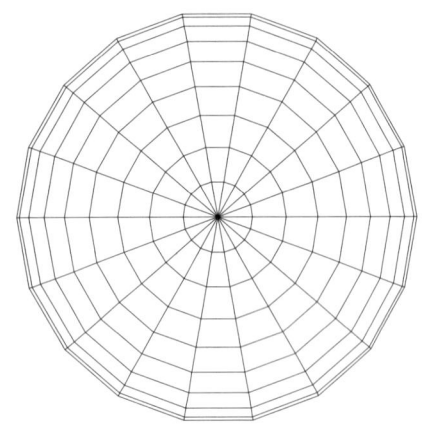

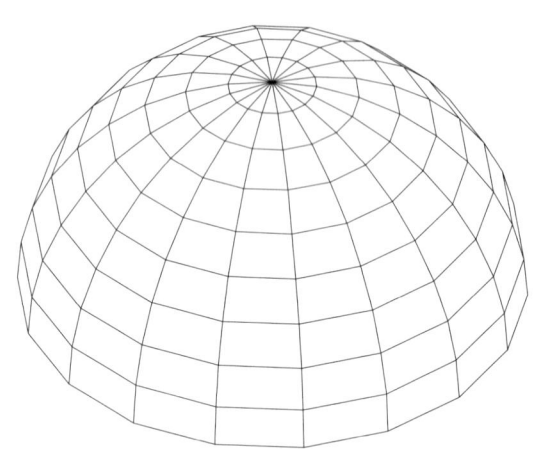

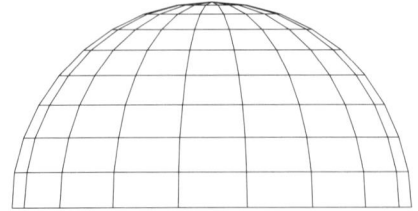

VISUALIZING WITH CAD

DOMES

Domes are vaults characterized by a circular plan and double curvature in the same sense in all directions at any point. By *in the same sense* we mean that the curvature in every direction is either concave or convex; both curvatures cannot occur at the same point, as they do in saddle-shaped forms.

The Hemispherical Dome

The simplest geometric shape associated with a dome is a hemisphere (figure V-10a). The CAD implementation is usually quite trivial if the system provides commands to generate hemispherical shells. Otherwise, the geometric model can be defined as a double rotation. A point on the x-axis is repeated and rotated about the y-axis n times at an angle of $90°/n$. The set of points so generated is contained in the x-z plane and is repeated and rotated about the z-axis at an angle α. All the points thus far created are boundaries for a set of surfaces, which are grouped as symbol and repeated and rotated m-1 times about the z-axis, where m is equal to $360°/\alpha$.

The Domical Dome

Most domes in the history of architecture are not hemispherical, but instead take the form of a repetition, in a rotational symmetry, of a characteristic component or bay.

One of the simplest examples is a domical or cloister vault (figure V-10b), which rises from a polygonal plan. The bay is generated as a **ruled surface**, the paths for which are given by two 90° vertical arches defining an angle α in the x y plane. The bay is replicated and rotated n–1 times at the angle α, where $n = 360°/\alpha$.

The Ribbed Dome

The ribbed vault of figure V-10c is generated by using rotational symmetry twice. First, initial points A and B on the x-axis are both rotated about the y-axis by angles α and β in succession, generating the vertices of two quadrilaterals, one above the other. (These corre-

Domes.

a. Hemispherical.

hem_dome.dwg

dom_dome.dwg

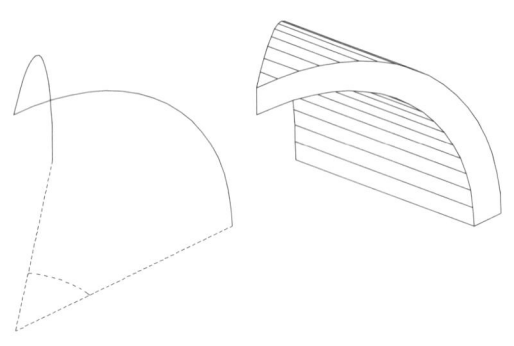

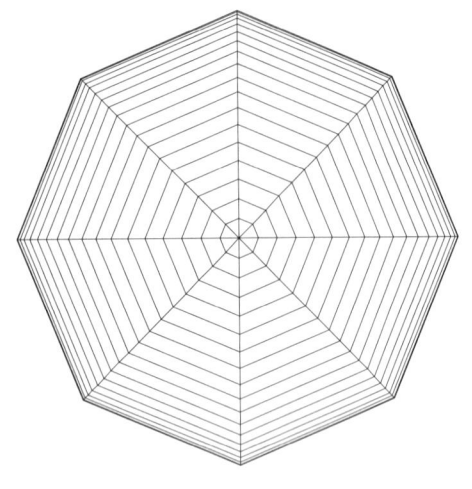

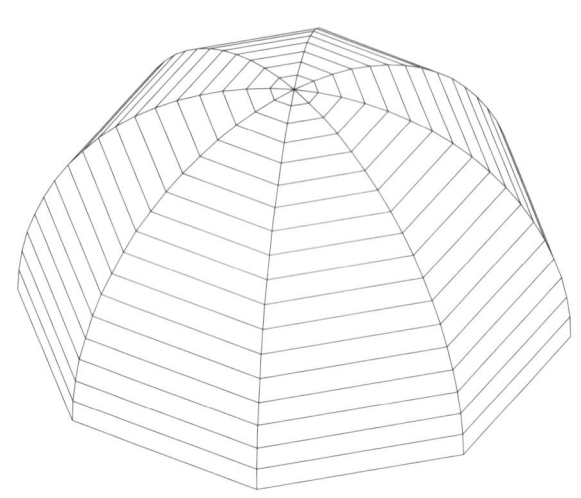

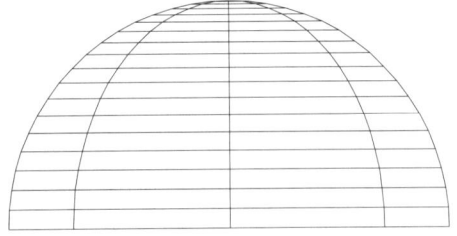

VISUALIZING WITH CAD

Domes.

b. Domical.

rib_dome.dwg

rib_dome.lsp

spond in *x-z* section to solid and void elements of the dome; see figure V-10c.) This cumulative rotation is repeated *n* times, where $n = 90°/\alpha+\beta$, producing the vertices (in the *x-z* plane) of alternating polyhedral forms. The other vertices of the polyhedra are defined by successive rotations of angles δ and γ about the *z*-axis. The angles δ and γ are determined in such a way that $\delta+\gamma = 360°/m$ and *m* is the number of bays in the dome. At this stage, all the vertices for the modeling of the polyhedral shapes forming the bay are determined. A symbol named *bay* is defined, then repeated and rotated *m*-1 times about the *z*-axis. The polyhedra that compose the bay establish a rhythm of solid and void in which, in *x-z* section, the quadrilaterals generated by, say, the α rotation correspond to the solid part of the dome, and those generated by the β rotation correspond to the void. Similarly, in *y-z* section, δ and γ also correspond to solid and void.

Domes and Boolean Operations

When a circular dome is erected from a square base, connective elements are needed between the spherical form of the dome and the square base. These connections are called *pendentives* and are geometrically represented by spherical triangles. A model of this type of dome can be constructed by using Boolean operations. Figure V-10d illustrates in detail the sequence of operations. The operations used to obtain each individual element of the dome are subtraction and intersection. The final dome is produced by the union of all the elements resulting from the sequence of operations.

Domes.

c. Ribbed vault.

d. Domes obtained from
Boolean operations:
sequence of construction.

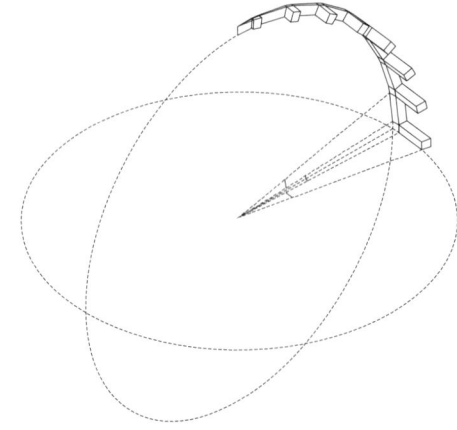

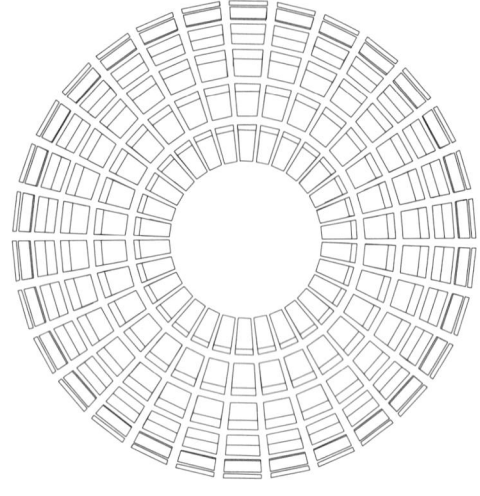

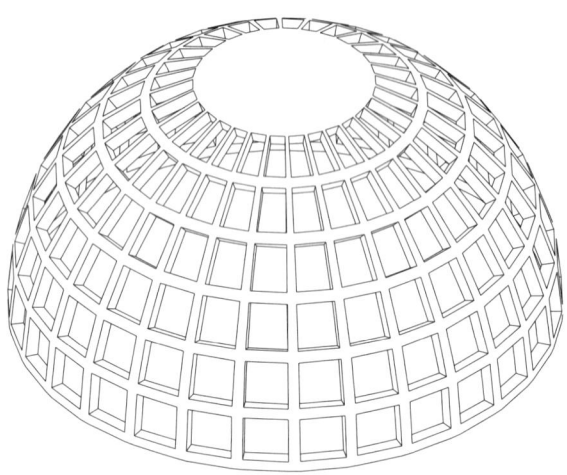

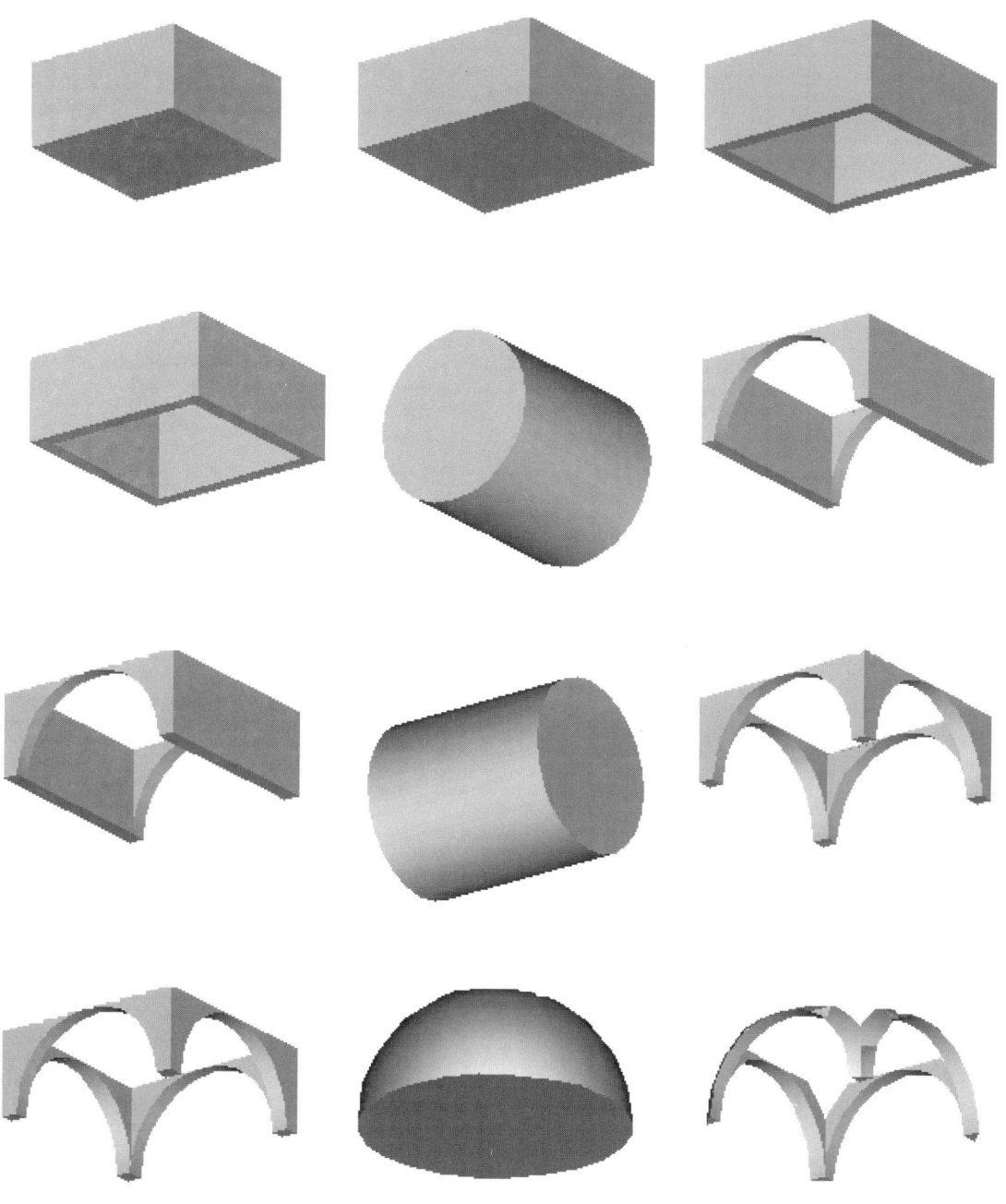

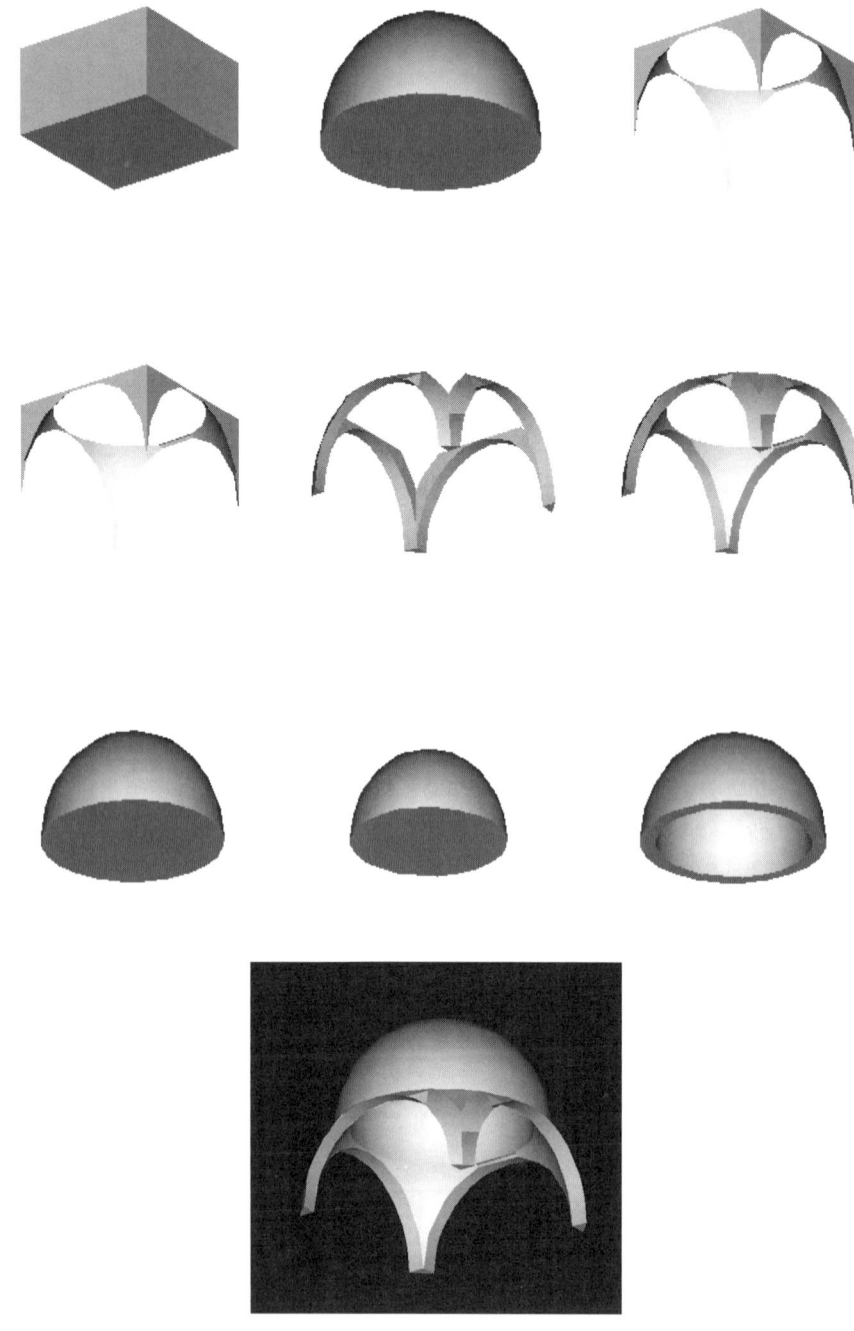

VISUALIZING WITH CAD

Domes.

d. Domes obtained from Boolean
operations: sequence of construction.

ANTICLASTIC SURFACES

Anticlastic surfaces have curvature in the opposite sense (both con-cave and convex) through any point. Their geometry and structural behavior are quite complex. Examples are found in tensile structures and reinforced concrete shells. An initial model can be obtained by defining the perimetral boundary and then generating the interme-diate surface. Often, CAD software provides procedures to generate these types of mathematical surfaces, which are usually defined by three-dimensional meshes of polygons that are automatically gener-ated from the user-defined perimeters. A typical example is the saddle-like surface in figure V-11a, generated as a polygon mesh from four adjoining lines in three-dimensional space.

All the CAD models of anticlastic surfaces in figures V-11b through V-11d can be generated as **ruled surfaces**, defined by lines connecting two points translated at uniform increments along paths given by two nonintersecting lines or curves. The two curves defining the boundaries of the membrane roof of figure V-11b are contained in two orthogonal planes and provide the paths for the translation of the line generating a ruled surface. The membrane in figure V-11c is also geometrically modeled as a ruled surface; two straight nonintersecting lines provide the paths along which the gen-erating lines are translated. In the shape of figure V-11d, the paths are provided by two curves contained in two different planes, both parallel to the x-y plane.

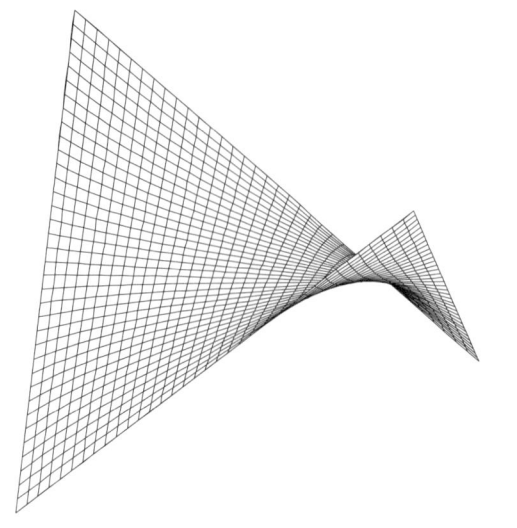

a

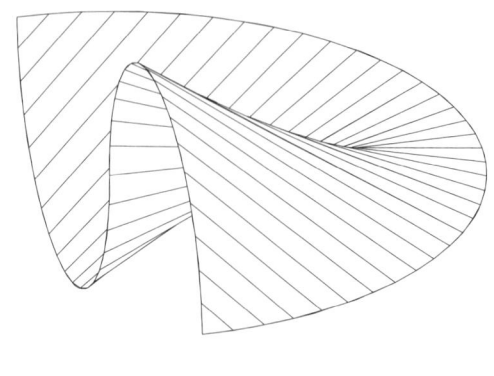

b

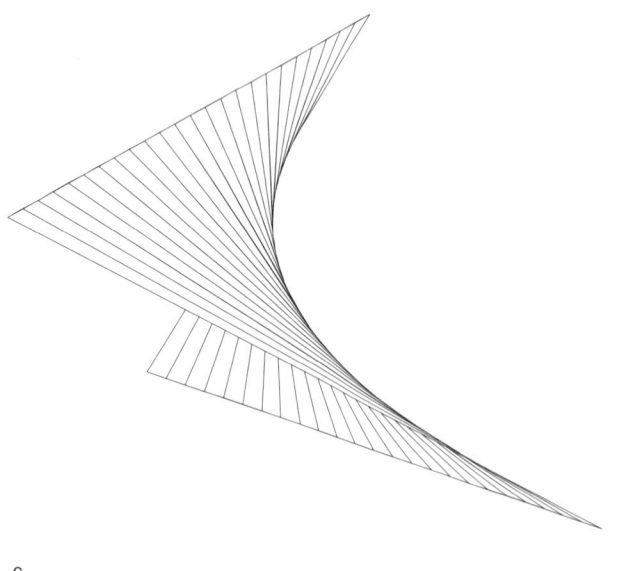

c

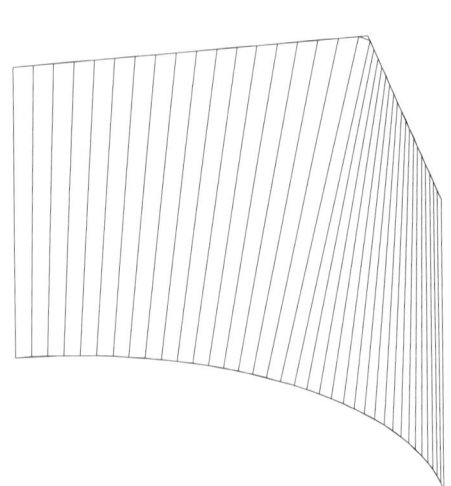

d

VISUALIZING WITH CAD

Anticlastic surfaces.

Stairs and ramps are elements in between horizontal planes; they provide the connection between two floors at different heights.

A ramp (figure V-12a) is geometrically represented by a sloping plane connecting two floors; the parameters are the **angle of slope**, the **length**, given as projection on the x-y plane, the structural **thickness** (height), and the **width**. The slope angle is often determined by the different elevations to be connected rather than by design.

Stairs are a functional element that can be stylistically interesting. Stairs are not simple elements, since they are composed of subparts or modules: the steps. The geometric model of a step, defined as symbol, is a parallelepiped defined by length, width, and height, where length and height correspond to the tread and rise, respectively. The two initial parameters that determine the geometric model of a stair are its shape and the vertical distance between floors. Once the shape is designed, the vertical distance determines the number of steps and the dimensions of the rise and tread of the steps, according to building code standards. The width is a design choice, independent of the standards. The shape of stairs can be defined as a sum of steps in a given geometric relation. The most common shape is that of a straight flight of stairs, which in projection on the floor plan appears as a rectangle (figure V-12b). The transformation involved is translation in both horizontal and vertical axes. Spiral stairs (figure V-12c) are shaped as a helicoid. The transformations used are rotation about the z-axis and translation, also about the z-axis. The other parameters are the inner and outer radius, which respectively represent the distance of each step from the center of the spiral and its width. The geometric principle behind spiral stairs can also be applied to double stairs, shown in figure V-12d.

Stairs represented a very important element in Baroque architecture. Fan stairs (figure V-12h) are typical of the Baroque style. They are formally based on concentric circles or semicircles. The model can be generated as the rotation of a profile representing a vertical

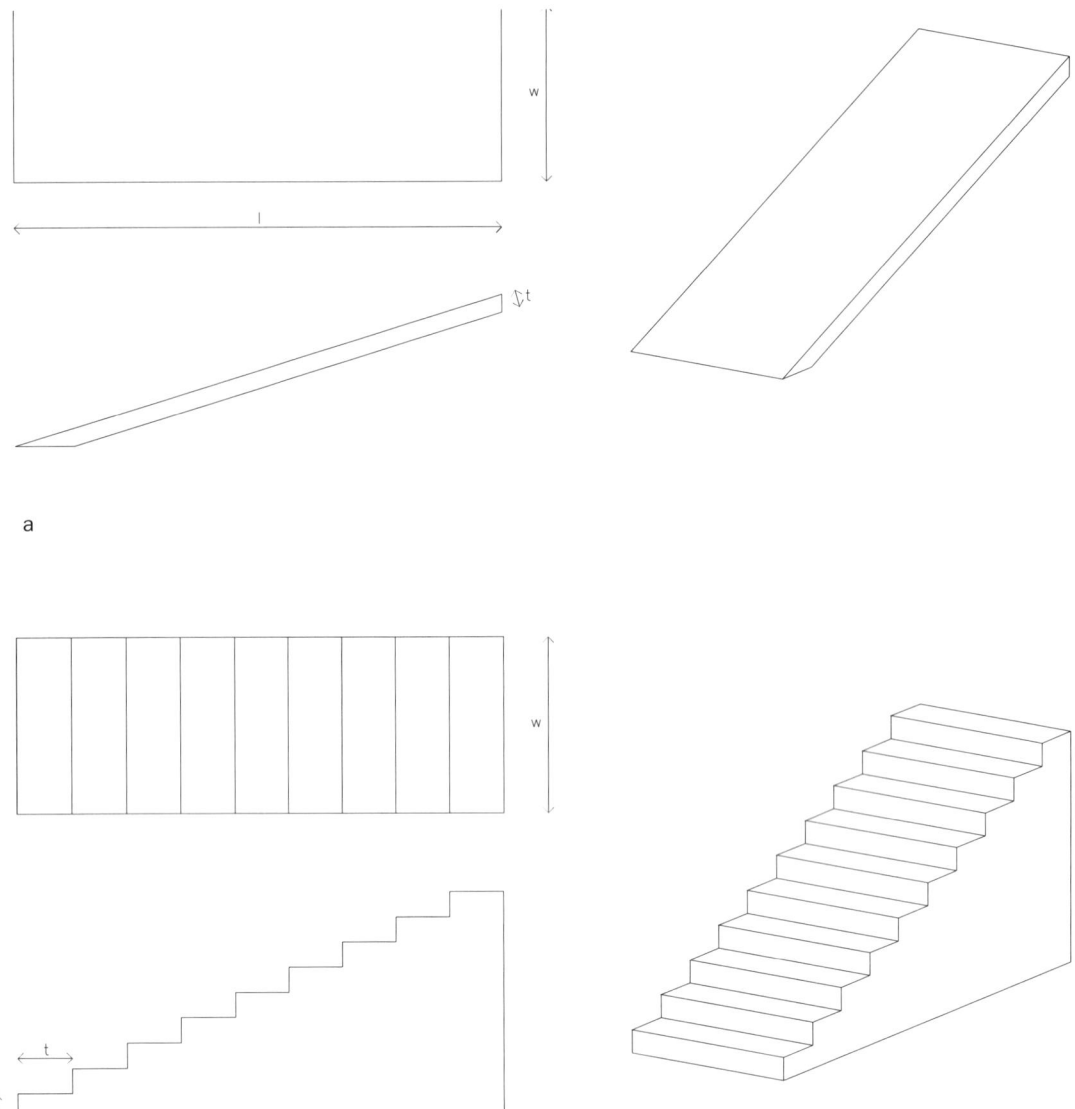

a

b

VISUALIZING WITH CAD

a. Ramps.   b. Stairs.   c. Spiral stairs.
d. Double stairs.

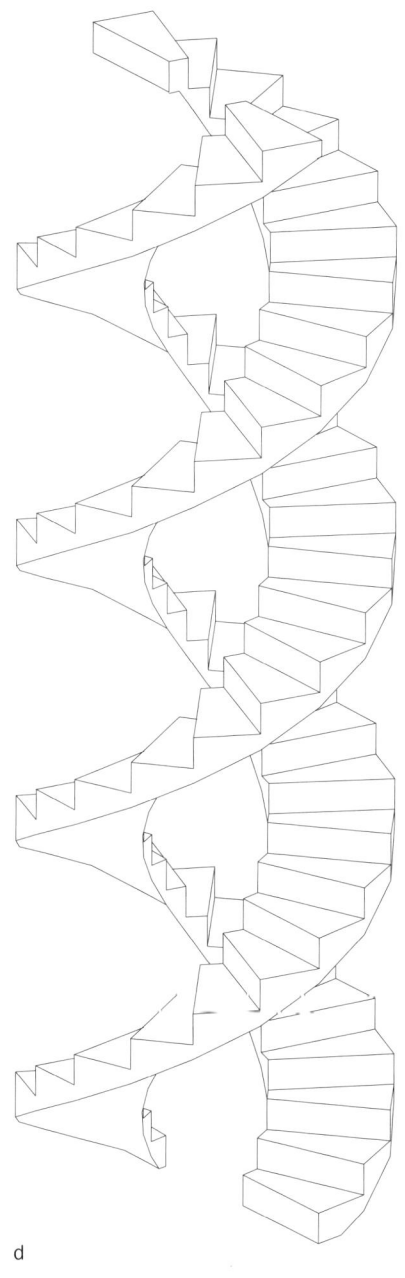

c

d

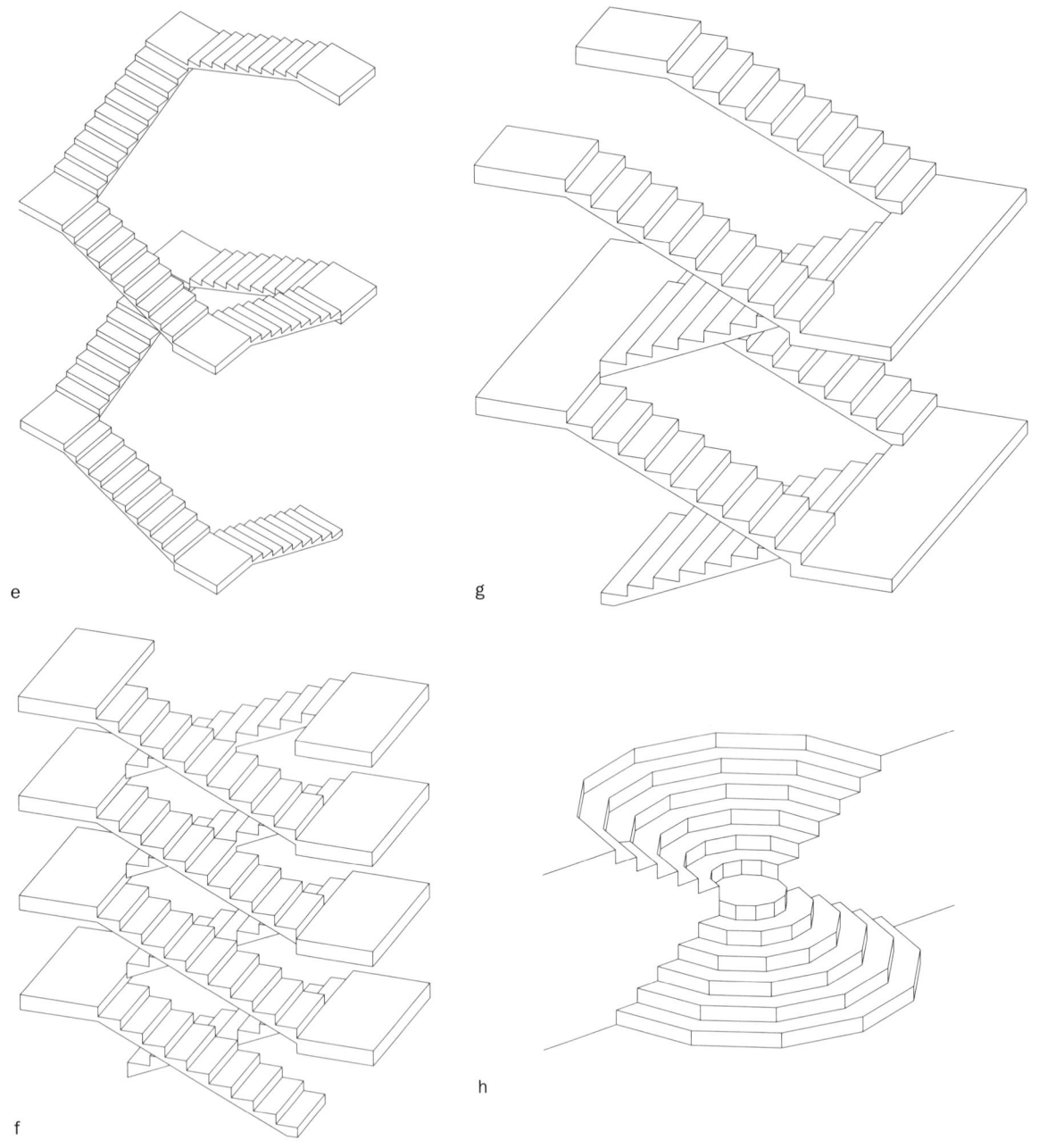

e

g

f

h

VISUALIZING WITH CAD

e. Divided stairs.   f. Fan stairs.
g. h. More stairs.

section of the stair. The other parameter is the rotation angle and the number of rotational elements to be included in it. A simple model of divided stairs is shown in figure V-12e. More models of stairs are shown in figures V-12g-h.

The form of stairs can be applicable to ramps as well. The model of a stair can become a ramp by substituting for each step, defined as symbol, a sloping portion of a plane.

BIBLIOGRAPHY

Alberti, Leon Battista, *The Ten Books of Architecture*, Dover, New York, 1986

Norberg-Schulz, Christian, *Genius Loci*, Rizzoli, New York, 1980

Norberg-Schulz, Christian, *Intentions in Architecture*, MIT Press, Cambridge, 1965

Palladio, Andrea, *The Four Books of Architecture*, Dover, New York, 1965

Pevsner, Nikolaus, J. Fleming, and H. Honour, *A Dictionary of Architecture*, Overlook Press, Woodstock, New York, 1976

Serlio, Sebastiano, *The Five Books of Architecture*, Dover, New York, 1982

Vitruvius, *The Ten Books on Architecture*, Dover, New York, 1960

VI

Cut out doors and windows in order

to make a room. Adapt the nothing

therein to the purpose in hand,

and you will have the use

of the room.

Lao Tzu

In the previous chapter we examined the generation of elementary forms for an architectural vocabulary. In this chapter we present the final series of operations in the construction of a model of an architectural form. In the recurrent analogy with human language, the process of creating an architectural composition from the simple elements previously described is similar to the articulation of sentences using nouns and verbs. In an architectural context, the "nouns" are represented by the elements in the architectural vocabulary, while the "verbs" are the relations between them.

Previous chapters have investigated architectural elements mainly as geometric models, even though for each element a semantic description associated with the geometrical definition was provided. A knowledge of the vocabulary of a language, however, does not imply the ability to articulate sentences, without a corresponding knowledge of grammar. In the same way, the creation of a work of architecture involves aspects, such as architectural context and style, that are irrelevant to the generation of architectural primitives. The principal focus of this chapter will be examples of classic architecture, in which the grammar of the composition is very clear. The concluding remarks, however, consider possible approaches to modern and contemporary architecture, where the "deconstructive"[1] character of a composition often does not allow its underlying syntactic structures to be clearly identified.

The compositional analysis that follows will be confined to formal aspects. The use of the term *architectural composition* instead of *building* is indicative of this approach.

ARCHITECTURE AS BOUNDARY CREATION

According to Christian Norberg-Schulz, a built space is defined by the boundaries of floor, wall, and roof, in the same way as the boundaries of a landscape are given by ground, horizon, and sky. Boundaries create perceptible distinctions by virtue of the elements they are composed of and the void space they enclose. The relations

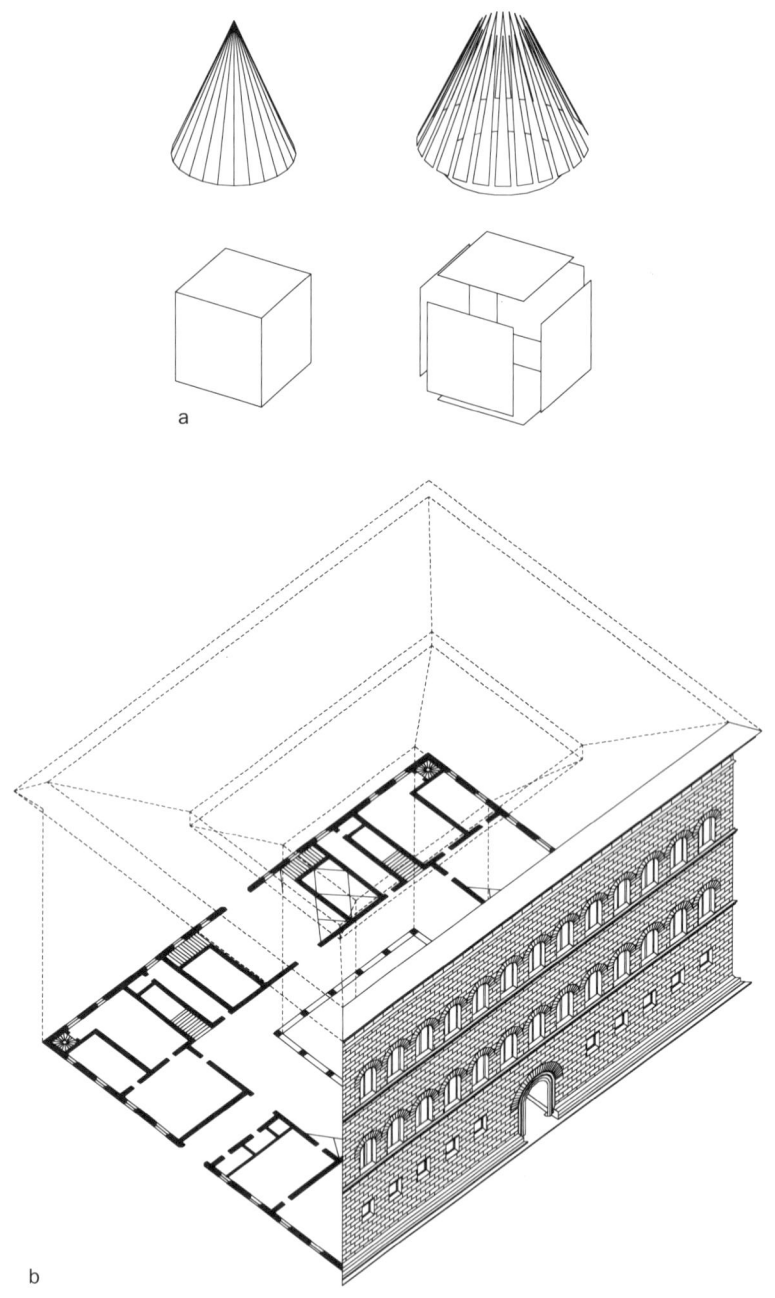

a

b

VISUALIZING WITH CAD

among the solid elements, defined as architectural primitives in the previous chapter, can be seen in the architectural boundaries they generate.

This interpretation of architectural forms in terms of boundaries does not preclude the consideration of volumes in design; *volumes* and the *boundaries* that delimit volumes are not mutually exclusive terms, but interrelated and complementary (figure VI-1). A design can start with a definition of boundaries and end up with a determination of volumes, or vice versa; ultimately, however, design is concerned with the definition of boundaries.

SOLID AND VOID MODELING

The discussion of Chapter III regarding a syntax of form generation can also have a semantic interpretation in architecture. The definition of geometric primitives and their grouping into part/symbols was done according to their spatial-formal characteristics of solid and void. This criterion is even more meaningful when we proceed from geometry to an architectural context. According to the Italian architect and theoretician Bruno Zevi, "Architecture . . . does not consist in the sum of the width, length and height of the structural elements which enclose space, but in the void itself, the enclosed space in which man lives and moves . . . The facade and walls of a house, church, or palace, no matter how beautiful they may be, are only the container, the box formed by the walls; the content is the internal space." The relevance of void is not to be undervalued in architecture or, for that matter, in computer-aided design.

In several areas of computer graphics modeling, the term *solid-modeling* is used; however, the concept of void-modeling [Yessios 1986] has a greater meaning in an architectural context. Conversely, void does not exist without the solid boundaries enclosing it. Thus, computer-aided design of architectural forms is properly concerned with the distinction between solid and void—the separation of inside from outside—established by boundaries. The solid architectural

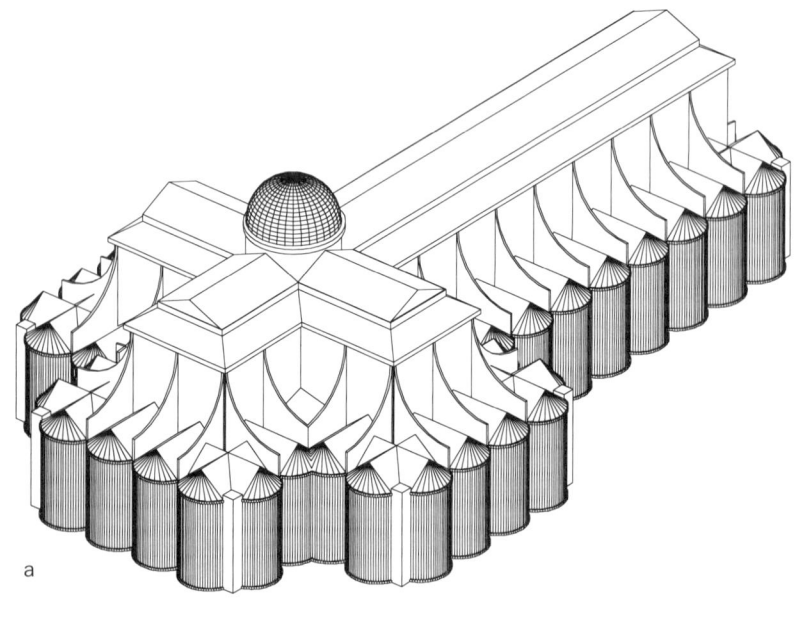

a

b

VISUALIZING WITH CAD

primitives discussed in Chapter V can be combined to enclose a void, and the way in which they do so is the essence of an architectural composition. An example is provided by the schematic model of the Santo Spirito Church, by Filippo Brunelleschi, shown in figure VI-2. The two images are views of the same model: Figure VI-2a shows the church from outside as a solid container, while VI-2b is a perspective view of the inside as a contained void.

SPATIAL MODELS FOR ARCHITECTURAL COMPOSITIONS

An architectural model can be defined in four ways, corresponding to different semantic descriptions: geometric, architectural, construction, and urban. These definitions are applicable in CAD, but they also apply generally to any architectural model. The model of Palazzo Strozzi, described later in this chapter, provides a clear example of these four definitions.

The boundary elements of a building create differentiations in the continuity of space. In the **geometric** definition, these boundaries can be defined as an organization of simple geometric elements, such as lines, points, and surfaces. Geometry is concerned with the spatial description of forms in terms of position and extension; this definition is therefore essential, even for designs generated with a nongeometric approach, since architecture is made of physical materials that have extension and position. There are several descriptive levels to the geometric definition itself. Starting from a diagram showing the spatial relations, such as symmetry and modularity (figure VI-3a1), the geometric definition can evolve into a volumetric diagram showing the solid-void composition (figure VI-3a2).

In the **architectural** definition (figure VI-3b), the boundaries are defined as walls, floors, and ceiling, each carrying attributes of style, function, or structure that go beyond geometric characteristics. The representations of the architectural boundaries become plans, eleva-

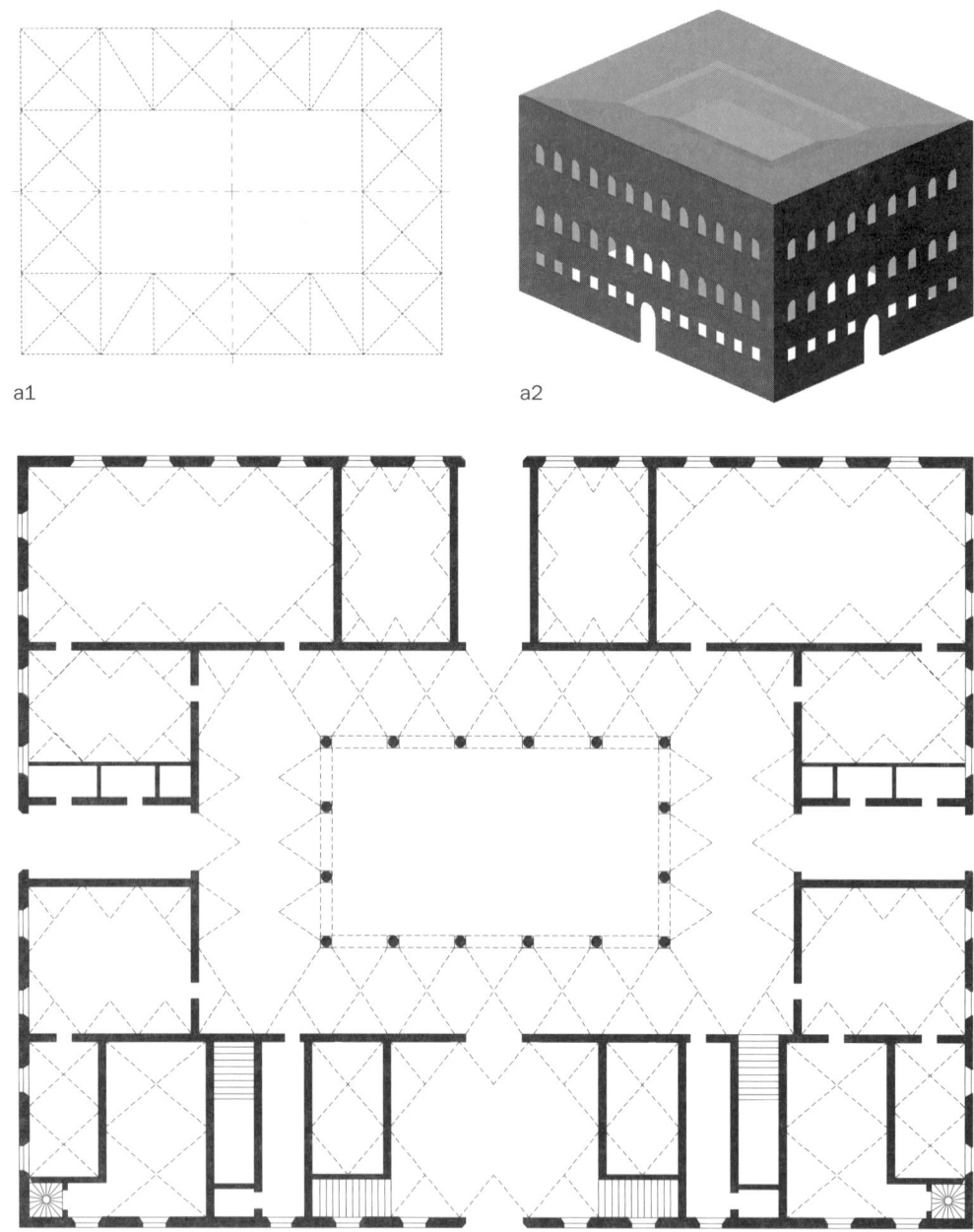

a1

a2

b

VISUALIZING WITH CAD

Palazzo Strozzi.

a. b. geometric and architectural
definition.

tions, and sections. The present chapter is concerned with this level of definition.

The **construction** definition is concerned with the material substance of the boundaries. In this phase, the designer is concerned only with the solid components of the composition, which are described individually and in greater detail than its spatial characteristics. Therefore, a purely visual representation is usually not sufficient to completely describe the construction, and additional written notes are required.

In the **urban** definition, the enclosed space is the public space defined by the outside boundaries of buildings. A clear example of the different definitions is provided by maps of different scales, from architectural to territorial.

These various definitions often are not distinct, however, but overlap. A geometric model defining in detail a solid-void relation can be sufficient for the architectural description; or, an urban design can relate several architectural enclosures. Nevertheless, for the purpose of creating a complete (not just electronic) architectural model, these definitions should be differentiated, as a general rule, in the data structure. Due to the physical limitations of the machine in memory and computing time, it would be extremely difficult and redundant to display simultaneously all the elements belonging to each definition, to say nothing of the visual confusion this would create.

In order to distinguish conceptually and visually the information relevant to each definition, we must focus on the data structure of the **dynamic model**. In each definition, the model as a whole has a hierarchical structure. Each part, defined as a symbol, can be a simple element or can be composed of subparts, also defined as symbols, so that when the hierarchical process is completed, the model is entirely defined as a set of symbols. Each symbol, however, has a different content for each of the four definitions. Thus, the designer can create four different files, each containing a model made of the same symbols. In addition, all the elements belonging to one defini-

a

b

VISUALIZING WITH CAD

Subtraction of regular shapes
in Baroque architecture:
Sant'Ivo alla Sapienza.

tion can be grouped in the same layer (or in a set of layers) that can be clearly distinguished.

Addition and Subtraction

Addition and subtraction are two different approaches to architectural composition; both can be recognized in works of architecture. Classic architecture is based on addition methods: The spatial composition is given by the organized hierarchical sum of elements. Each element can be considered part of the whole, but can also assume its own identity, independent of the whole. The Greek temple, which will be discussed and implemented later in this chapter, is a typical example of the additive approach. Each part of the temple, such as columns, can be considered separately from the rest. The Renaissance period provides many examples in which the whole is clearly a hierarchical sum of parts. In the Santo Spirito Church, by Filippo Brunelleschi, the total architectural composition is composed of external walls, columns, domes, vaults, and floors (figure VI-4). Each of these elements can be separately defined and implemented as a CAD model; the total compositional model represents the addition of architectural elements.

In Baroque architecture, however, the complex formal articulation is obtained by subtraction of regular geometrical shapes. The alternating convex and concave masses, which are typical of the Baroque language, can be easily derived from the subtraction of elemental curved shapes. In the Sant'Ivo alla Sapienza, by Francesco Borromini, the geometric model of the plan is clearly based on forms derived by the subtraction of a circle and equilateral triangle (figure VI-5a). The subtractive process between straight and curvilinear forms is repeated at several different hierarchical levels, as shown in the schematic plan (figure VI-5b).

For the CAD implementation of this second type of formally derived composition, the Boolean operations provide a first approximated geometric scheme, which integrates the elements created from the subtraction of the two initial shapes. From this scheme, elements can be grouped in parts, and the composition can then revert to the addition of parts. In this way we can go back to the dynamic model approach, which allows more flexibility for the modeling of each part according to the four definitions.

A Library of Architectural Elements

The CAD models of the basic architectural elements generated in Chapter V can be stored as a database in the computer memory in separate directories or subdirectories. Each directory or subdirectory represents a library organized according to subject. Libraries can consist of such elements as columns, walls, windows, floors, and roofs. According to type and level of detail, each library/directory can be divided into subdirectories containing parts of each element. For instance, the library of columns can be divided into three subdirectories for the base, capital, and shaft. The library can be organized in terms of style, chronology, or materials, or according to the number and complexity of the elements. The examples implemented here borrow the basic elements from the following libraries: column, vault, dome, floor, and wall.

The actual dimensions of the elements in the libraries are irrelevant since, for particular compositions, the desired dimensions can be obtained by scaling transformations. Each library element represents a dynamic model, identified as a symbol, with associated parameters corresponding to its extension (height, length, and width). Dynamic models are prototypes of architectural elements, based on the geometric description of an element as determined by its structural and functional characteristics. The designer can modify the prototype to fit specific design requirements. The simplest variation is a change of parameters. If the model is composed of part/symbols, it can be modified by replacing one or more parts, while the syntactic relations remain intact. The designer can also incorporate additional information in the modified prototype, which then becomes a new prototype.

A Second Library of Spatial Relations

An architectural composition can be viewed not only in terms of its basic architectural elements, but also in terms of spatial organization. As Norberg-Schulz emphasizes, there are three basic configurations: centrality, axiality, and grid. These configurations are organizing principles for the plan of the composition, but they can also sometimes be recognized in section. Even if none of these three principles is immediately recognizable in a composition, almost any example of classic architecture can ultimately be attributed to at least one of them.

Centrality (figure VI-6) refers to buildings based on a circular plan, or to any configuration determined by rotational symmetry about one point, or element, at the center of the composition. In Greek architecture, amphitheaters were organized according to a rotational symmetry. In Renaissance and Baroque architecture as well, centrality represents the organizing principle for many buildings; illustrious examples include: the Pantheon, the Tempietto of San Pietro in Montorio by Bramante, and Villa Rotonda by Palladio (figure VI-6b,c). In Baroque architecture, the centrality is often obscured by the complex geometry of the composition, based on subtractive operations. Rotation is the geometric operation associated with a central spatial organization. A CAD model (prototype) of this organization is given by a point/symbol replicated and rotated at an angle α about the center. Points can be replaced by simple elements such as columns or walls. The parameters are the rotation angle α between two point/symbols and the distance d of each point/symbol from the center.

Axiality (figure VI-7) denotes a spatial organization based on a line, or axis. Axial compositions are found in some Greek and Roman temples, Romanesque and Gothic cathedrals, and Italian Renaissance churches. The geometric transformation that generates an axial composition is translation; the parameters are represented by the distance d between two successive point/symbols and the total number n of point/symbols, which also gives the total length of the axis.

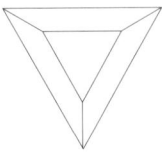

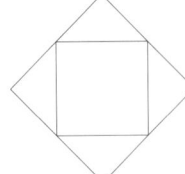

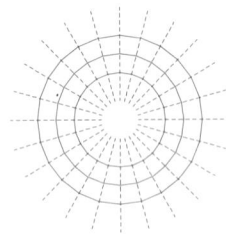

a

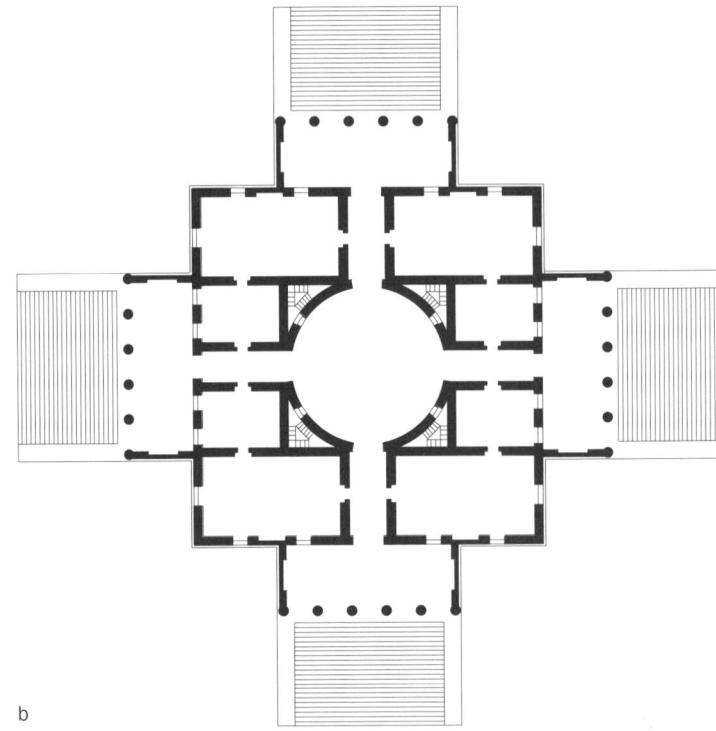

b

Centrality.

a. Diagrams.

b. Villa Rotonda: plan.

c. Villa Rotonda: schematic
   axonometric view.

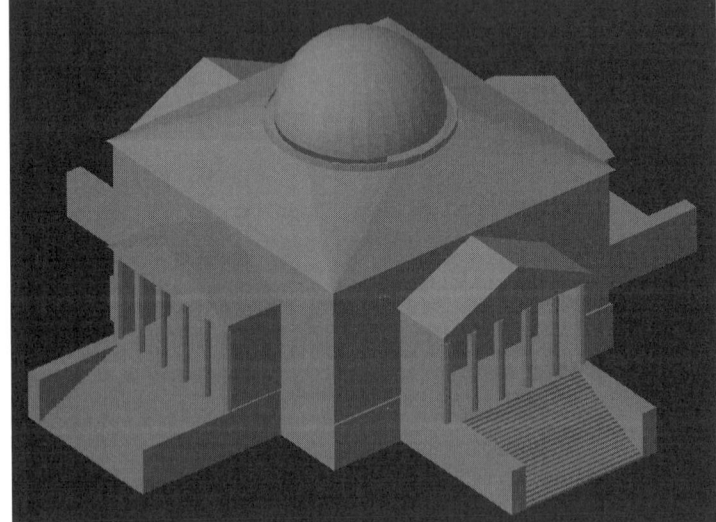

c

VI-7

Axiality.

a. Diagrams.

b. Temple of Hera Argiva
   in Paestum: plan.

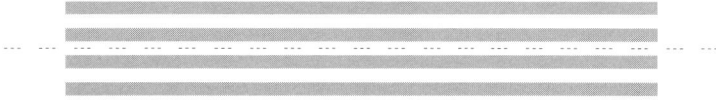
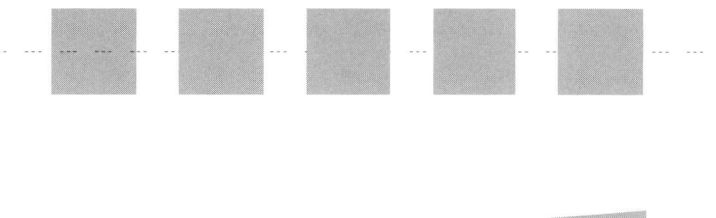

a

b

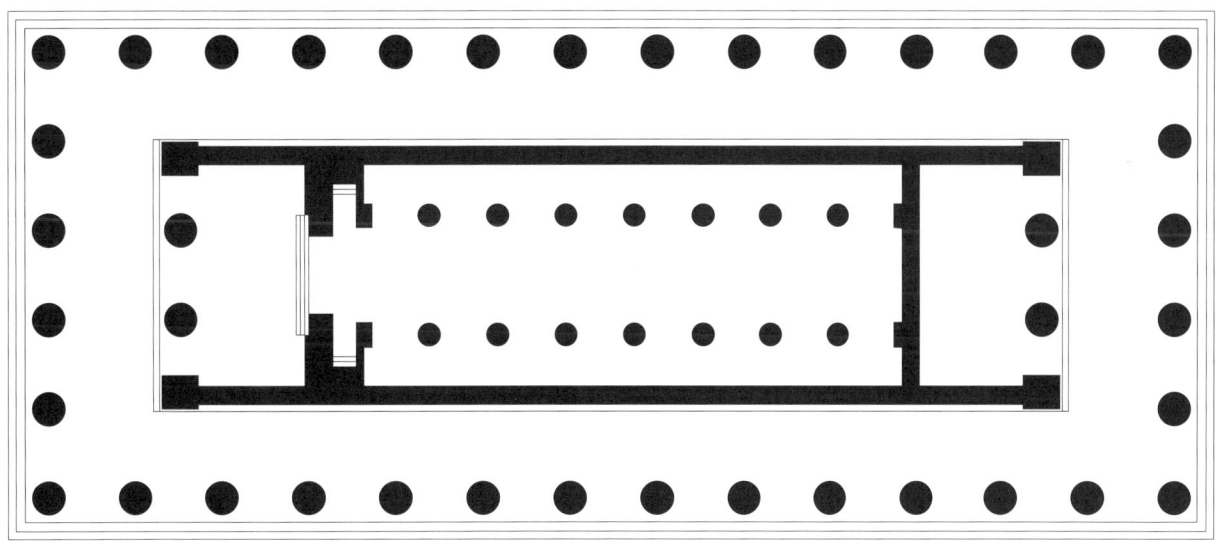

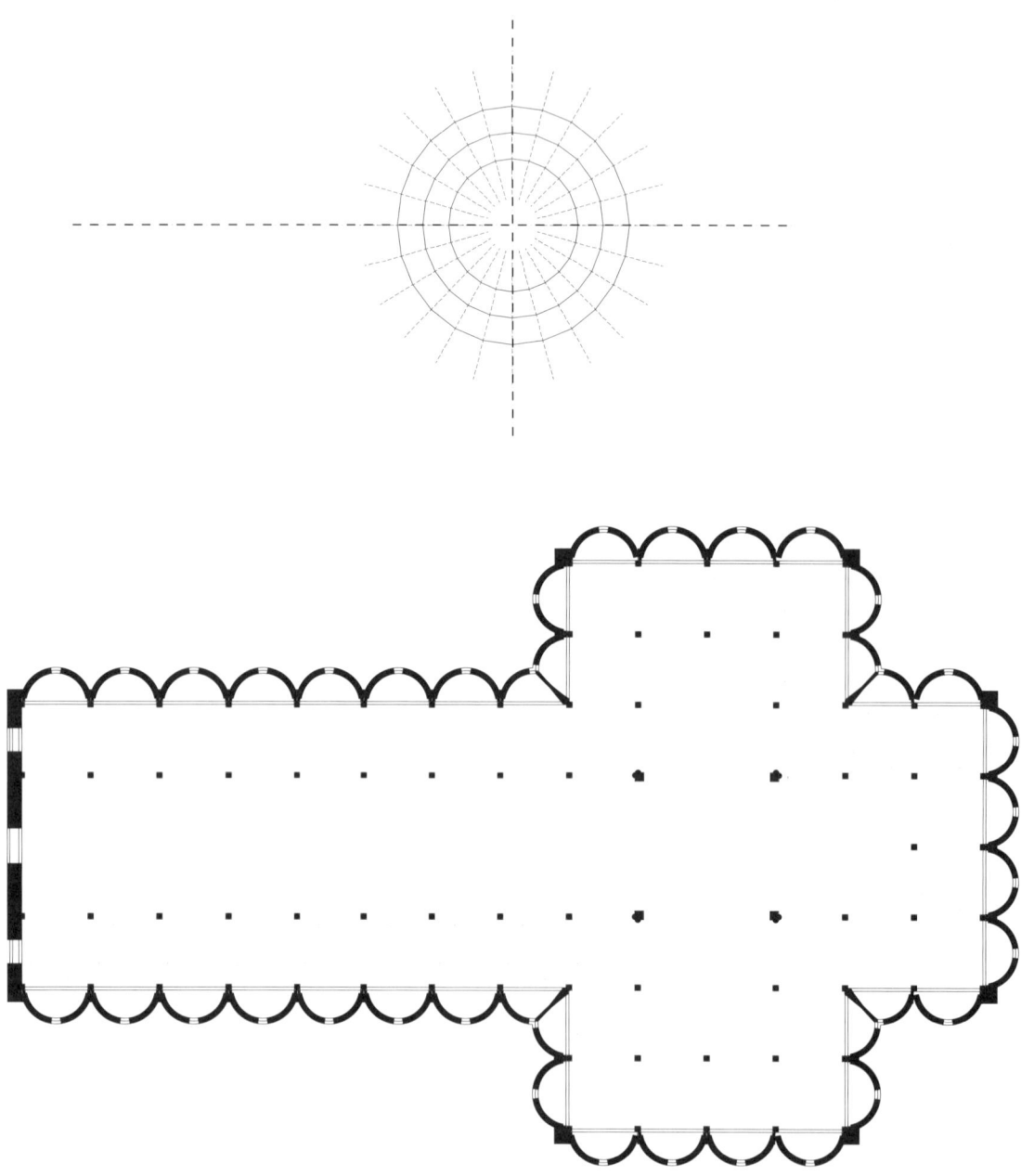

Axiality combined with a central
composition: Santo Spirito by
Filippo Brunelleschi.

The **grid** configuration, the third mode of spatial organization, is based on the intersection of a set of parallel lines with another set of parallel lines, at a certain prescribed angle. Grids deserve a more detailed investigation, since they are often the basis for elevations as well as floor plans and have structural significance. They will be discussed later in this chapter.

Although an architectural composition may exhibit only one of these three spatial organizations, more often a composition is generated by the combination of these configurations, either as relations of equal importance or in a hierarchical structure. The schematic example in figure VI-8a shows a central composition of points imposed upon an axial relation. Santo Spirito by Filippo Brunelleschi (figure VI-8b) is an example of this organization.

Hierarchical Composition of Elements

In some compositions, the combination of the three primary spatial organizations can be richly articulated. In this case, as in the implementation of models of existing designs, it may be easier to generate a model by the addition of architectural elements according to spatial relations, without recurring to the three basic organizational modes. Since, in CAD, the relations between the different architectural elements composing a building are defined in terms of geometric transformations, the same geometric operations used in Chapter V to generate the elements of the architectural vocabulary from geometric primitives can be used once again to generate a composition.

A graphic diagram can be helpful in clarifying the hierarchical classification of all the elements composing a building since it provides immediate information about the spatial relations between them. The diagram can take the form of a tree structure (as shown in figures VI-9a and VI-11a), in which each element or set of elements is represented by a node and the complete building is the root of the tree. The lowest level contains information about the geometric primitives of the simplest architectural elements. In the tree, different semantic levels are combined.

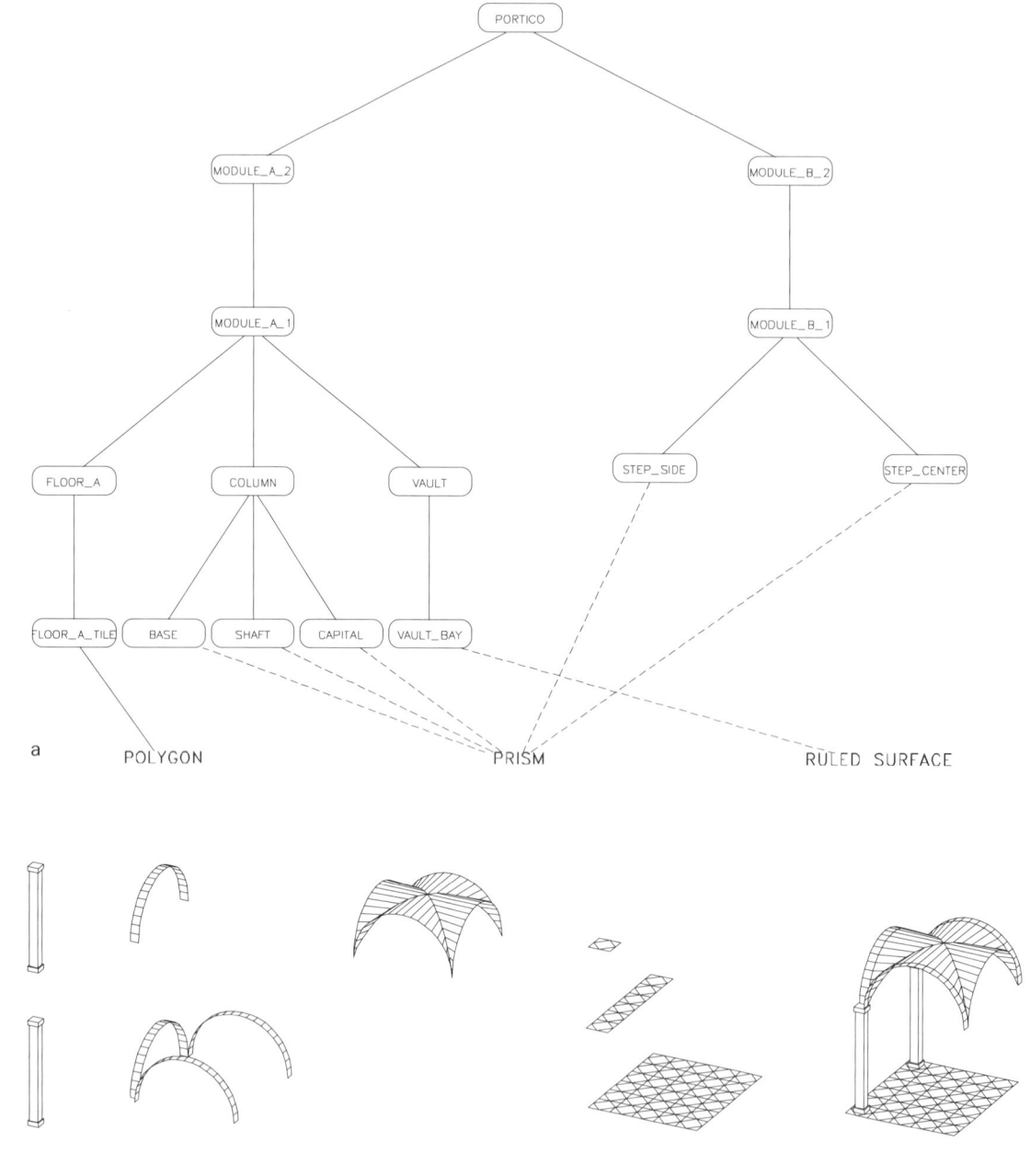

a

POLYGON   PRISM   RULED SURFACE

b

VISUALIZING WITH CAD

Portico.

a. Diagram.

b. Hierarchy of forms.

Ordered spatial relations are easily identified in historical examples that follow rigorous symmetric compositions, but they may not be apparent, at first glance, in a modernist building. This does not mean that definable relations are completely absent in such buildings, but it shows how, in a different vocabulary, elements can be ordered by relations less obvious than symmetry and evident geometric operations.

**TWO EXAMPLES OF
IMPLEMENTED SPATIAL MODELS**

The two dynamic models that follow represent examples of architectural compositions derived formally by the principles of spatial organization discussed above. They also illustrate how the spatial configurations previously discussed can be combined to generate new designs.

Portico

porti_2d.dwg

portico.dwg

The composition in figure VI-9 recalls some of the porticos found in Renaissance iconography. It is based on a hierarchical spatial organization, in which an axial configuration (at the lower hierarchical level) is arranged according to a central plan.

The geometric transformations generating the composition are given by repeated translations (for the axial organization) and rotations (for the creation of the central plan). The modular element is contained in a square in plan projection and is composed of four columns, one groin vault, and a floor pattern. The most basic elements, at the lower hierarchical levels, are **column**, **vault**, **arc** (given parameters) and **floor tile** (derived) (figure VI-9a,b). A groin vault is placed, together with three arches along its western, southern, and eastern edges. Two columns are placed at the two southern corners of the vault. A square floor tile is dimensioned according to a number n of divisions of the distance between the two columns. The floor tile is placed with its center corresponding to the center of the southwest column and then repeated and translated n times along the x-axis.

C

VISUALIZING WITH CAD

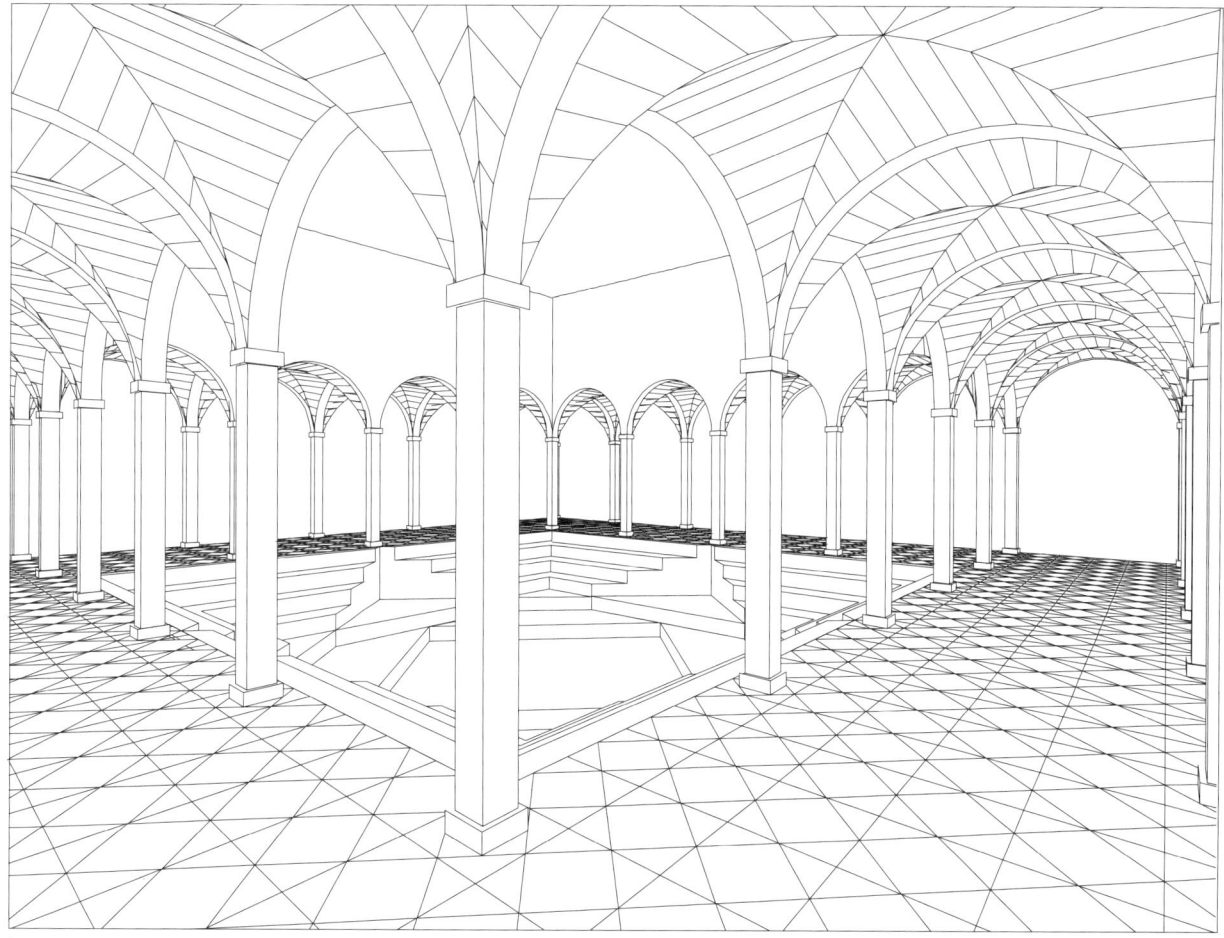

d

VI-9

Portico.

c. Plan and section.

d. Perspective view.

The row of tiles so obtained is replicated and translated n-1 times along the y-axis. All the floor tiles can be grouped into a symbol, which may be called *3D-floor*. All the elements created thus far, including the newly created symbol, are then grouped into another symbol, which can be called *3D-module*, with the coordinates of the insertion point given by the x and y coordinates of the center of the vault and the z coordinate corresponding to that of the floor. From the library of spatial organizations, a central configuration is retrieved. The parameters are represented by a rotation angle of 90° (90° = 360°/4) and a distance from the center equal to some multiple m of the sum of the width of the edge of the vault plus the depth of the arch. The four point/symbols of the central configuration are each replaced with an axial configuration, given by a number of point/symbols equal to $2m$+1 and separated by an interval equal to the sum of the edge width of the vault plus the depth of the arch. This creates a square grid of point/symbols, each of which is replaced with the symbol *3D-module*.

The dynamic model generated in this way can evolve through the replacement (figure VI-10a1) of the basic architectural elements—column, vault, arc, floor tile—with other stylistically different elements (figure VI-10). In the model illustrated in figure VI-10a, the groin vault has been replaced with an angled roof and the simple cylindrical columns with classical columns.

Vaulted Room

The vaulted room in figure VI-11a–d is characterized by bilateral symmetry combined with an axial organization. The diagram of the composition is shown in figure VI-11a; in the following description the given parameters are the **position of the columns**, their **distance from the y-axis,** and the **number of tiles** (figure VI-11b). The dimensions of the coffered vault and the tiles and the distance between two successive columns along the y-axis are derived.

A column is placed with its center at point $P(x,y)$, where x is given and y is arbitrary. The column is then replicated and mirrored about the y-axis; the centers of the two columns establish a distance a. A square floor tile is inserted so that its center corresponds to the center of one column. (The sides of the tile have length a/n-1, where n is the total number of tiles in the x direction.) The repetition of the tiles according to a translatory movement along the x direction creates the first row of tiles, which are grouped into a symbol called *3D-floor*. This will be repeated and translated along the y direction. All the elements created thus far, including the newly created symbol, are then grouped again into a symbol called *3D-module*. An axial configuration is retrieved with point/symbols at intervals equal to a fixed multiple of the y dimension of the floor tile. These point/symbols are then replaced with the symbol *3D-module*. An arbitrary point on the y-axis determines the center of the rotation for a coffered vault (see Chapter V), which will be scaled in the y direction according to a value dependent on the module of the floor tile. (These operations can be inverted, so that the module in the coffered dome functions is a given parameter, and all the other proportions are derived from it.)

An evolution of this model is shown in figure VI-12. The coffered vault and columns have been replaced with metal frame arches.

THE CLASSICAL APPROACH: EXAMPLES FROM ARCHITECTURAL HISTORY

The methods used to create the two models above can also be used to model a building that is already designed or built. In this case, a CAD model is not used as an exploratory tool as in new designs, but as a means by which the designed building can be analyzed in terms of geometric transformations and elements. Thus, the process is inverted: While in the former case the geometric definition could eventually be transposed into physical (built) forms, in the latter case the existing forms (built or designed) with material characteristics are reduced to geometric shapes. Architectural compositions may need to be simplified at different levels since the richness of description for each element can obscure the more basic compositional

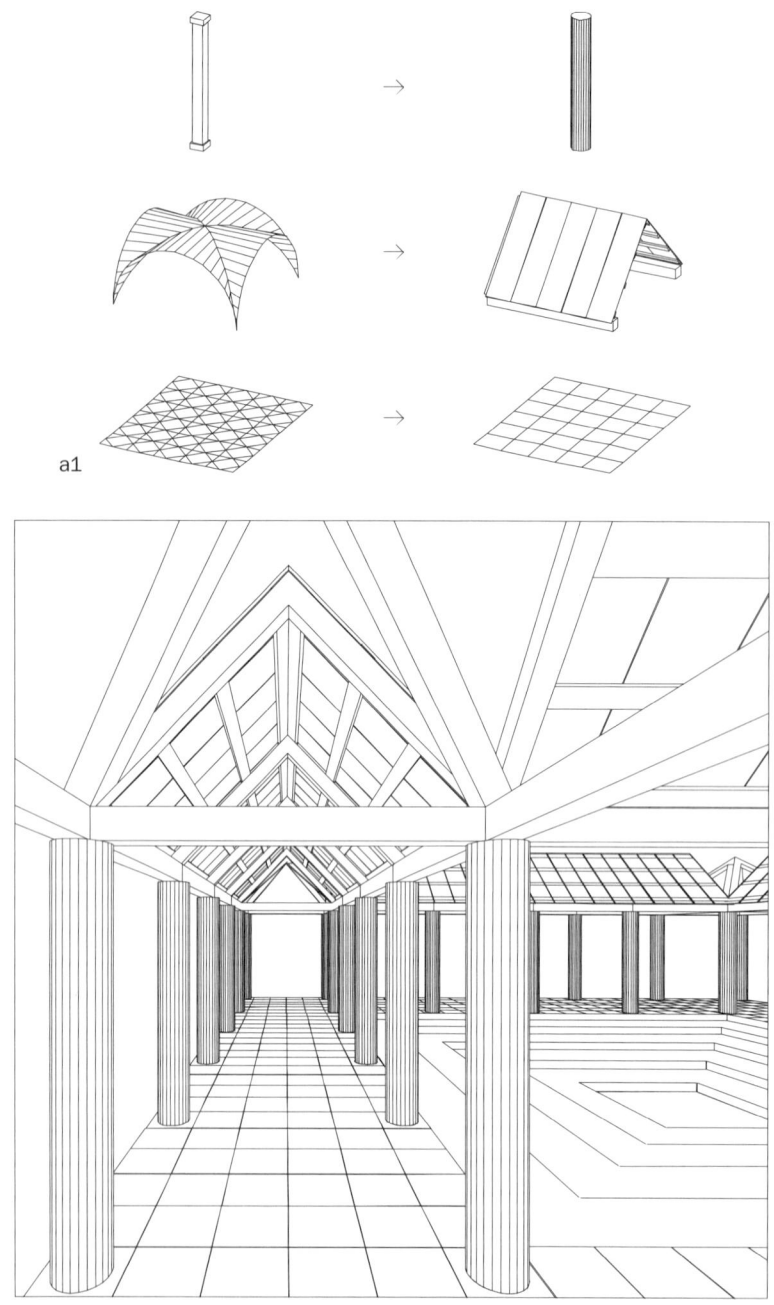

a1

a2

VISUALIZING WITH CAD

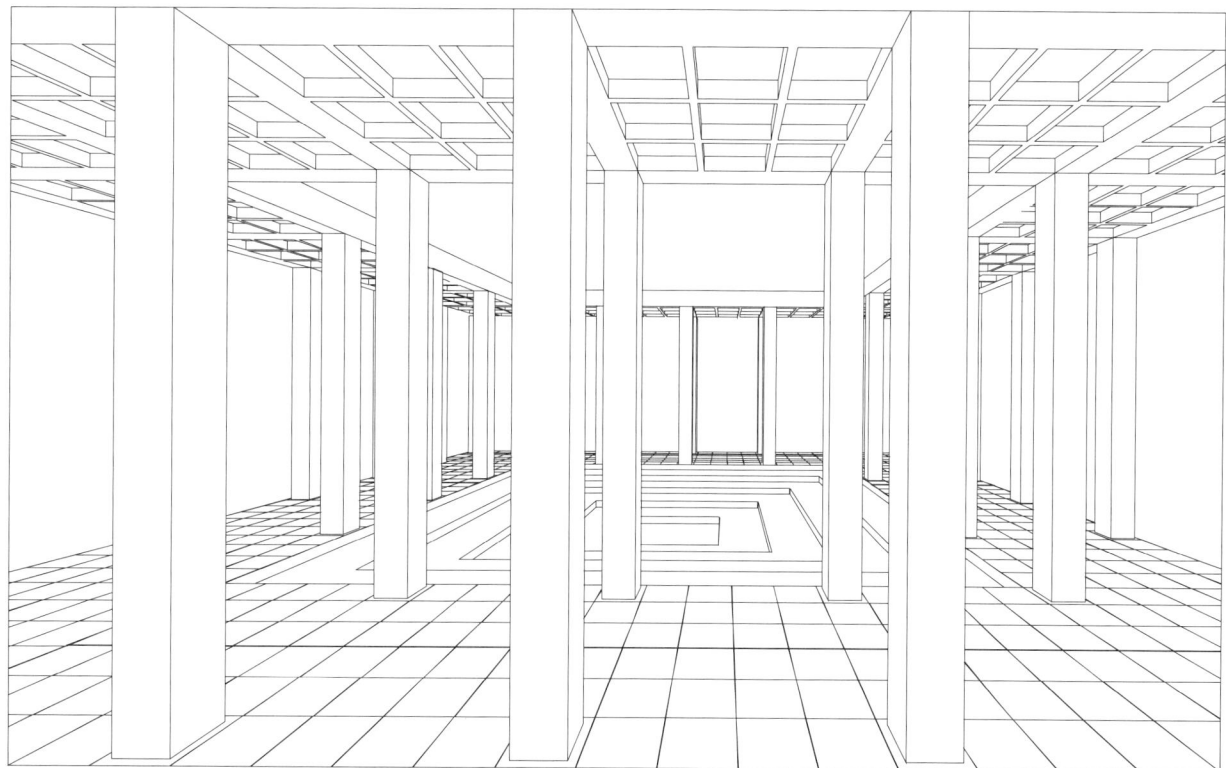

b

VI-10

Evolution of the portico in
different architectural styles.

Vaulted room.

a. Diagram.   b. Hierarchy of forms.

c. Plan and section.

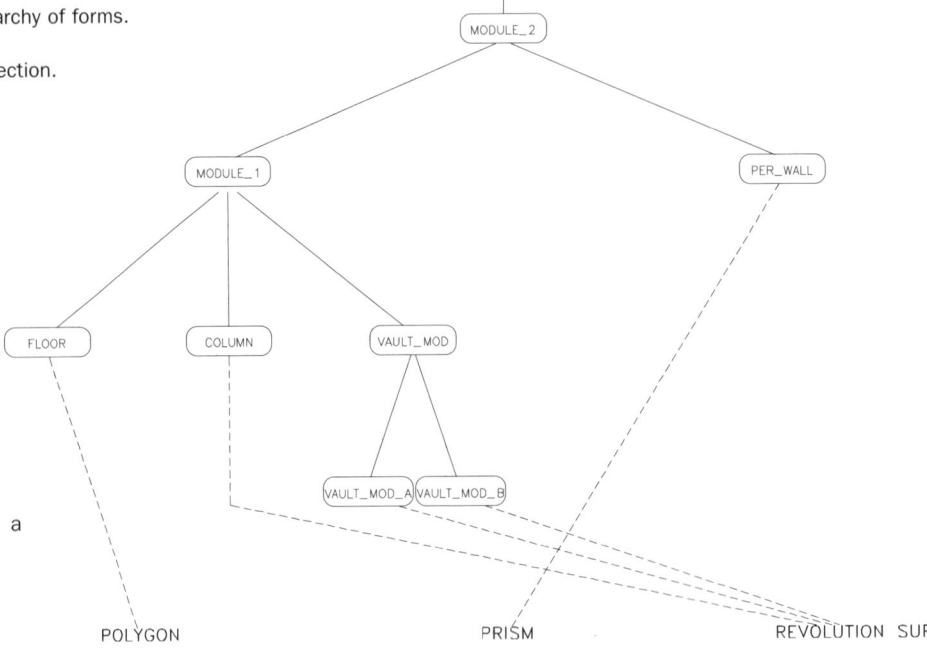

a

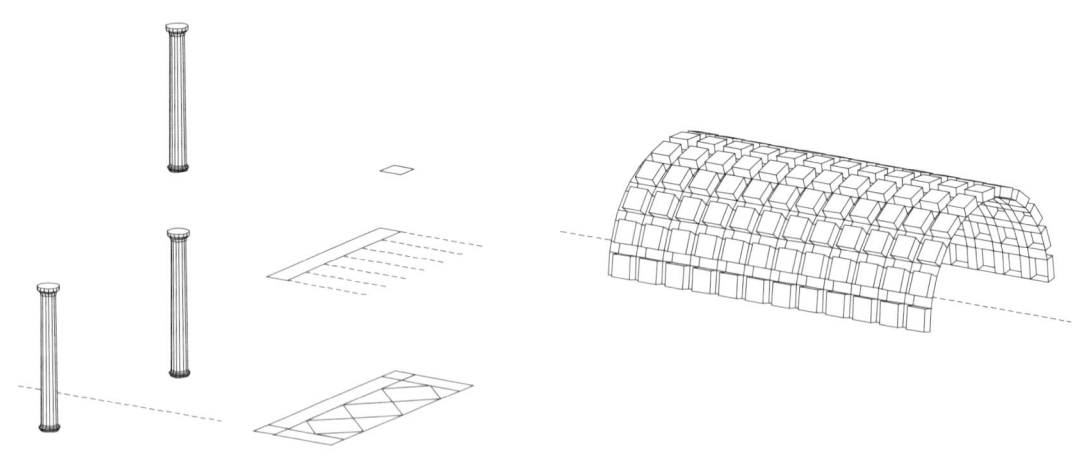

b

VISUALIZING WITH CAD

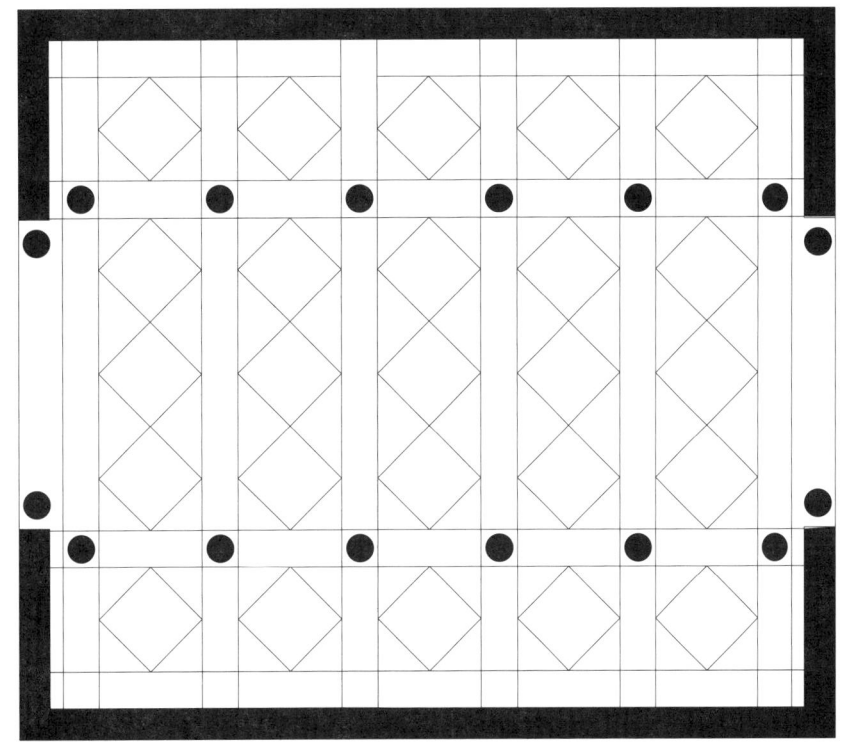

c

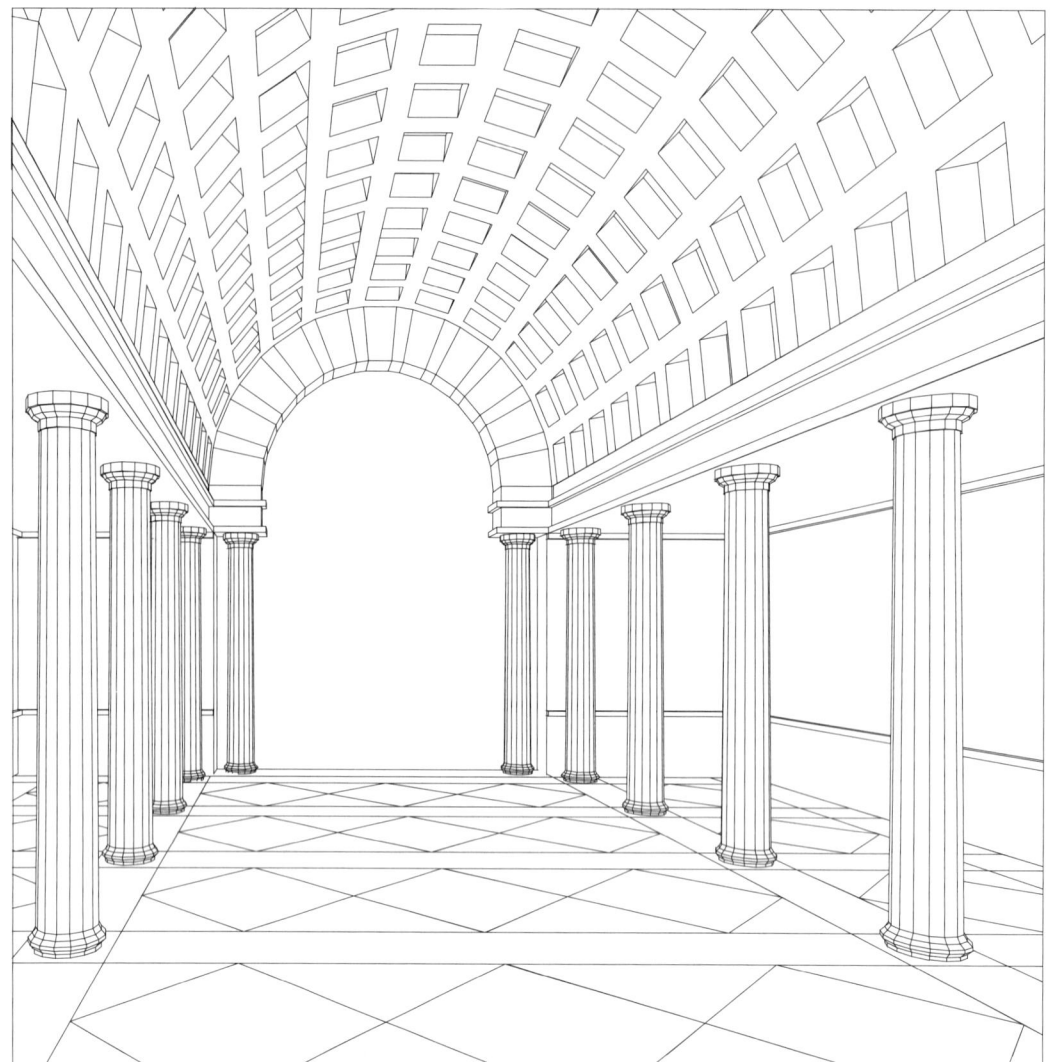

VI-11

Vaulted room.

d. Perspective view.

VISUALIZING WITH CAD

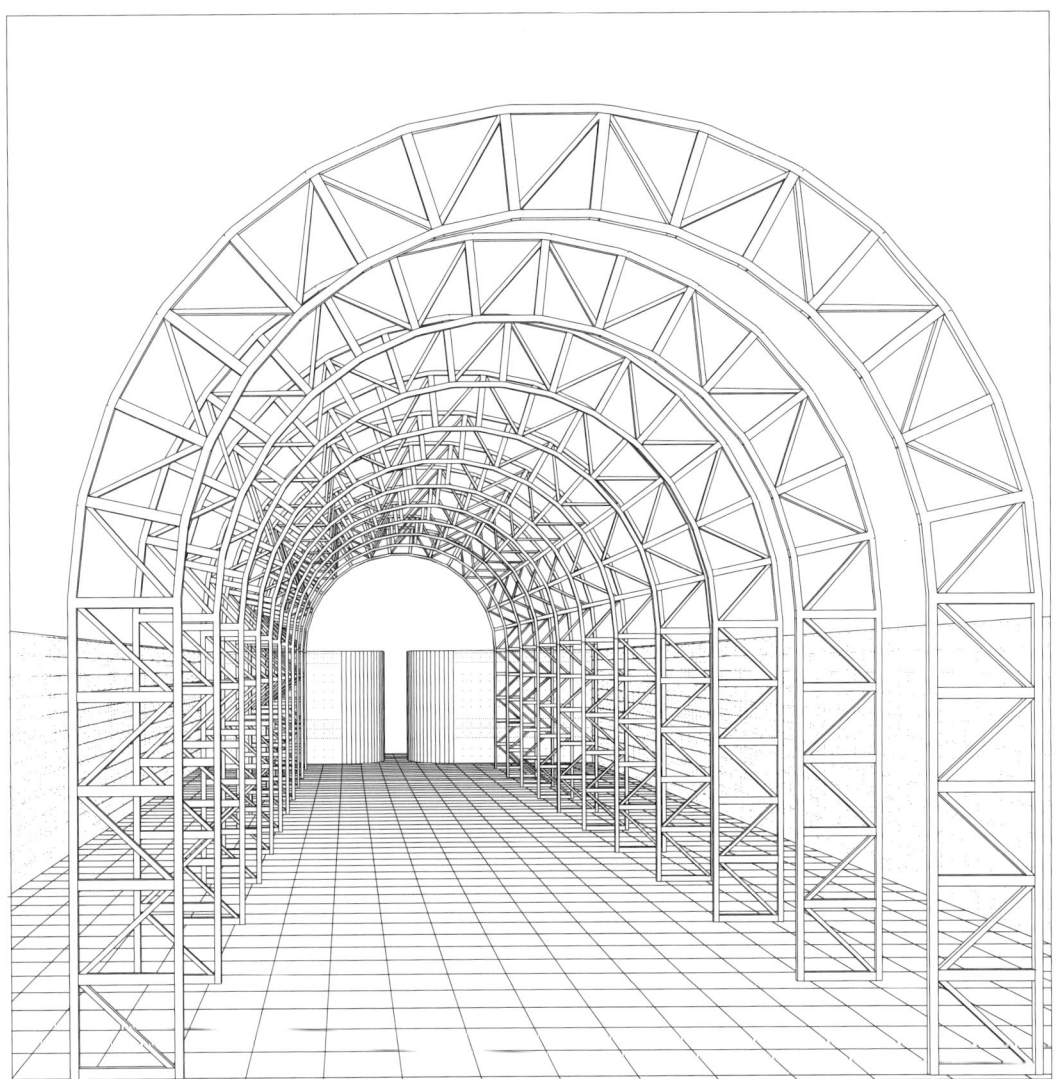

VI-12

Evolution of the vaulted room in
a different architectural style.

relations. The level of detail to be used in the model is a matter of choice. The following models attempt to distill the essential characteristics of the buildings by focusing on the composition as a "whole" made of "parts." The parts are identified in the CAD model as symbols to allow different evolutions in which each part may change in its characteristic style or proportions. This shows how spatial relations are what truly identify an architectural composition, which keeps its identity beyond the modifications in style. The models are intended as interpretative tools, rather than photorealistic simulations. Nevertheless, the architectural elements in each of the following compositions, although simplified, represent the correct proportional values, and the spatial relations between the elements are the same as those of the real buildings. The buildings below have been chosen as CAD models on the basis of two criteria: their popularity, which makes them recognizable to most readers, and their susceptibility to geometric interpretations.

The Second Temple of Hera

The Second Temple of Hera (figure VI-13) is one of the best-preserved examples of Greek architecture. This temple of the Doric order dates from about 450 B.C. and is located in Paestum, a town in southern Italy originally founded as a Greek colony.

A Greek temple is a clear example of an architectural whole made of parts. In plan, the temple is characterized by an axial composition and bilateral symmetry. By taking advantage of this symmetry, the model construction requires implementation only of the bottom left corner of the composition. The dominant element in the composition is the massive Doric column, inserted as a symbol in the lower left corner, and repeated and translated twice in the x direction and six times in the y direction. The last column in the x direction is then repeated and translated once along the y direction to frame the entrance to the cella. The elements generated thus far can be grouped and mirrored about the y-axis and then, together with the new elements produced by reflection, mirrored again about the x-axis to

paestu_2d.dwg

paestum.dwg

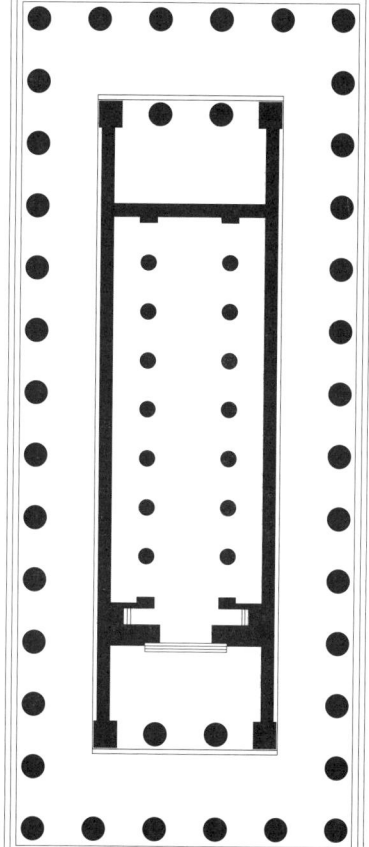

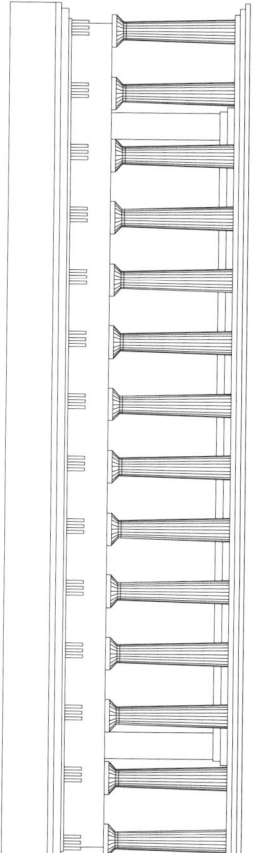

VI-13 ▶

Second Temple of Hera, Paestum.

a. Plan and elevation views.

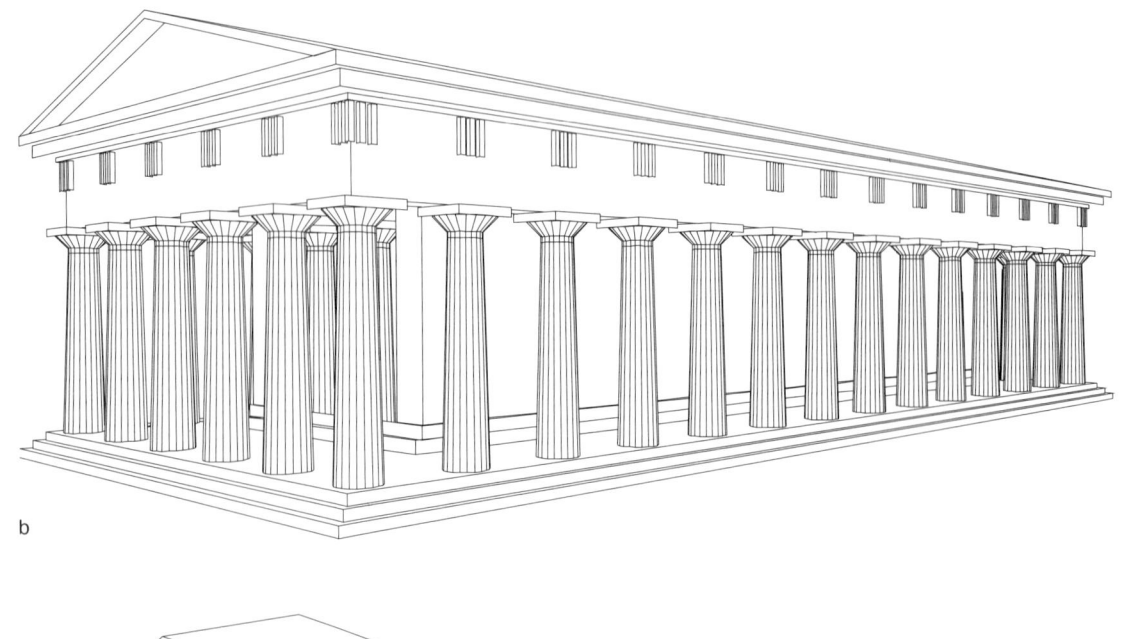

b

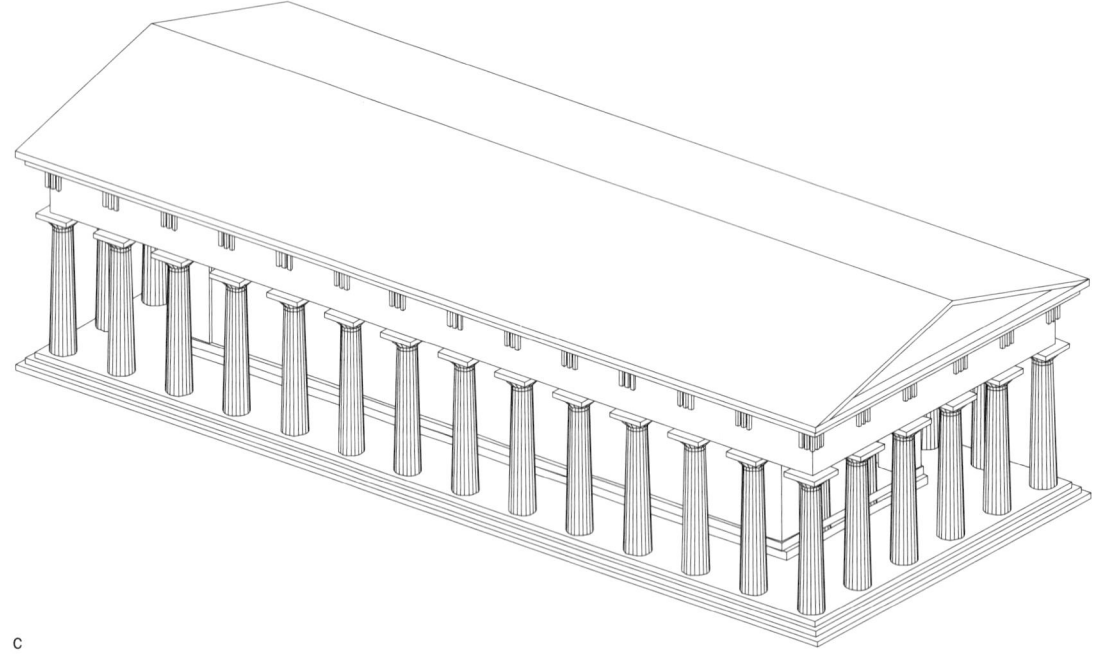

c

VISUALIZING WITH CAD

Second Temple of Hera, Paestum.

b. Perspective view.   c. Axonometric view.
d. e. Perspective views.

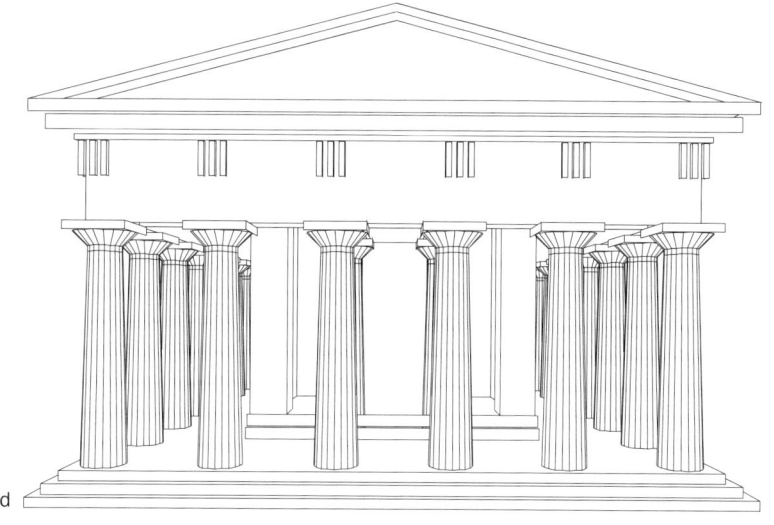

d

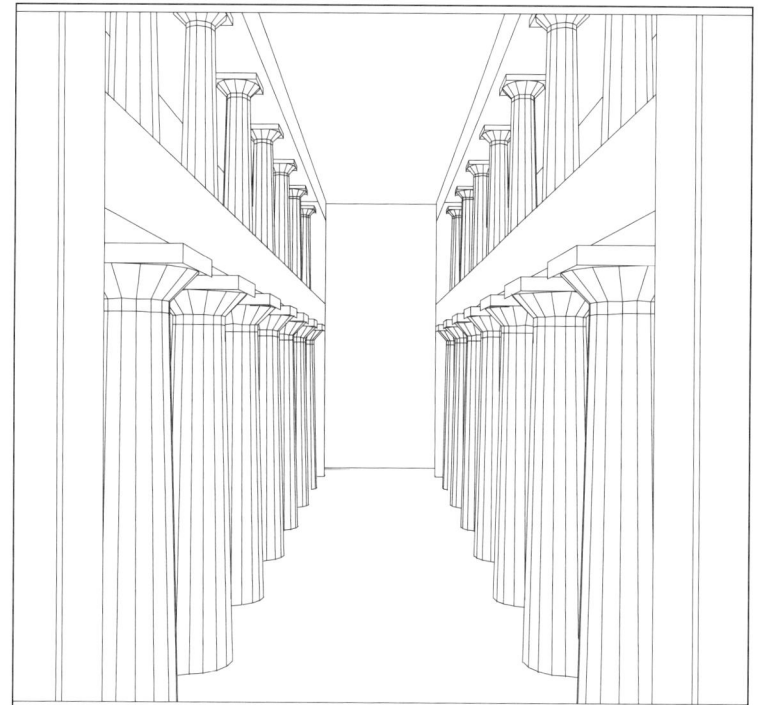

e

create the full complement of large columns. The cella is again symmetrical about the y-axis and is composed of plain rectangular perimetral walls enclosing seven double rows of a new type of column, which is a scaled version of the initial column. The roof is made by the translation and extrusion of a triangular element. All the elements in the pediment and the base, including the decorative ones, can be defined geometrically as parallelepipeds of various dimensions.

Figures VI-13f1 and VI-13f2 show examples of two different evolutions of the longitudinal elevation of the temple, obtained by replacing the column and the frieze with different elements. In the composition of figure VI-13f2, the visual rythmn of the facade is maintained while the architectural character is modified by replacing

Second Temple of Hera, Paestum.

f. "Evolutions"of the longitudinal elevation
of the temple.

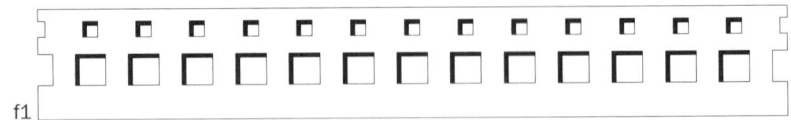

f1

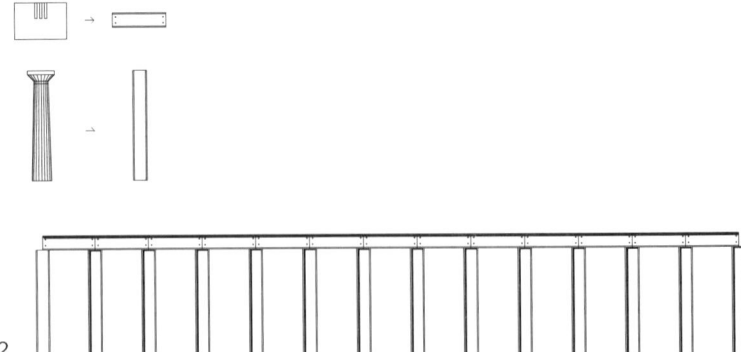

f2

VISUALIZING WITH CAD

the doric column with a steel one. In figure VI-13f1, the column is replaced with wall-like elements, changing completely the solid-void relations established in the original temple model.

Palazzo Strozzi

Palazzo Strozzi, by Giuliano da Sangallo, Il Cronaca, and Benedetto da Maiano, was built in Florence in the late fifteenth century. The composition of the floor plan is almost entirely bilaterally symmetrical. This regularity is also apparent in the composition of the three-story facades, interrupted only at the center by the entrance doors. The main facade presents thirteen columns of windows; the other facades are identical and are articulated by nine columns of windows.

For the model of Palazzo Strozzi (figure VI-14), the elements retrieved from the architectural library are walls, windows, floors, ceilings, stairs, and roofs. These elements are composed according to the symmetry relations cited above. The composition of the facade deserves particular attention since it is typical of many classical palaces. The geometric element organizing the facade is a connected orthogonal grid. The two minor facades are ordered by a 9x3 rectangular grid while the main one has a 13x3 grid. The connecting line/symbols are replaced by massive rusticated wall elements. Arched windows are placed at the openings for the second and third floor while simpler and smaller square windows form the openings of the ground floor.

The dynamic model of Palazzo Strozzi exemplifies the four definitions of an architectural model—geometric, architectural, construction, and urban—which were described earlier in this chapter and identified in figure VI-3. Again, the respective definitions of the model are derived from information that, for each part/symbol of the composition, has been differentiated according to four layers or sets of layers.

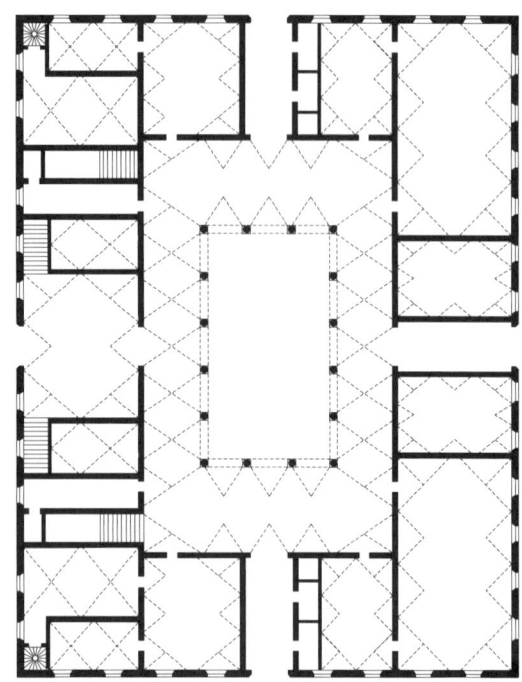

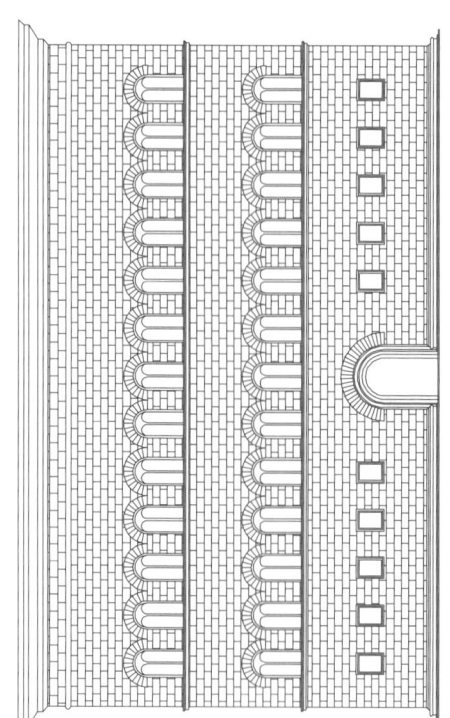

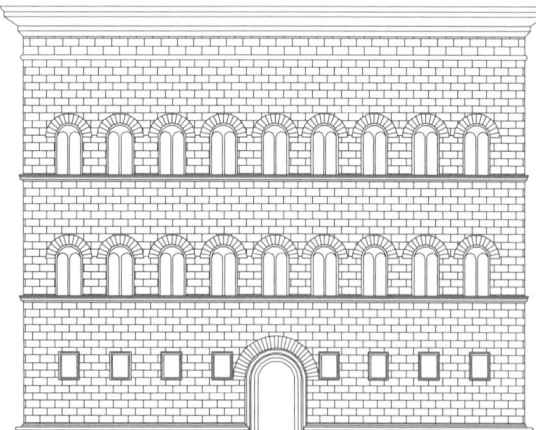

VISUALIZING WITH CAD

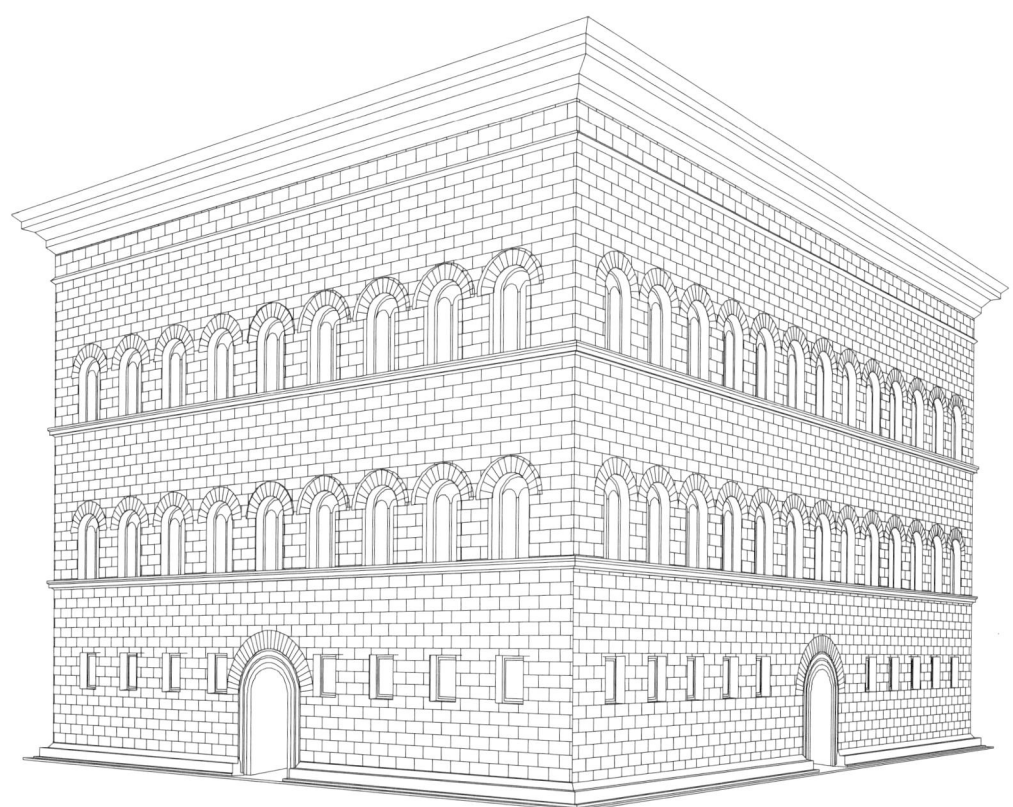

VI-14

Palazzo Strozzi.

a. Plan and elevation views.

b. Perspective view from outside.

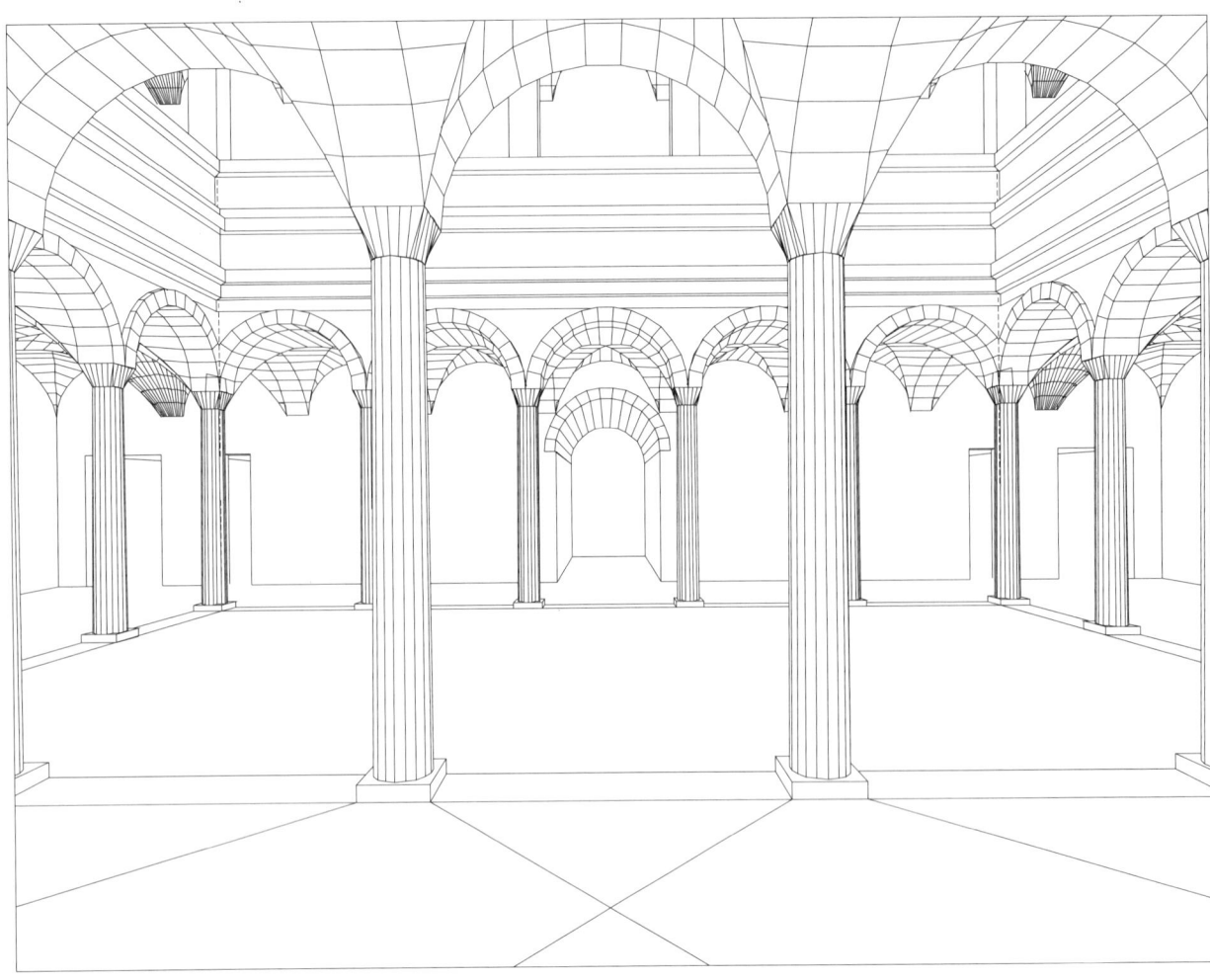

c

VI-14

Palazzo Strozzi.

c, d. Perspective views of the court.

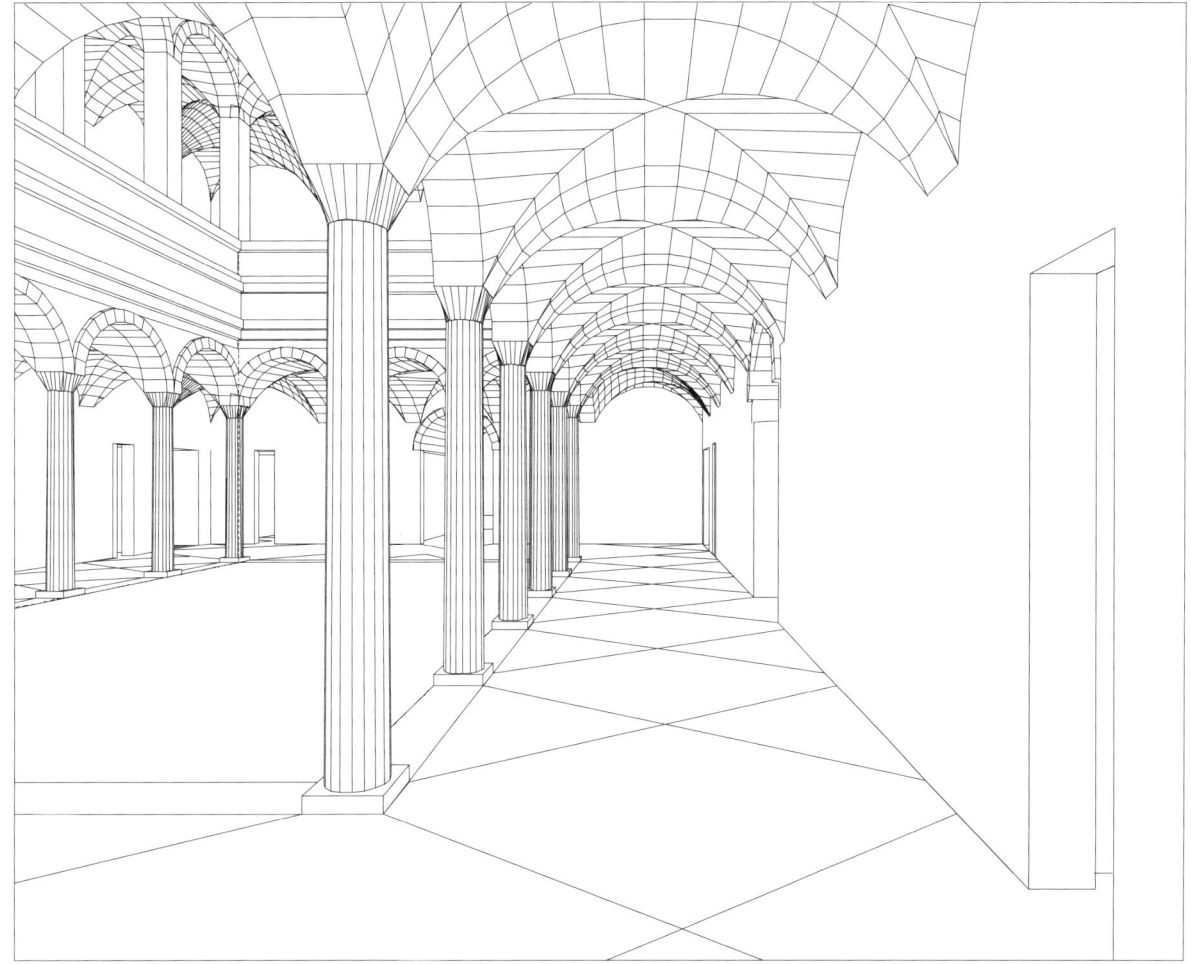

d

Palazzetto dello Sport

Palazzetto dello Sport (Small Sport Palace) in Rome, by the Italian engineer Pier Luigi Nervi, is a twentieth-century example of composition based on classical principles (figure VI-15). The geometry interacts with the structural characteristics, which in a large-span building, greatly influence the formal composition. The spatial organization is rigorously based on centrality.

The implementation of a CAD model for the Palazzetto shown here focuses on the essential characteristics of the structure, since that represents the most interesting formal aspect of the project. The entire structure of Palazzetto dello Sport is generated by the circular rotation of thirty-six parts, which form a perfect structural and formal continuity from the ground to the top of the roof. In contrast with other rotational elements, such as domes, there is no separation between vertical supporting elements and the roof. For this reason, the CAD model is constructed by starting from the implementation of the complete rotational module (figure VI-15a), which will then be replicated and rotated thirty-six times at a 10° angle. The final composition of the ribbed structure cannot be anticipated from the individual module; it is produced only after the rotational composition of all the modules. This case is typical of formal composition that perfectly expresses the structural behavior.

Palazzetto dello Sport.

a. Rotational module.

b. Plan and elevation view.

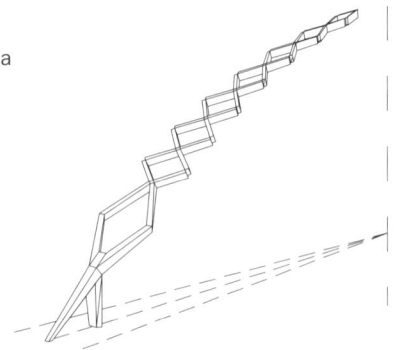

a

b

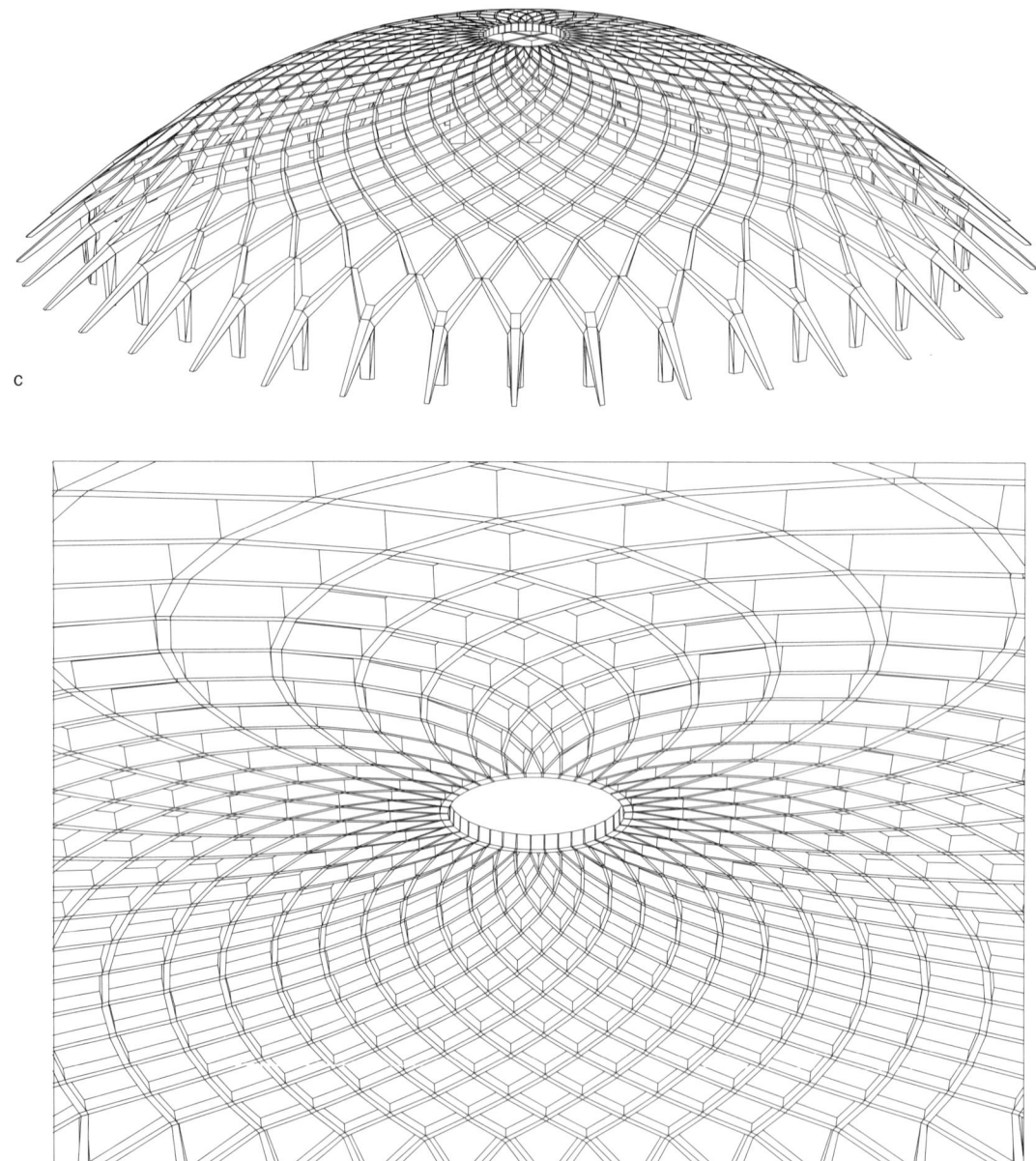

c

d

VISUALIZING WITH CAD

FROM CLASSICAL COMPOSITION
TO MODERN ARCHITECTURE

In classic architecture the rules of composition are very clearly defined. Modern architecture instead has different hierarchical and geometrical principles, which are often obsured by the emphasis on functional motivations that determine the design. Notwithstanding this different approach, a library of basic architectural elements can provide a basis for design alternatives.

Villa Savoye at Poissy

The model for Villa Savoye (figure VI-16) represents the transition to a modern architectural vocabulary and grammar. Villa Savoye in Poissy, France, was designed in the late 1920s by Le Corbusier. Its vocabulary is typical of Functionalist architecture.

The basic elements retrieved from the architectural library are columns, floors, ceilings, walls, windows, curved walls, ramps, and flat roofs. A CAD model can be created following the additive approach. The structural skeleton is shaped by an approximately square grid of pillars. Another essential component of this example is the ramp, which is a formal design element as well as a functional connection between floors. The Villa Savoye demonstrates clearly the use of modern structural technologies, which allowed the architect to separate the enclosure from the structure, thereby permitting a free-form composition. Just as modernist architecture found greater design freedom in the independence of the structural and enclosure systems, the segregation of these components into various layers of the CAD database (figure VI-16c) gives the designer greater facility in the creative process.

Villa Savoye at Poissy.

a. Roof plan, first floor plan
and west elevation.

b. Perspective view.

c. Independence between enclosure
and structural system.

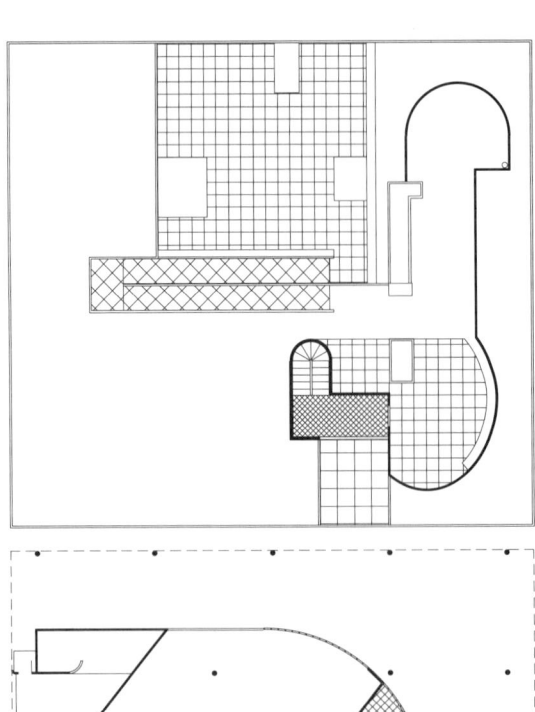

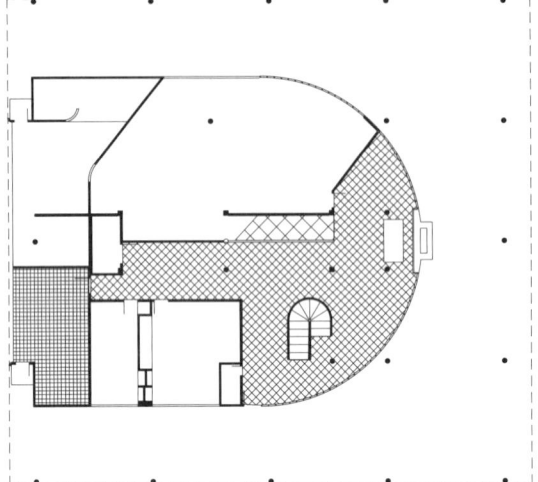

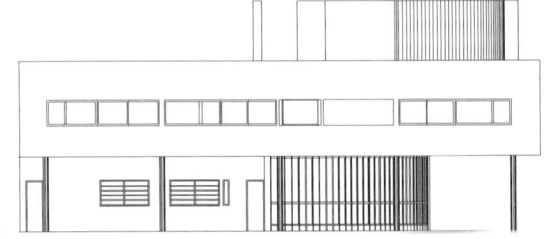

a

VISUALIZING WITH CAD

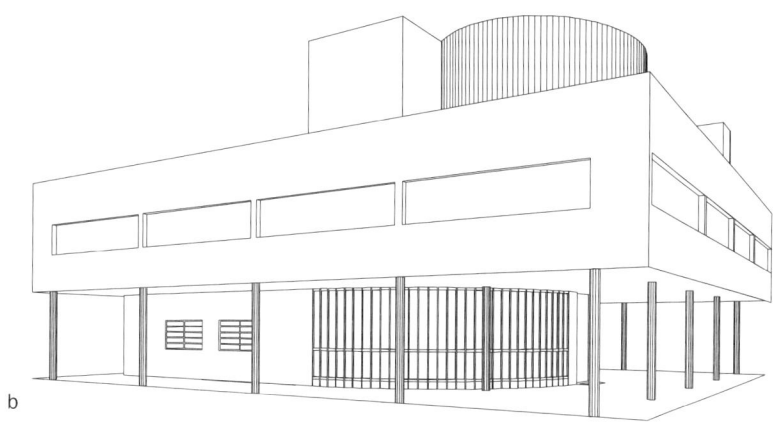

b

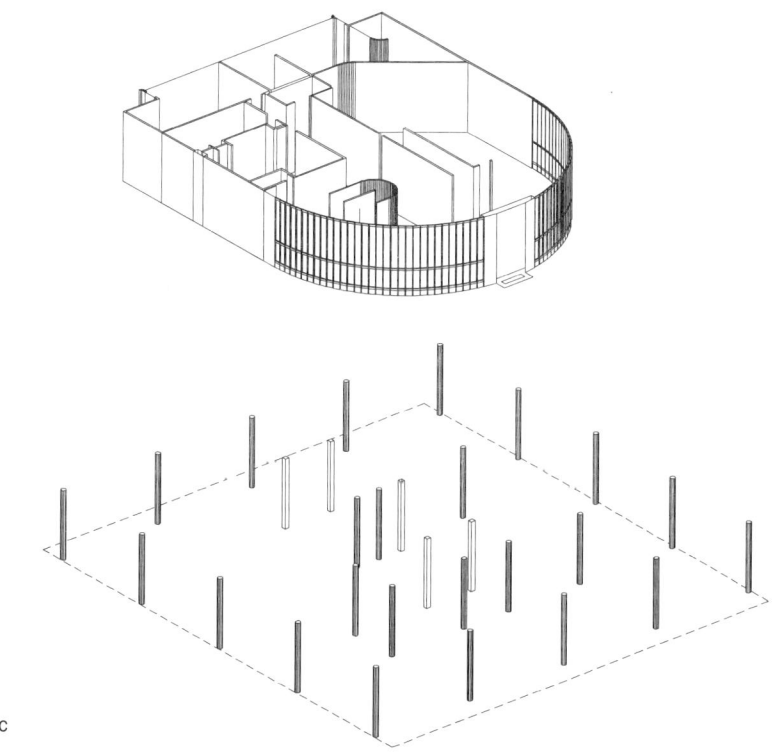

c

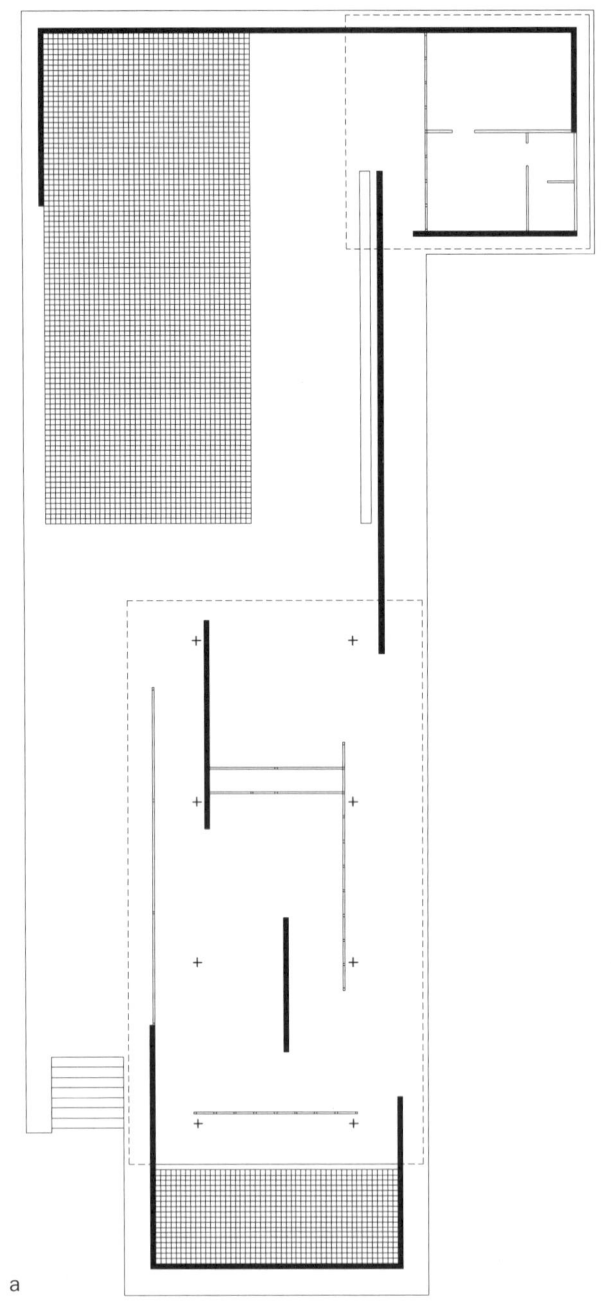

a

VISUALIZING WITH CAD

Barcelona Pavilion.

a. Floor plan.

b. Perspective view.

Barcelona Pavilion

The Barcelona Pavilion (figure VI-17), designed by Mies Van der Rohe, was built for the World's Exposition of 1929 and destroyed eight months later. It represents another landmark of modern architecture and a copy has recently been reconstructed at the same site. As with the Villa Savoye, the Pavilion makes a strong statement for the open plan implanted on a regular structure. The architectural composition derives from the intersection of horizontal and vertical planes, but this spatial conception goes beyond the traditional box, where planes are faces bounding a volume which strictly separate the inside from the outside. In the Pavilion, instead, inside and outside fluidly interact: The inside is more properly regarded as the space resulting from the intersection of portions of orthogonal planes, rather than the contained volume. The CAD model focuses on these aspects; each portion of plan can be defined as symbol. It is left to the reader to experiment by changing the model of the symbols identifying each element of the Pavilion, and allowing evolution from the basic spatial composition of the original design.

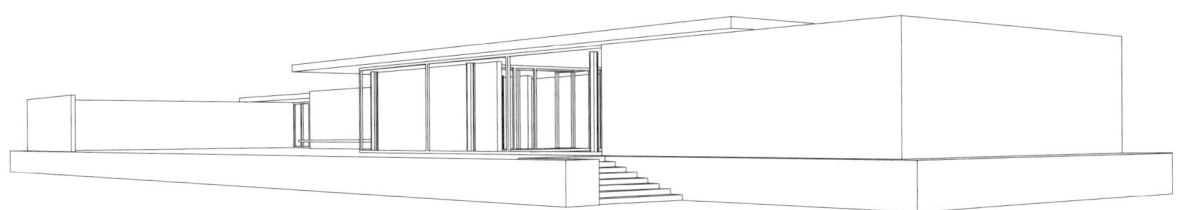

b

Barcelona Pavilion.

c.–e. Perspective views.

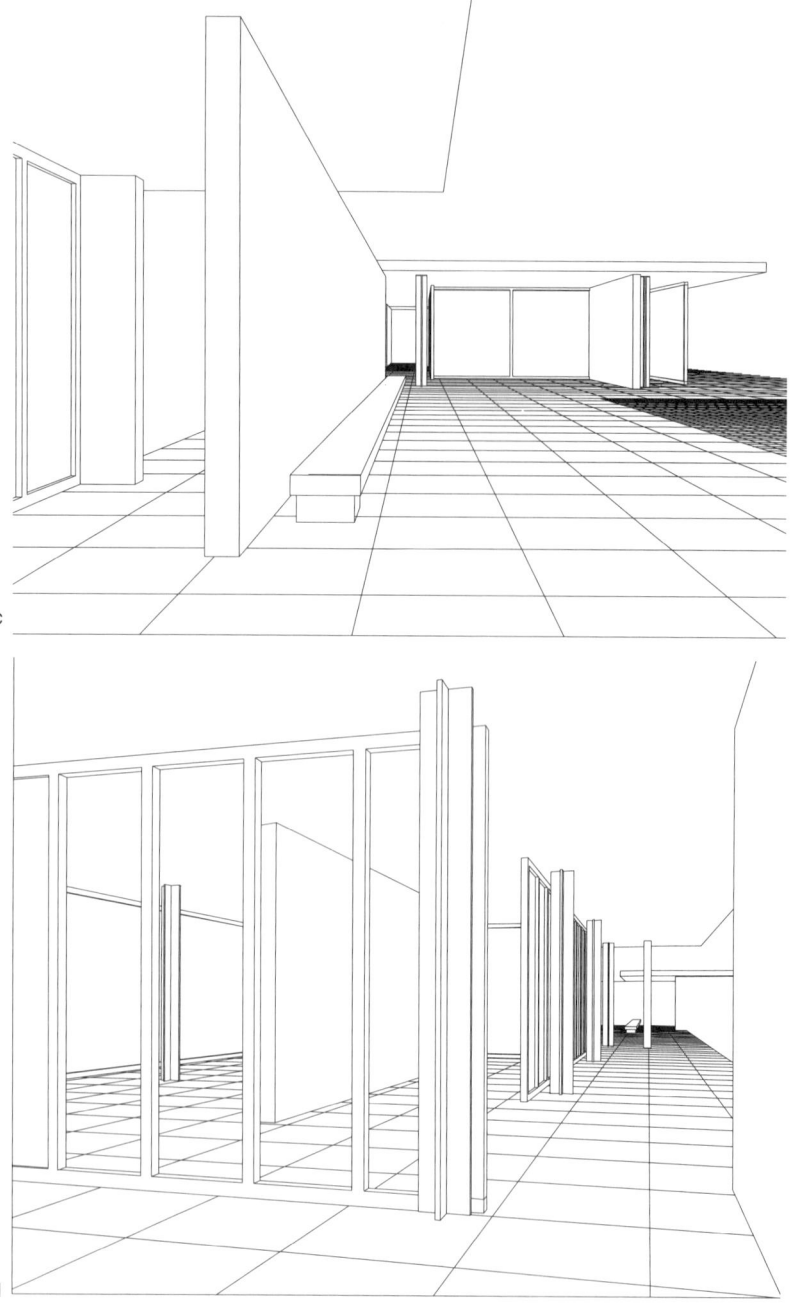

c

d

VISUALIZING WITH CAD

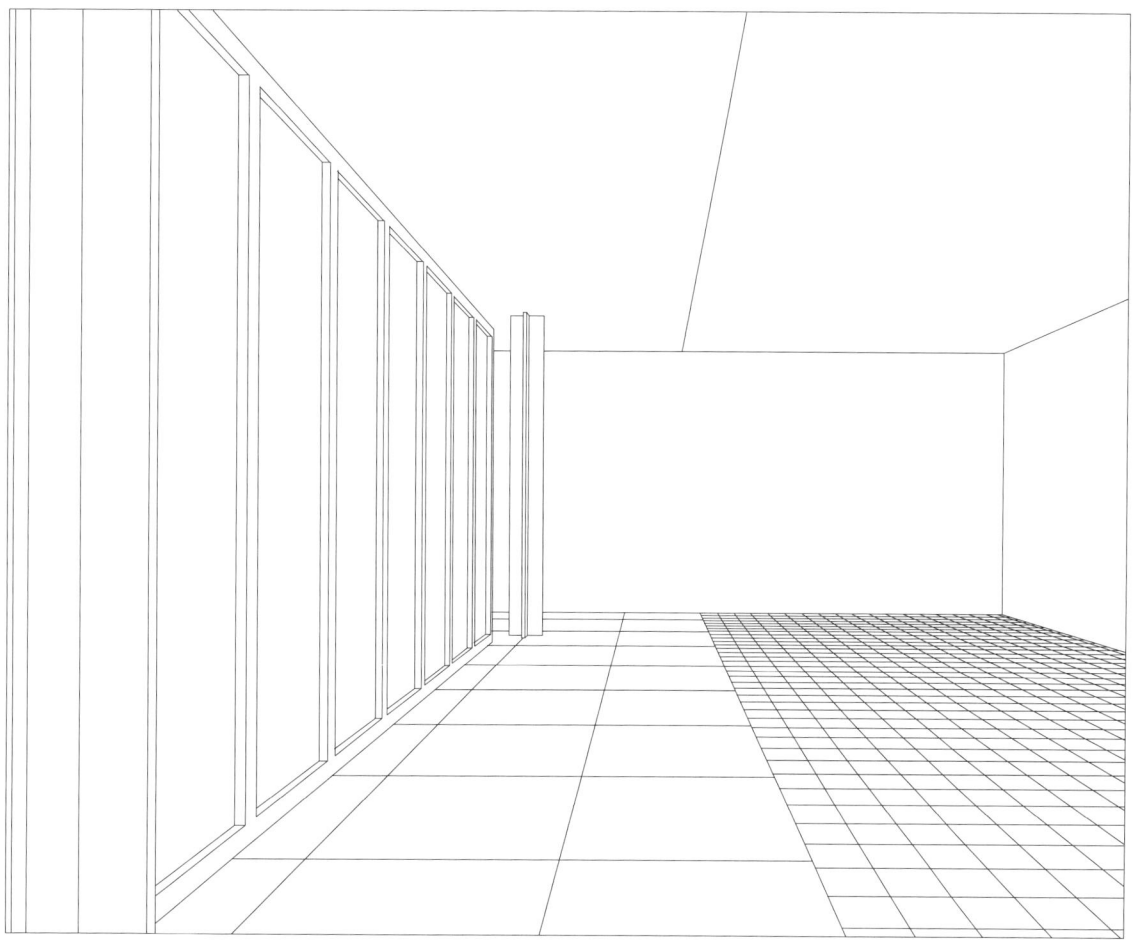

e

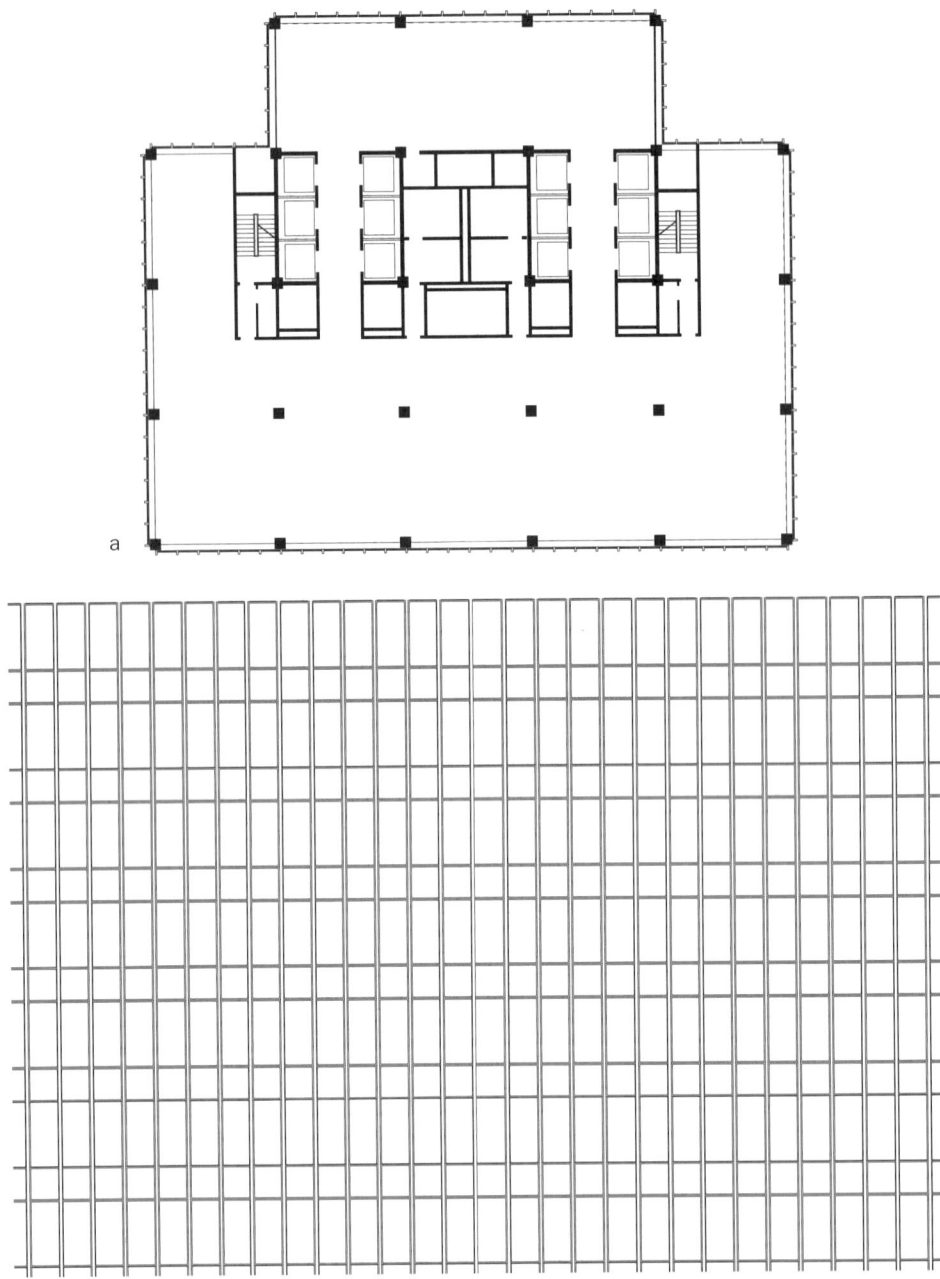

a

b

VISUALIZING WITH CAD

Seagram Building

Seagram Building.

a. Typical floor plan.

b. Partial elevation.

The Seagram Building (figure VI-18), designed by Mies Van der Rohe and built in New York in the late 1950s, is a milestone in modern architecture, especially in the design of tall buildings. The plaza created by the setback of the building and the "weighty" characteristics of the composition have been emphasized in many texts on modern architecture [Benevolo 1977, Jencks 1973].

The CAD model is based on the repetition of a few basic elements, such as columns and curtain wall modules. Repetition and translation are the generative motifs in both plan and elevation. A grid is the geometric model from which the building evolves. The floor plan is based on a five-bay by four-bay grid of square columns, with the two east corners carved out (figure VI-18d); the core is characterized by a bilateral symmetry. The regularity of the floor plan is followed in elevation, where the generative grid is based on six mullions per structural bay, repeated in elevation for thirty-eight floors. The geometric repetition of curtain wall elements in the facade (figure VI-18b,c) contrasts sharply with the traditional facades of the premodern period. Similarly, the sense of scale is modified not by changes in dimension but by the repetition of small elements.

VI-18

Seagram Building.

c. Perspective view

VISUALIZING WITH CAD

VI-18

Seagram Building.

d. Northeast corner: perspective view.

We now return to a more detailed discussion of the third basic form of spatial organization. The use of grids, both conceptually and physically, as a way to measure and appropriate space can be found in architecture dating back to the earliest civilizations. In Western civilizations, a partition by grids often marks the transformation of a natural space into an urban settlement. Grids, in either three-dimensional volumes or two-dimensional surfaces, represent an interpretation of space in terms of a rational framework, and can be an investigative tool for different types of compositions.

We have already considered the possible geometrical compositions derived from grids and connected grids. Those geometrical relations can now be used to generate architectural compositions. The dynamic models of grids and connected grids examined in Chapter IV were based on geometric elements, such as points and lines. These grid models can now be evolved into different architectural and even urbanistic models. For this purpose we distinguish two-dimensional grids from three-dimensional.

The implementation of a dynamic model of a grid can be generalized as follows. A particular type of grid is retrieved from the geometric library (for the geometric description of different types of grids, see Chapter IV). The point/symbols or the connecting line/symbols of the retrieved grid can then be replaced according to the function of the grid in the architectural composition.

Two-Dimensional Grids

Two-dimensional grids are often used in floor plans as both an organizing principle and a structural element. Recall that the geometry of grids is not limited to orthogonal relations (figure VI-19). They can also be generated by rotation, in the case of polar grids, or translation, which can generate skew grids in which the angles are not orthogonal. Such grids can become structural elements if the grid points are replaced by columns (figure VI-19a,b).

Evolution of two-dimensional grids
in floor plans: points evolve
into columns.

a. Orthogonal grid.

b. Polar grid.

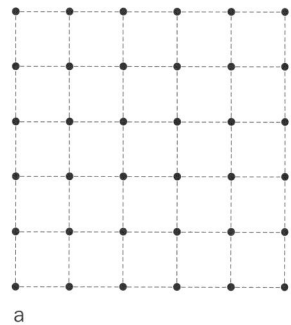
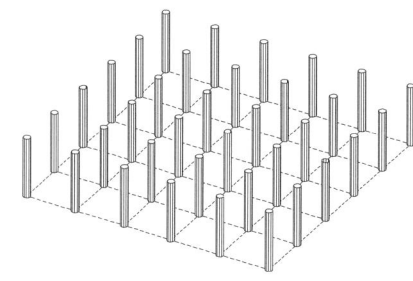

a

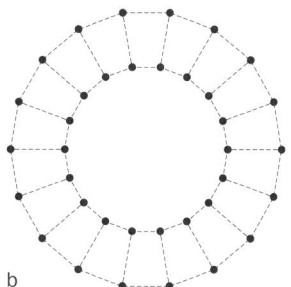
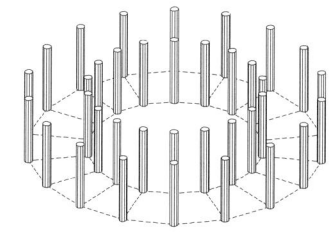

b

A two-dimensional connected grid, usually orthogonal, can also be used in elevation for compositional purposes. The line/symbols of a connected grid are differentiated as horizontal and vertical and can be replaced by architectural elements such as walls (figure VI-20a) or, in the case of a curtain wall, by mullions (figure VI-20b). The void space between the connecting elements is replaced by windows.

Three-Dimensional Grids

Three-dimensional grids are often used in architectural compositions as a space-frame structure (figure VI -21a). The line/symbols defining connected grids are replaced by structural elements such as steel bars, and structural joints replace the point/symbols of the same grid. Grids based on triangular configurations (figure VI-21b) are

Evolution of two-dimensional
connected grids in elevations.

a. Points and lines evolve into walls.

b. Points and lines evolve into mullions.

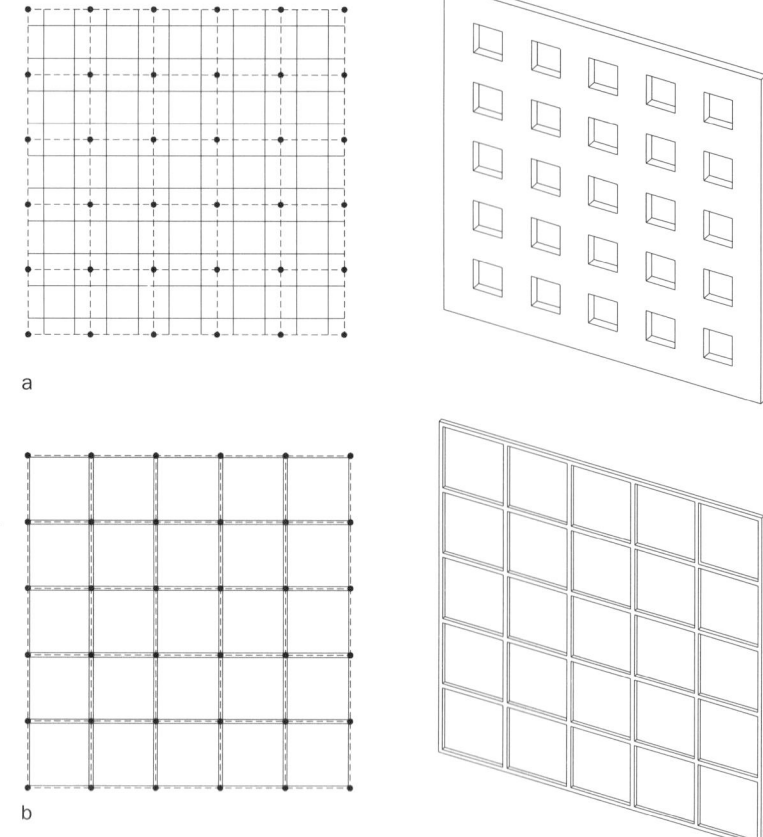

a

b

especially efficient in their structural properties.

Three-dimensional grids can also define a structural skeleton system. The vertical connecting lines represent columns while the horizontal lines are replaced by beams. Such grids are geometrically characterized by vertical elements orthogonal to the horizontal plane.

Evolution of three-dimensional
connected grids in space-frame
structures.

a. Cubic grid.

b. Grid based on tetrahedra.

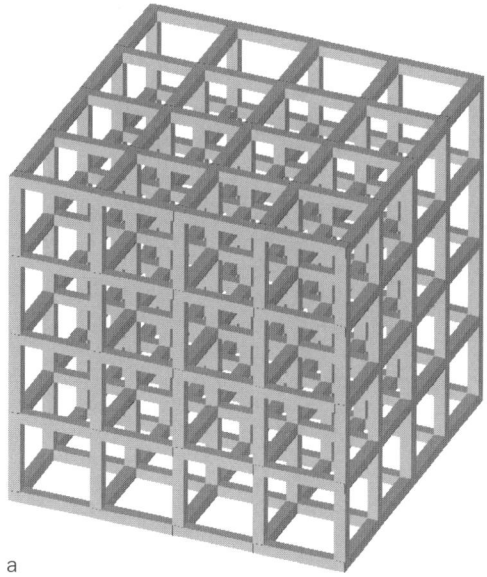

a

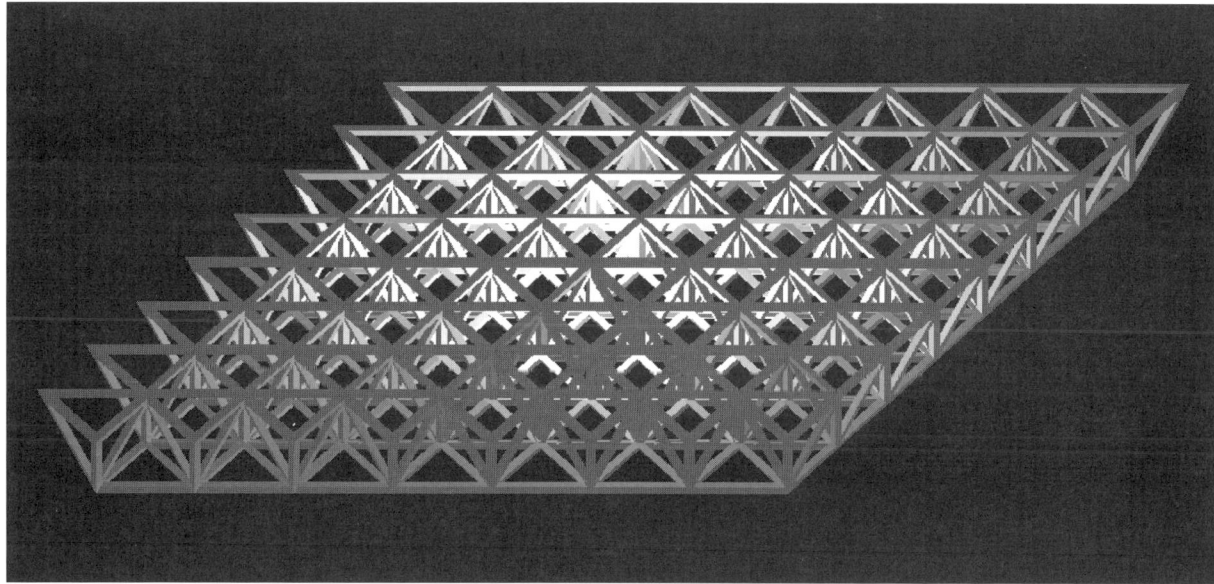

b

Urbanistic Hypothesis: Ideal Cities

Everybody is familiar with the orthogonal street grids of American cities. Urban grids need not be orthogonal, of course, and are not restricted to the modern tradition. In the early nineteenth century, the Utopian Socialists proposed models of cities based on regular grids, where a correspondence was established between the geometric configuration and the economic and social structure. Examples include the Phalanstery by Fourier, the model town Victoria by Buckingham, and the Happy Colony by Pemberton. Later in the century, the Garden City proposed by Ebenezer Howard was based on a hexagonal grid. Even earlier, the "ideal cities" of the Italian Renaissance were also based on grids, usually polar but sometimes orthogonal.[2] The city center is well defined by the main piazza and public buildings such as the cathedral or town hall. The boundaries of the city, functionally interpreted as fortifications, have a stellar polygonal shape. Unlike cities in the modern tradition, where the regularity of the two-dimensional plan grid contrasts with the choatic development of the third dimension, some ideal cities, such as Sabbioneta, Palmanova, and Sforzinda, are based on grids that are regular in both plan and elevation.

Since, as these examples demonstrate, regular geometric configurations are often used for urban proposals, it is worthwhile to consider the possibility of dynamic models of urban grids. Regular urban grids are geometrically represented by connected grids, but, in contrast to their usual interpretation, the connecting elements, which represent streets, are regarded as void instead of solid. The CAD generation for one of these typical urban forms is straightforward. A dynamic model of a connected grid defined in the *x-y* plane, polar or orthogonal, is retrieved from the geometric library; the model shown in figure VI-22 is based on an orthogonal grid. The point/symbols and line/symbols composing the grid can be replaced by vertical surfaces defined in three dimensions by lines of variable length and position. These surfaces in turn define solid enclosures in

A regular urban grid.

a. Diagram.   b. Evolution (plan).

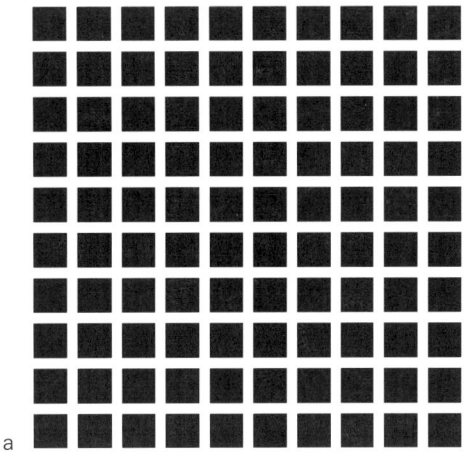

a

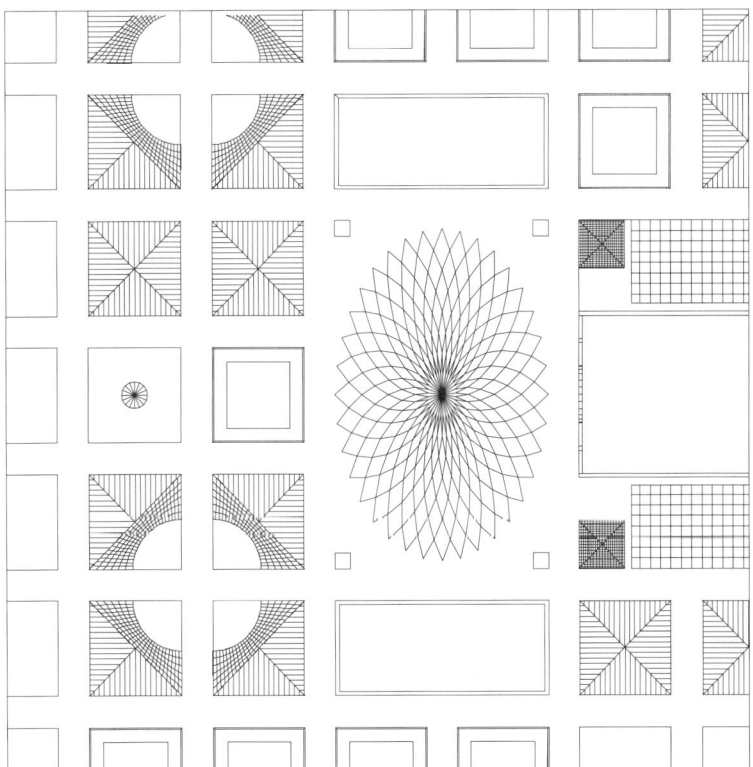

b

VI-22

A regular urban grid.

c. Evolution of a regular urban grid.

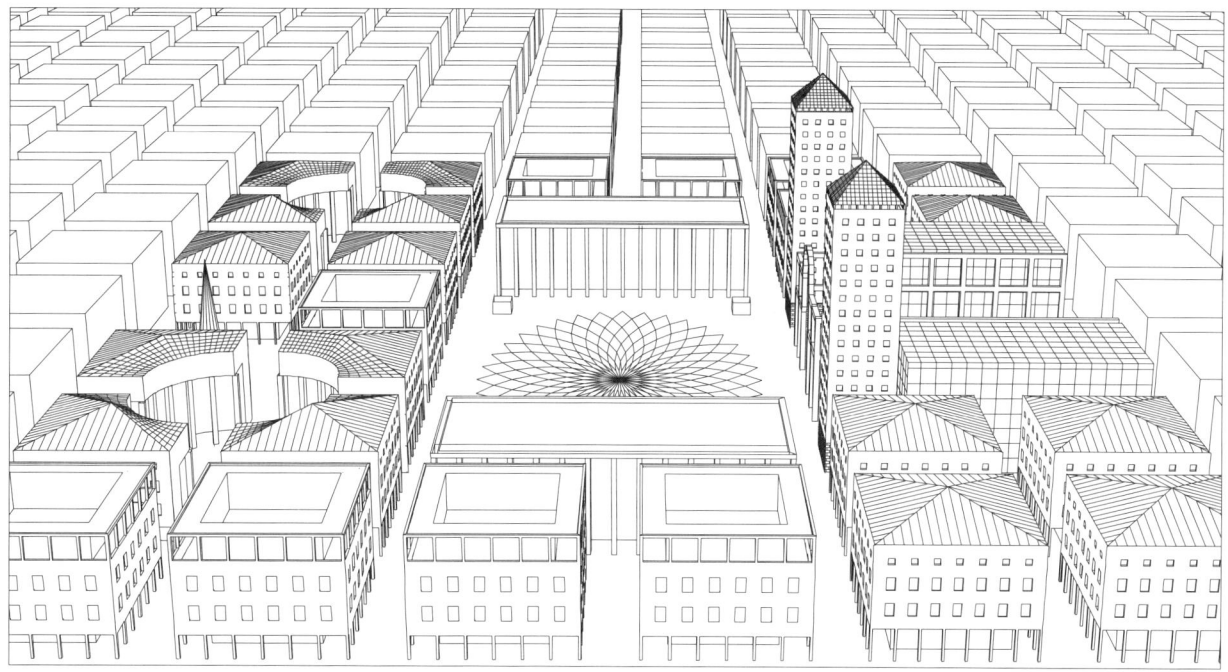

VI-22

A regular urban grid.

d. Evolution of a regular urban grid.

the areas between the grid connections, which were left void. The solid enclosures, which represent buildings, can assume different heights and varied geometric configurations. Each solid can be replaced by a more detailed architectural composition at a more detailed scale (figure VI-22b,c,d).

Palmanova, a schematic model of which is shown in figure VI-23, is one of the few ideal cities that was actually built (in 1593, from a design attributed to Scamozzi). The polar grid is based on a rotational angle of 40°, and composes nine repeated modules. Each module is symmetric with respect to an axis originating from the center and perpendicular to an edge of the nine-sided polygon. Note that, as the schematic model shows, the plan grid has several interwoven rotational symmetries.

Palmanova.

a. Geometric diagram.

b. Diagram showing streets as voids and buildings as solids.

c. Perspective view.

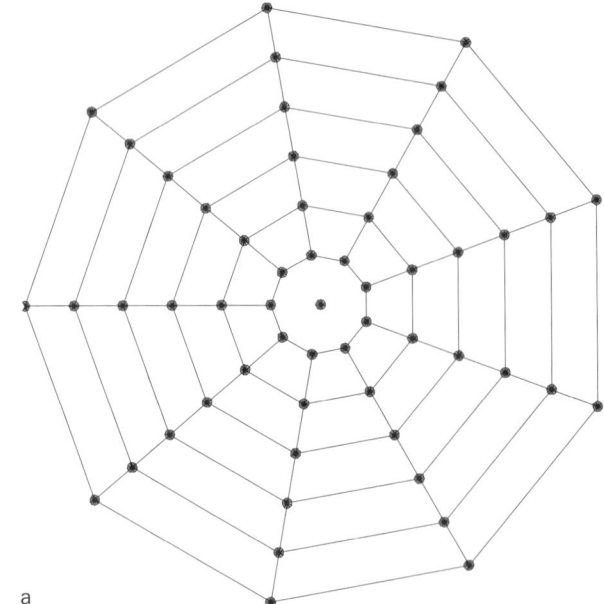

a

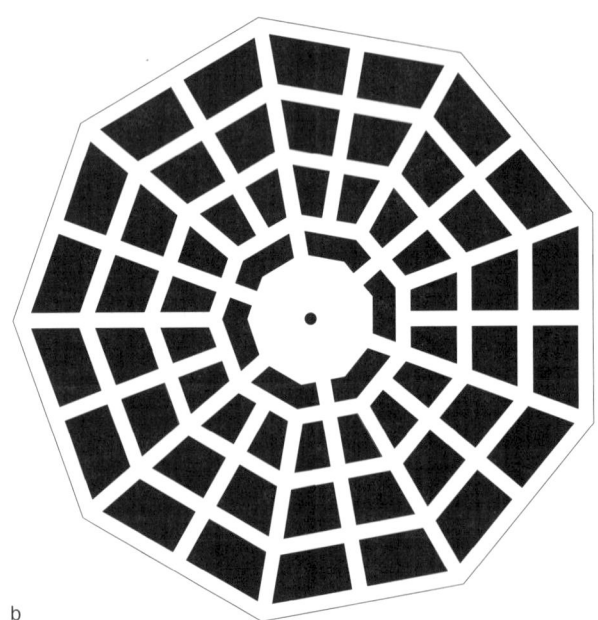

b

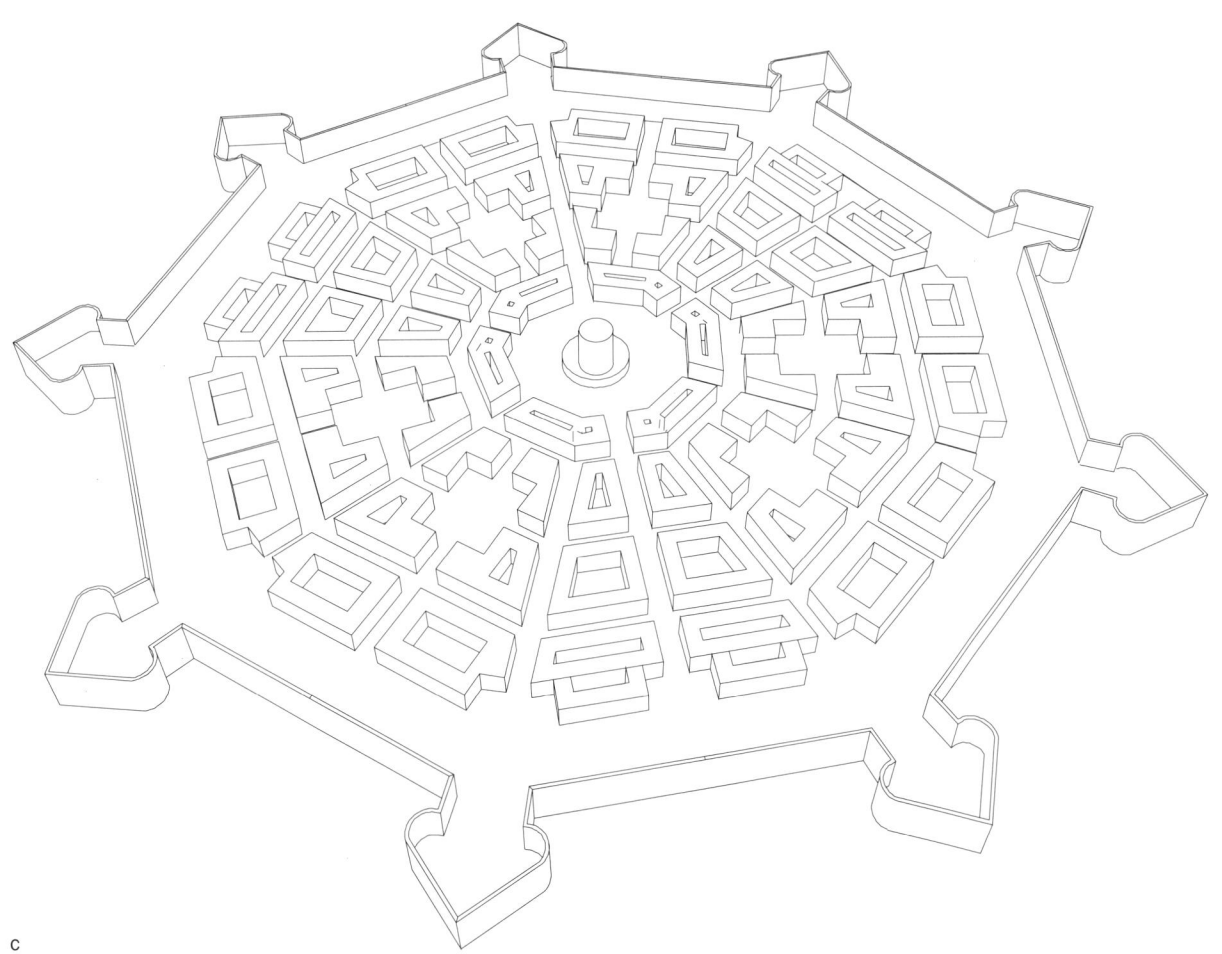

c

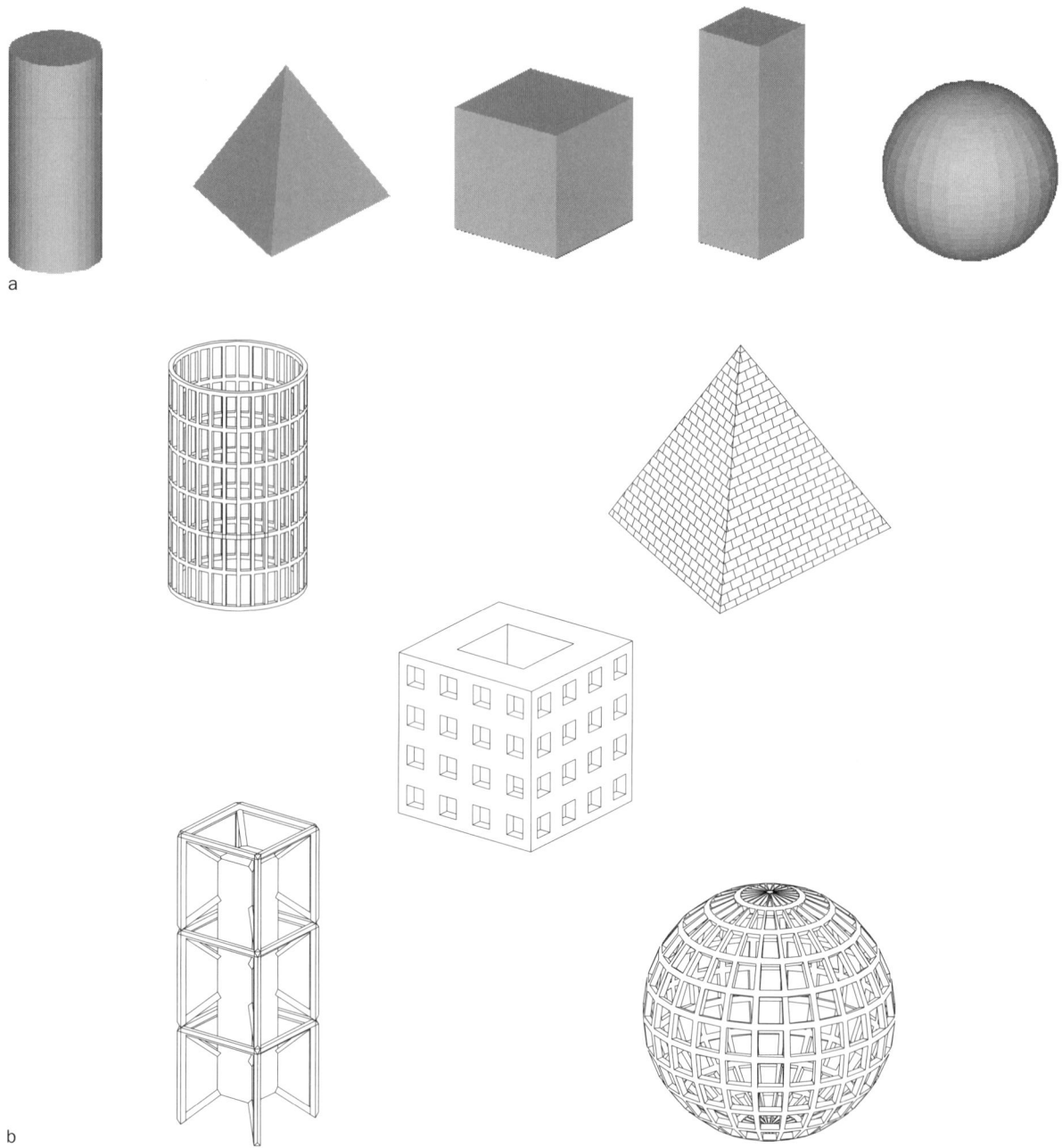

a

b

a. Volumes in architecture
(according to Le Corbusier).

b. Evolution of geometric volumes
into architectural forms.

Architectural compositions based on pure volumes find expression in the French revolutionary architecture of the eighteenth century. Typical of this period are the works of French architects Boullée and Ledoux, whose projects were widely published but rarely built. These two architects used cylinders, pyramids, and spheres as symbolic forms. A good example is the Cenotaph for Newton by Boullée, in which a sphere of superhuman scale represents the universe. Volumes have also been used in more recent architecture, especially in the Functionalist style. According to Le Corbusier, "Tout est sphères et cylindres," and the articulation of simple volumes is a characteristic of many Functionalist compositions.

In CAD, volumetric compositions can be explored by using volumes bounded by face/symbols (figure VI-24). The boundary faces, since they are defined as symbols in the computer data structure, can later be replaced by architectural elements. The relations among different interpenetrating volumes can be explored through Boolean operations. Although these operations traditionally deal with solids, in this case the solids are replaced by the boundary surfaces that enclose them; thus, the solids become containers and the operations are applied to the contained volumes.

THE SKYSCRAPER

The skyscraper affords one of the most powerful architectural examples of how a composition can be generated by geometric and Boolean transformations and can evolve architecturally through the replacement operation. In CAD, nested symbols aid in the analysis of the hierarchical composition of the floor plan. In elevation, the same geometric boundary elements can evolve in many different architectural shapes.

In the figures that illustrate the following discussion, different design stages are identified notationally. The suffix letters **a** through **e** represent the different generation rules; the suffix ".**P**" is used to indicate the various stages of evolution of floor plan; and the suffix ".E" denotes various elevation replacements. Different generation and evolution rules can be combined to produce alternative designs (denoted, *e.g.*, VI-25a.P2.E4), each to be evaluated in terms of the initial requirements. The models used in the following discussion are schematic, but still interpretative of the design concept.

Generative Models

The characteristics of skyscrapers, regardless of their architectural style, are governed by the repetition of a floor along the z axis (figure VI-25), which is a multiplication operation of a translation. This characteristic can be rigorously followed, as in skyscrapers by Mies Van der Rohe and Skidmore, Owings and Merrill. A floor is composed of floor plan and a floor elevation. A **floor** symbol is, for simplicity, defined by a constructed square. The sides of the square can also be identified as symbols, to be replaced by the elevations of the building.

Different transformations can be added to the basic translative multiplication of the core and the curtain wall to define several generative models. Figures VI-25a,b show models obtained by rotation and scaling, respectively; the core and the structure have to be excluded from the rotation and scaling transformations and will follow just the basic generation rule. In these cases—as in the generative models of figures VI-25c,d—the floor symbols are composed of two types of symbols, one subjected to complete transformations while the other is only translated and repeated along the z-axis. For example, in the model of figure VI-25b, the scaling transformation is applied to the x and y dimensions, while the z dimension is subject only to translation. The model of figure VI-25c is generated through rotation, combined with the subtraction of a floor module with each z-translation. In this case, the usual floor symbol made of a square cannot be used, since the subtractive operation changes the floor each time it is used. The transformation operations therefore become generative not only of the model as a whole but also of the floor plan; it may be practical to use a floor plan made of modules/symbols that are equal in number to the number of rotations. Evolutions of these geometric models into eventual architectural compositions are shown in figures VI-25c1,2.

All the schematic models thus far considered have a common characteristic: They are generated by geometric transformations of a basic module. Often, however, a skyscraper design is generated as much by the sculptural interpenetration and subtraction of masses (Boolean operations) as by a repetition of the same floor. The diagrams of figure VI-25d show a floor's structure and the core, which are the parts of the floor symbol to be conserved after the transformations are applied. If the Boolean transformations operate only in plan, the previous models apply, with the floor replaced by the product of the Boolean operations. More interesting models are obtained when the results of all the different Boolean operations are combined in elevation (figure VI-25d2): The monotony of the box is interrupted by the solid-void articulation, which allows the ground open spaces to be repeated at higher levels. A Menger sponge (see Chapter III) also offers a good example of combined Boolean operations; figure VI-25d1 shows a model that combines different portions of it.

A skyscraper has often been compared to a column and said to be composed of the three column parts [Huxtable 1984]. The base, capital, and shaft are surely emphasized, for example, by Adolf Loos in his entry for the Chicago Tribune competition, and often are recognizable in contemporary examples. A similar approach can be followed by using different generative models for the base, the shaft, and the capital, and combining these different parts in the whole building (figure VI-25e).

Evolution Models: Floor Plan and Elevations

The square/symbol used so far can be replaced by a floor plan (figure VI-25.P), and its side/symbols by curtain walls (figure VI-25.E). Numerous variations evolve from the same generative models (figures VI-25.P1.E2, VI-25.P1.E5, VI-25.P1.E6, VI-25.P2.E3, VI-25a.P1.E3, VI-25a.P1.E5, VI-25a.P1.E6, VI-25a.P2.E3, VI-25b.P1.E3, VI-25b.P1.E5, VI-25b.P1.E6, VI-25b.P2.E4).

Basic generation rule.

More generation rules:

a. Rotation.   b. Scaling.

c. Rotation and subtraction.

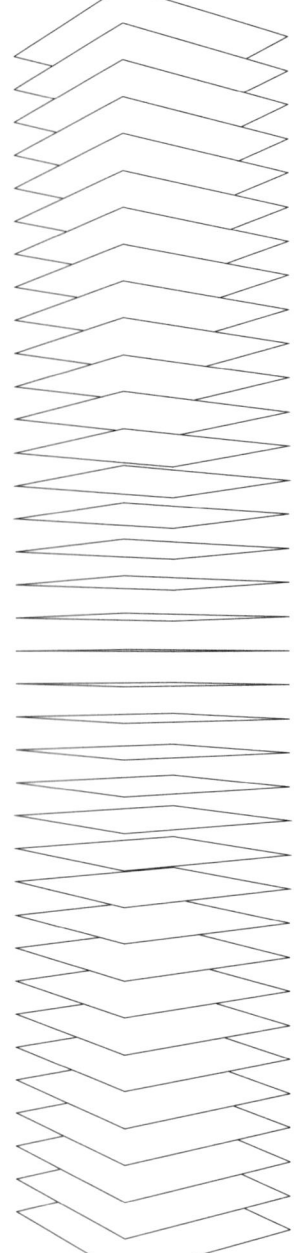

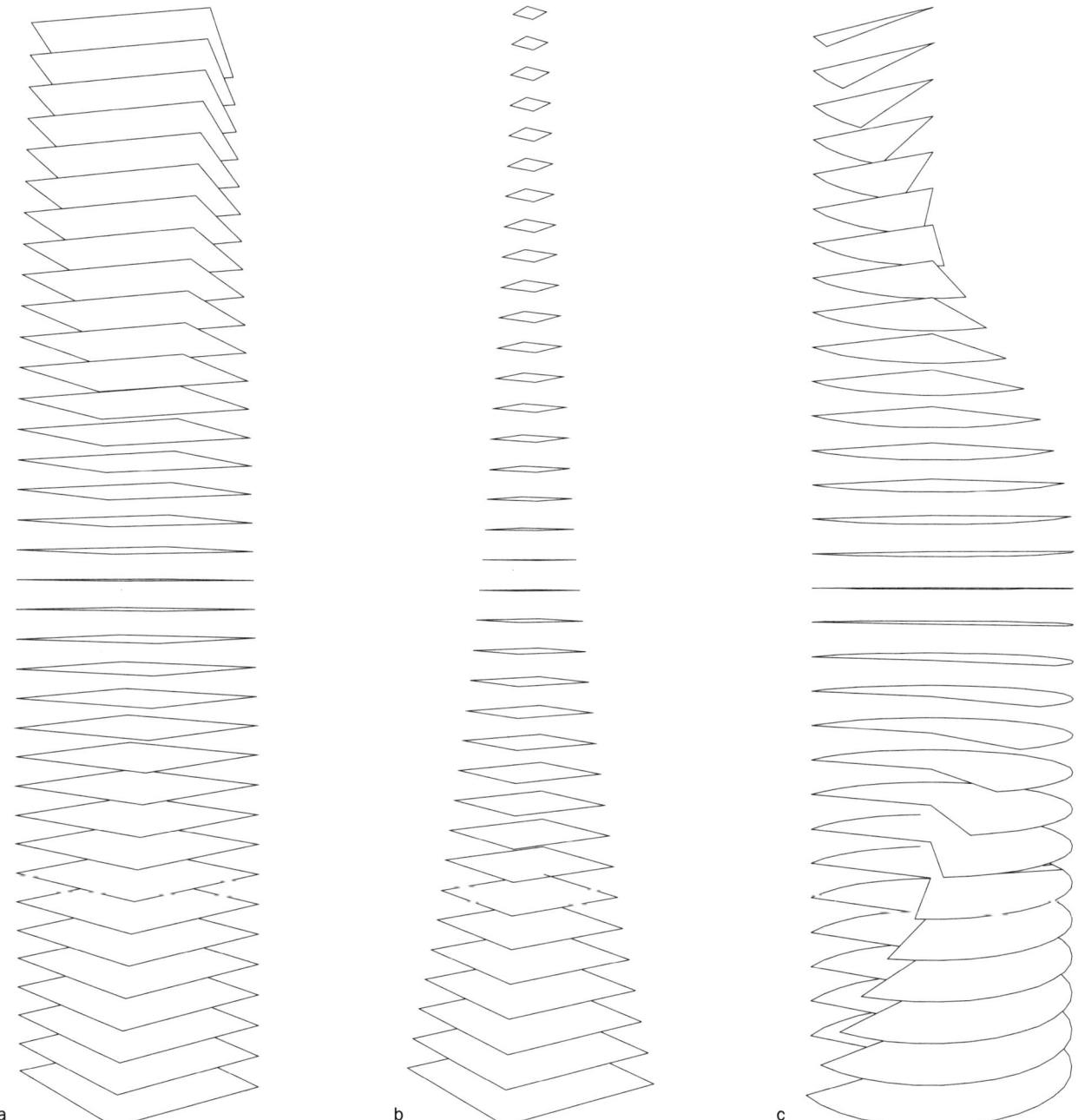

a

b

c

Perspective view.

Axometric view.

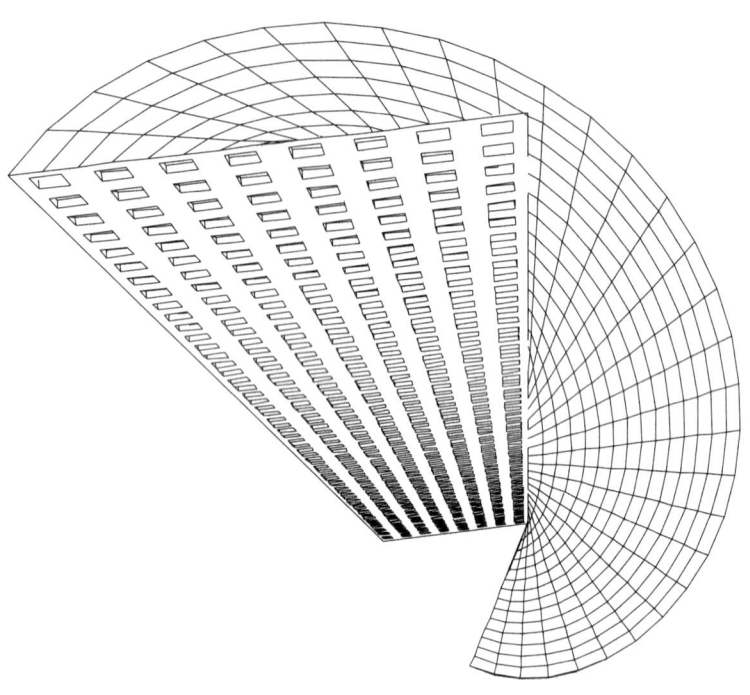

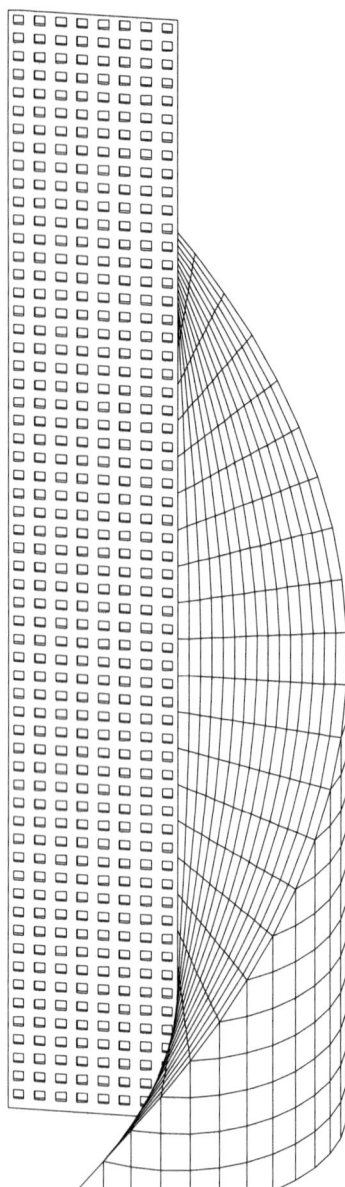

Perspective view.

Axometric view.

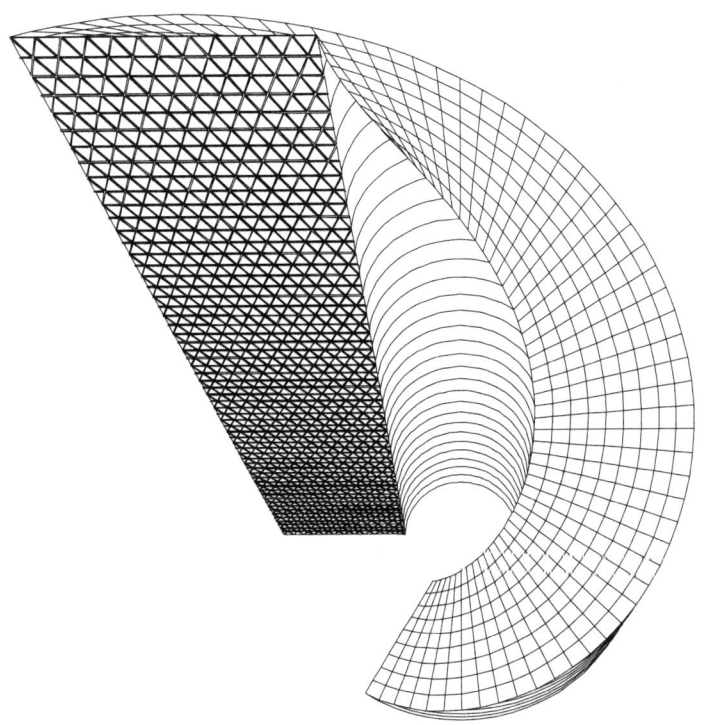

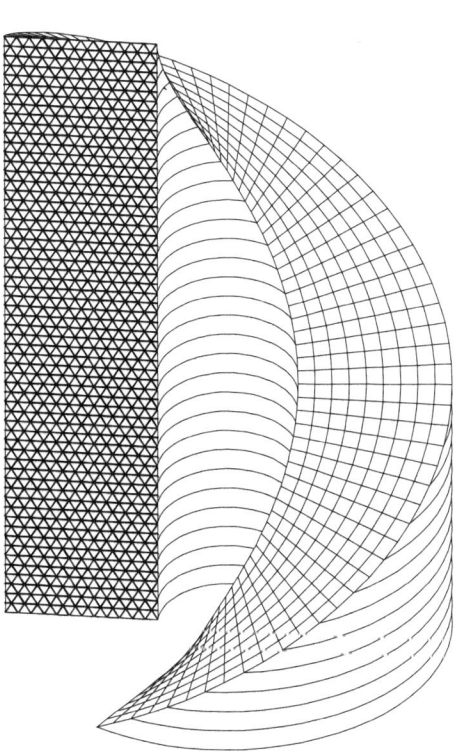

a. Boolean transformations.

b. Elements retained from
Boolean transformations.

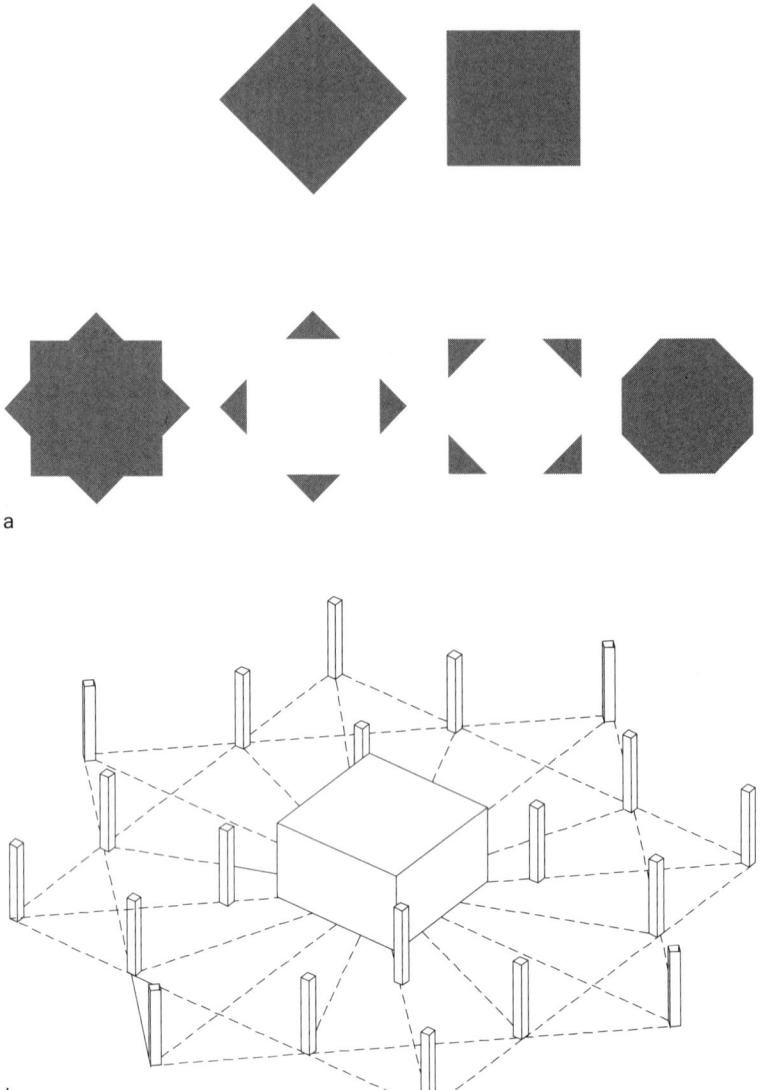

a

b

The Menger sponge applied
to a skyscraper.

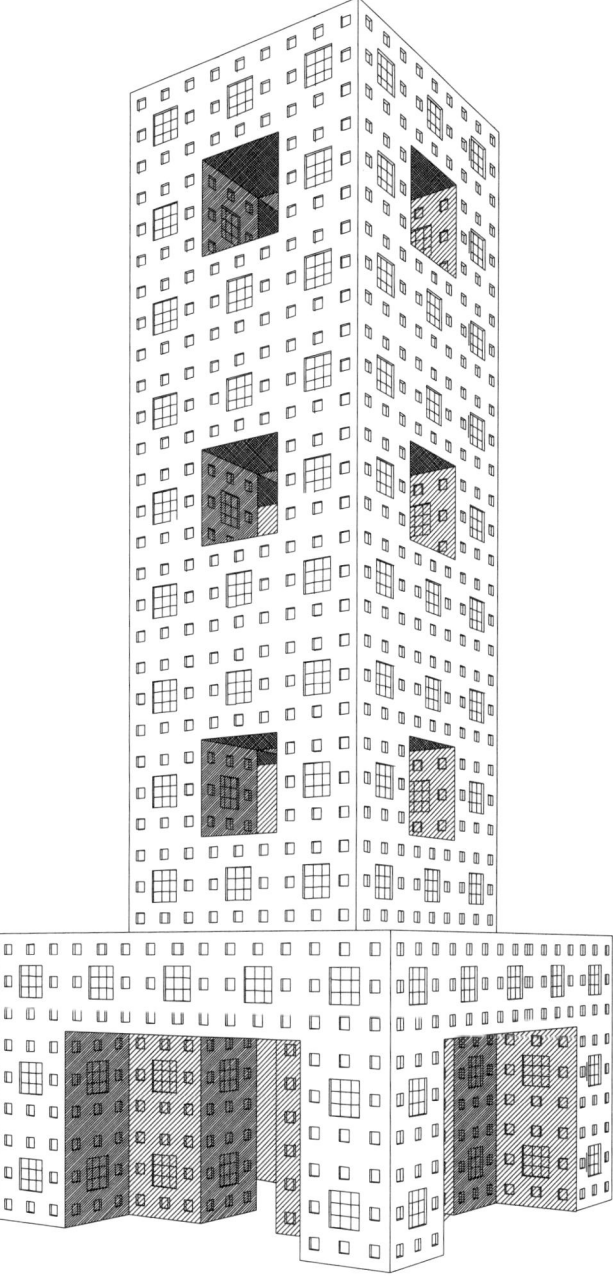

Elevation view.

Perspective view.

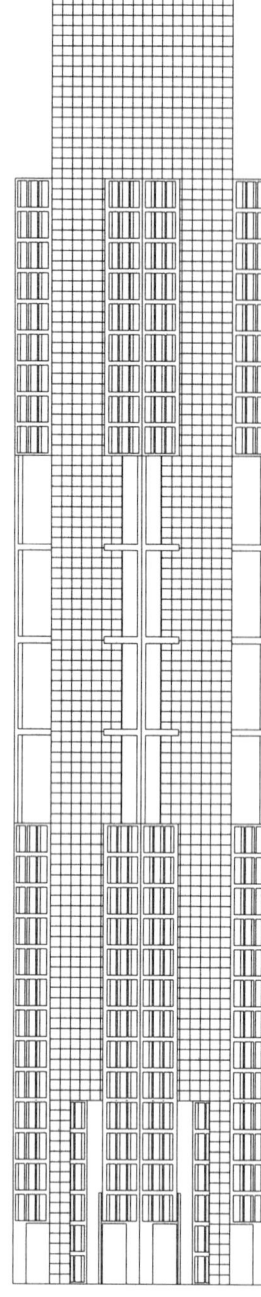

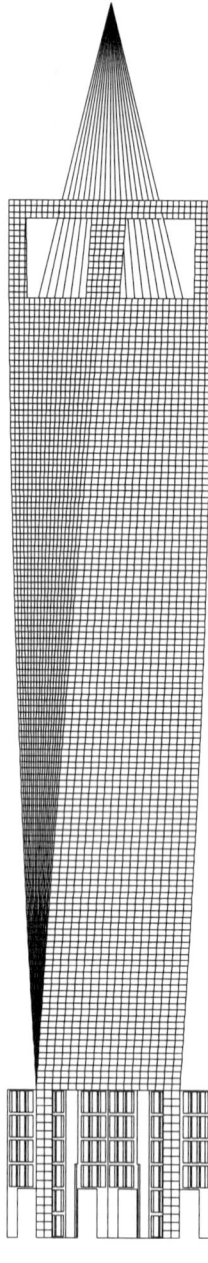

VI-25E

VISUALIZING WITH CAD

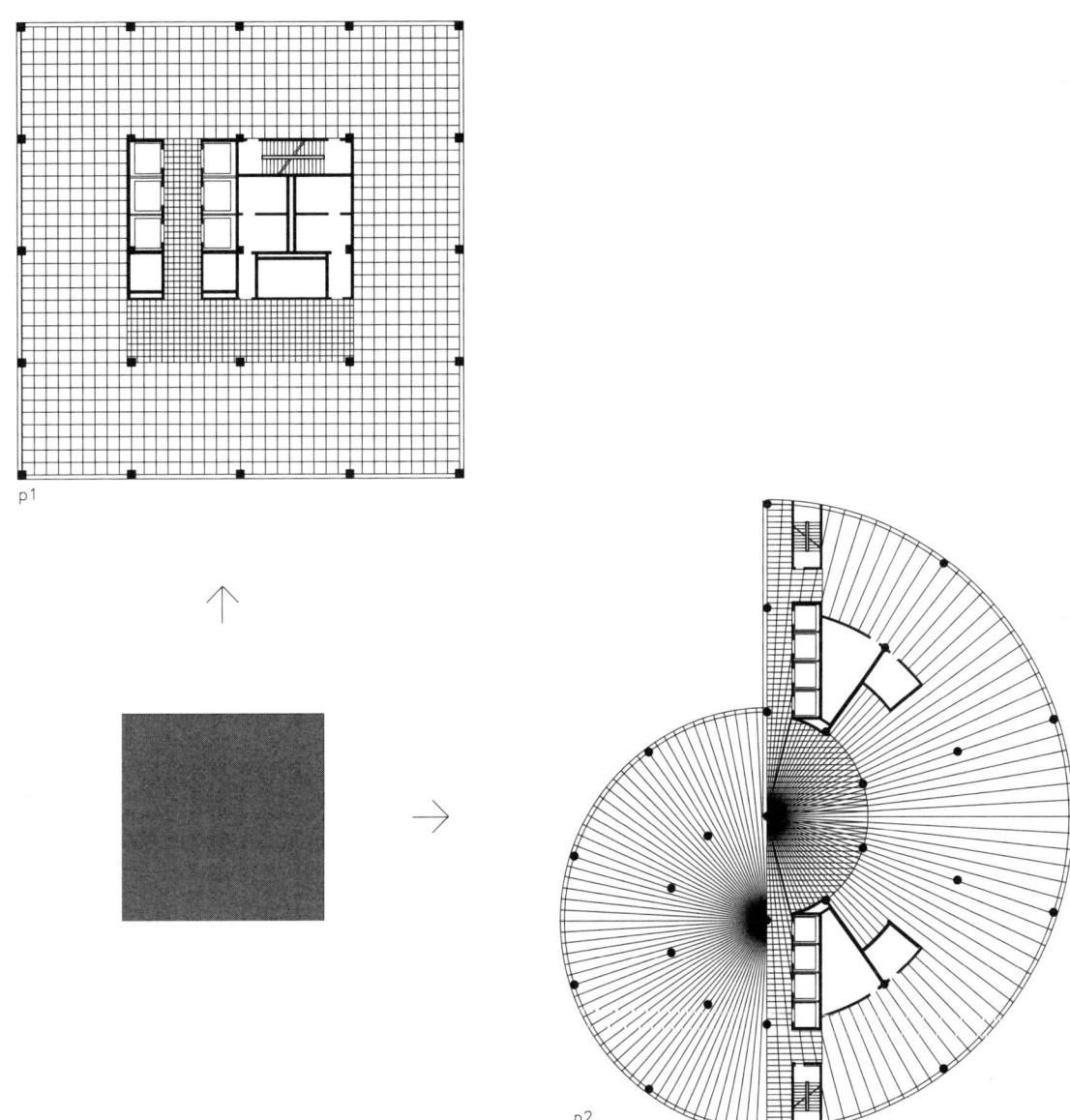

p1

p2

VI-25.P

Evolution from floor plans.

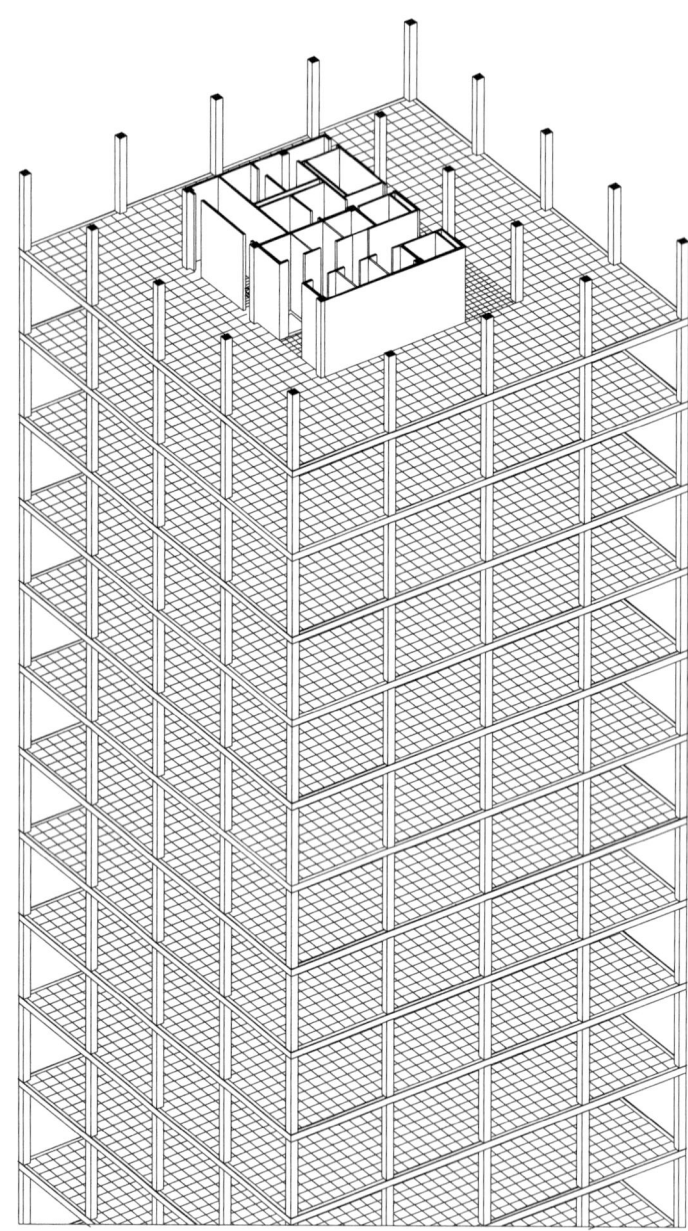

VI-25.P1

Evolution from floor plan p1.

VISUALIZING WITH CAD

VI-25.E

Curtain wall variations.

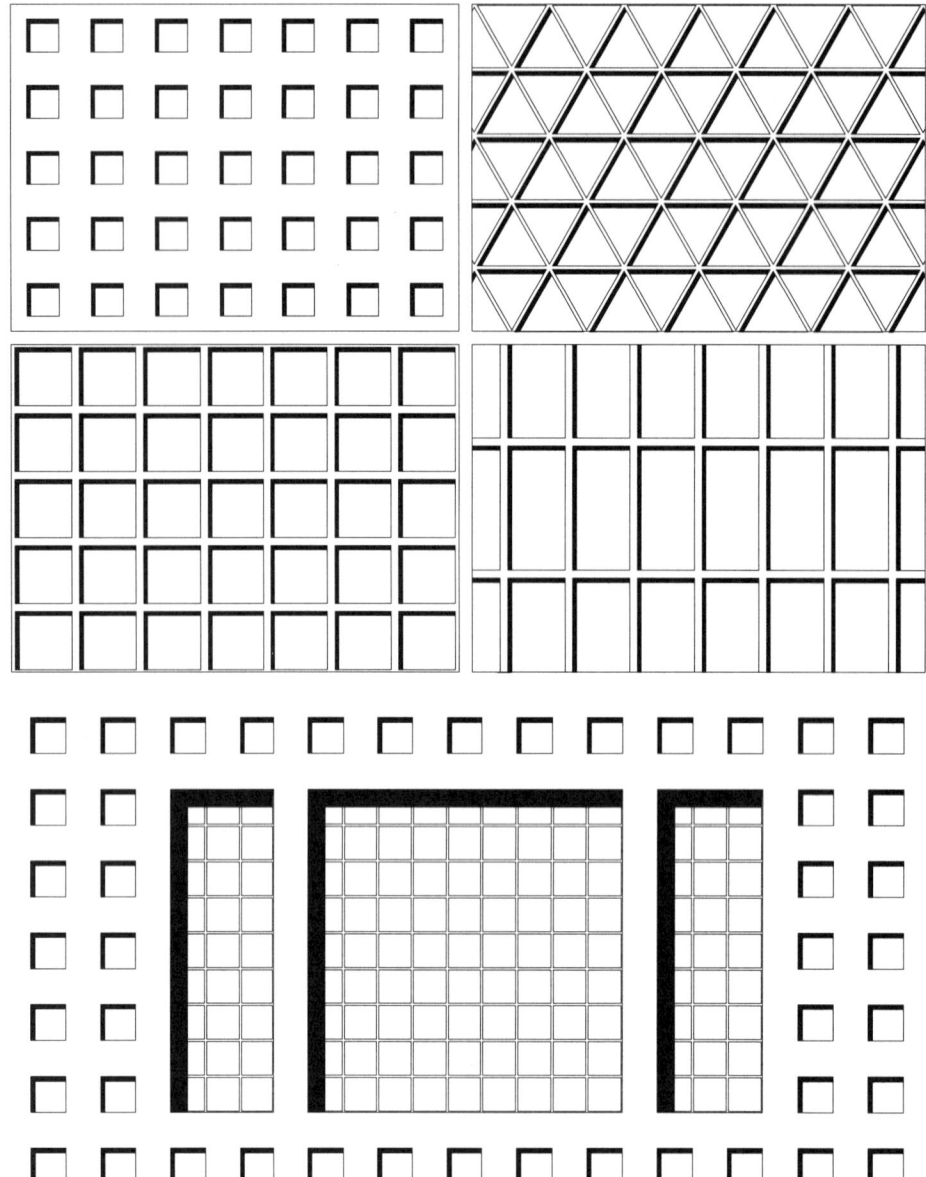

VI-25.E

Curtain wall variations.

VISUALIZING WITH CAD

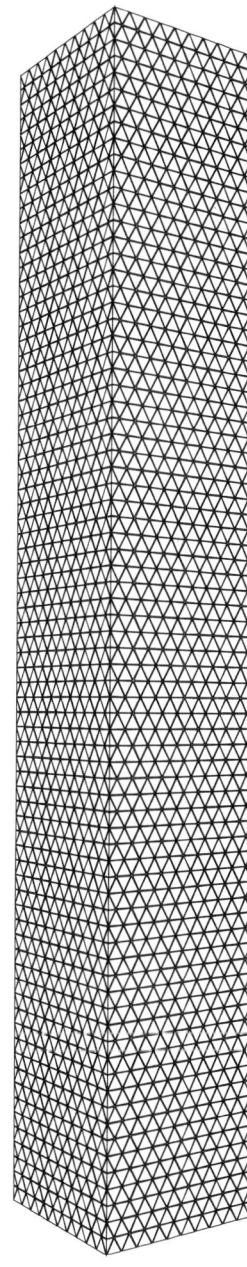

VI-25.P1.E2

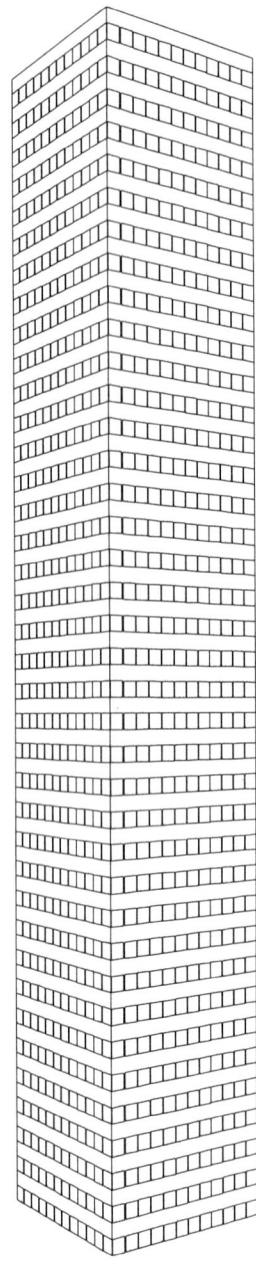

VI-25.P1.E5

VI-25.P1.E6

VISUALIZING WITH CAD

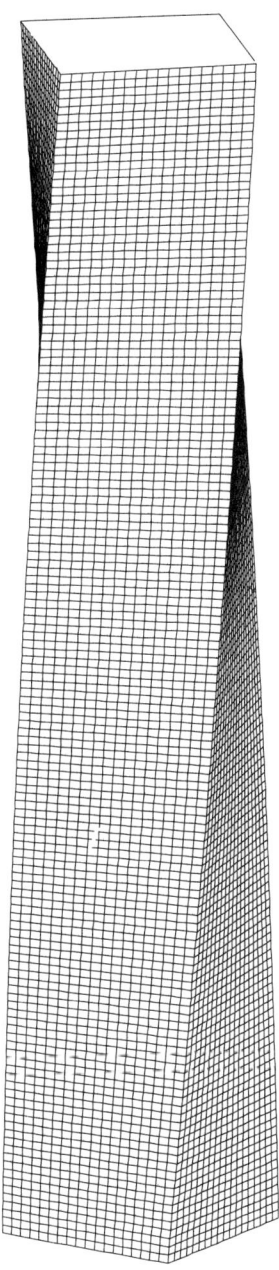

VI-25A.P1.E3

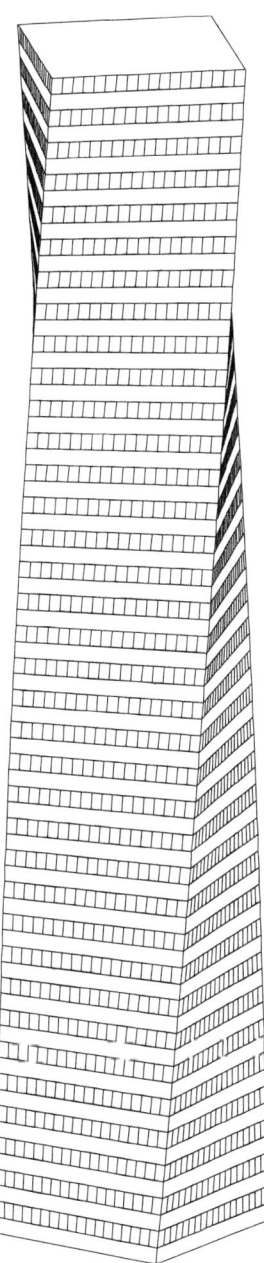

VI-25A.P1.E5

VI-25A.P1.E6

VISUALIZING WITH CAD

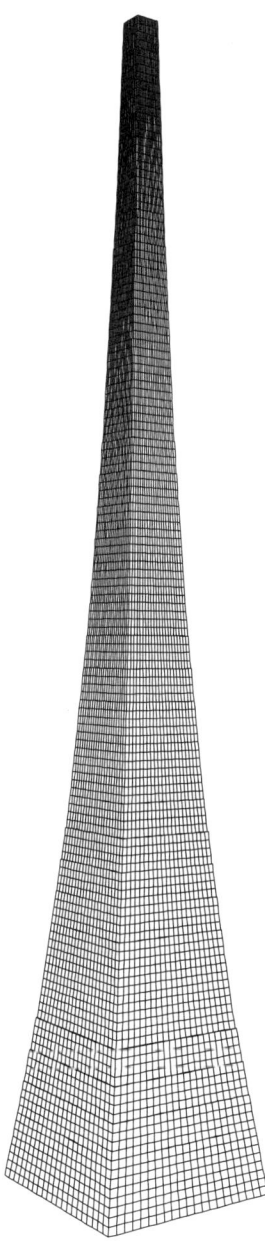

VI-25B.P1.E3

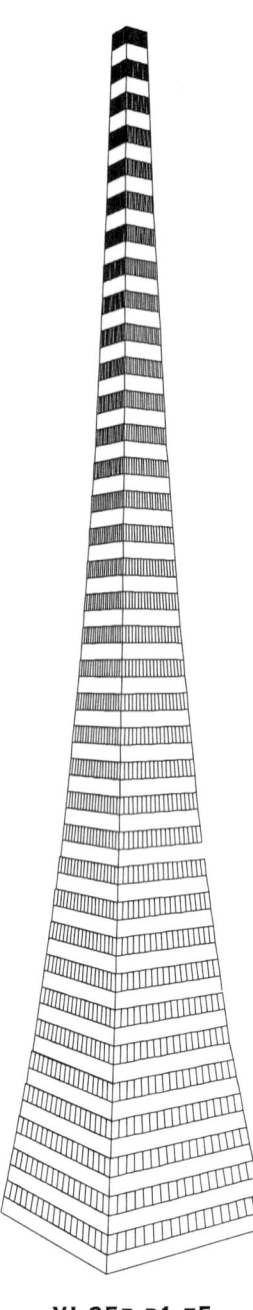

VI-25B.P1.E5

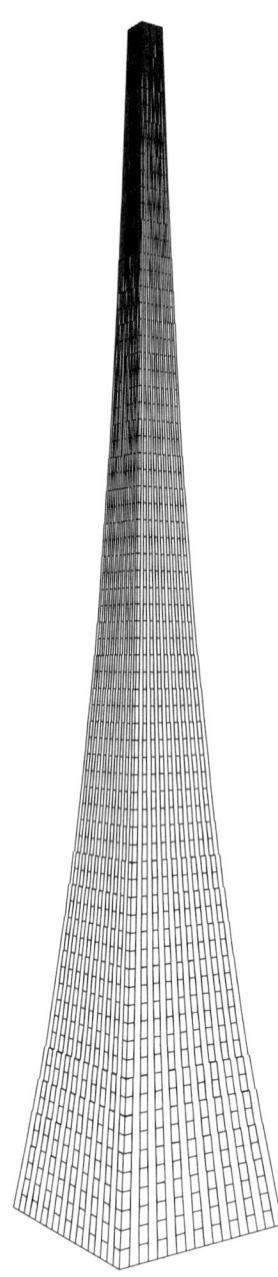

VI-25B.P1.E6

VISUALIZING WITH CAD

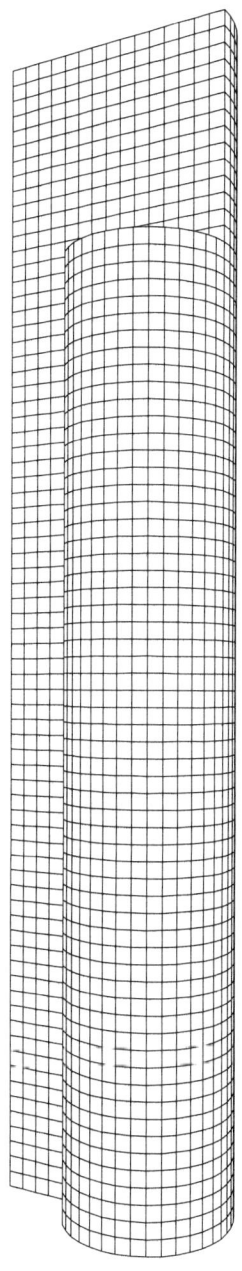

VI-25.P2.E3

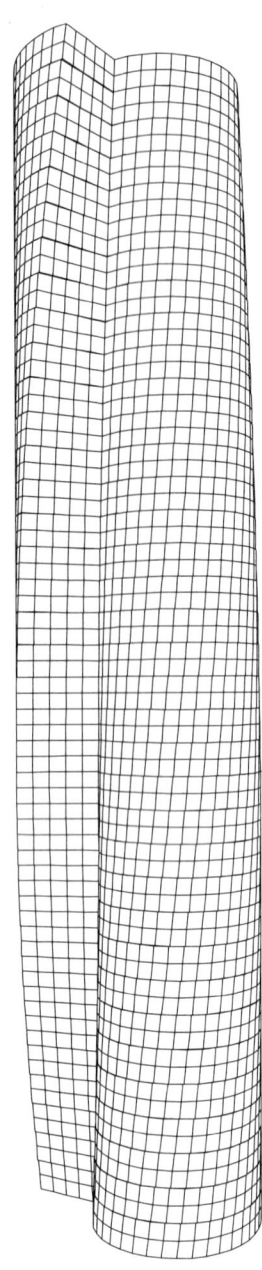

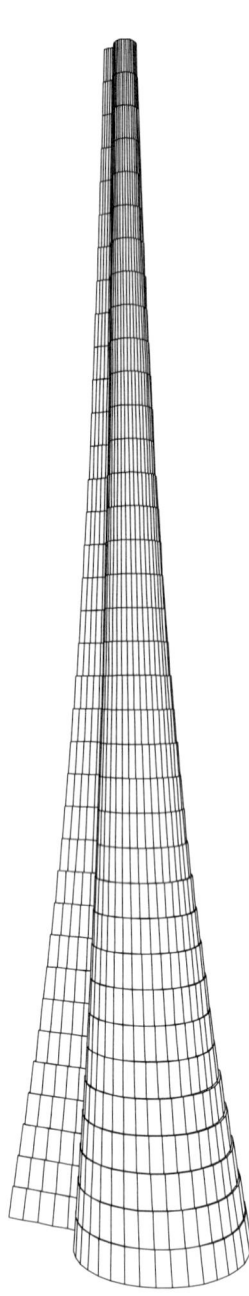

VI-25A.P2.E3   VI-25B.P2.E4

These simple compositional rules offer a variety of design solutions, governed by floor plan area and other functional requirements, to be developed from the basic modules.

TOWARD A GEOMETRY OF "DECONSTRUCTION"

In classic buildings, and in many examples of modern architecture, there is a very clear syntactic organization, revealed in the hierarchical composition of architectural elements, in the spatial organization, and in the volumetric articulation. This distinct organization of relations, which can be easily explored with CAD dynamic models, is less evident in contemporary architecture. The term *deconstruction*, which originated in philosophy and literary criticism, is often associated with contemporary buildings. Deconstructivism in architecture is often related to the Russian Constructivism of the early decades of this century, which began to emphasize the dynamic relations in forms, rather than the static balance and order of classic arrangements. The following discussion and models focus on the geometric aspect of architectural deconstructivism; that is, shape is generated from the "de-construction" of primary geometric forms, a process often achieved through the operations of shifting, rotating, stretching, tearing, superimposing, or scaling. Figures VI-26 and VI-27 show these operations for two different grids, in the plane and in three-dimensional space. The shape shown in figure VI-28 is generated by the "stretching" of a cubic grid. Another motif, the montage of two parts, obtained by the operations of superimposition, "cut and paste," and collage, is also typical of deconstructivist architecture (figures VI-29 and VI-30).

At first glance, it may seem that the CAD models discussed so far, structured according to ordered and hierarchical relations, are not applicable to deconstructivist compositions. A deeper analysis suggests, however, that the CAD methods can still be used; in fact, a deconstructivist composition often starts with two or more parts, clearly structured according to "classic" grammar. The model of a

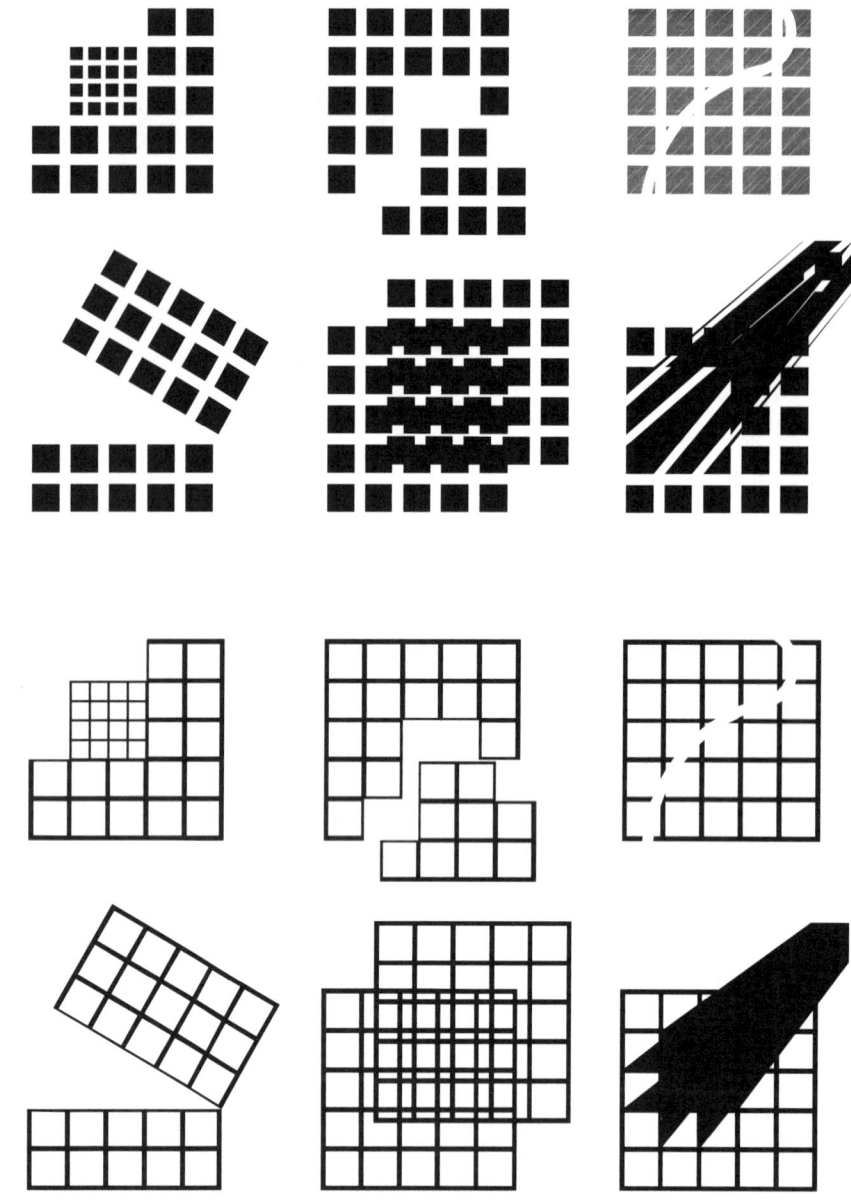

VI-26
"Deconstructivist" operations in the plane.

VISUALIZING WITH CAD

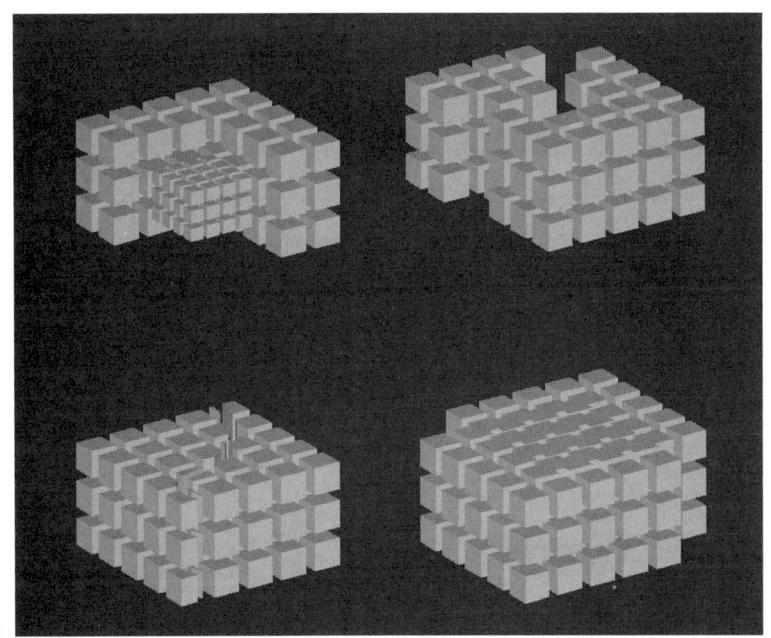

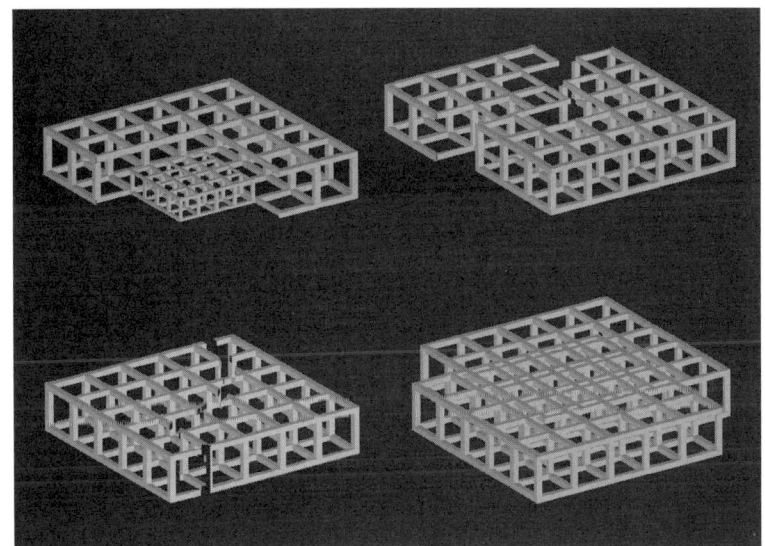

VI-27
"Deconstructivist" operations in three-dimensional space.

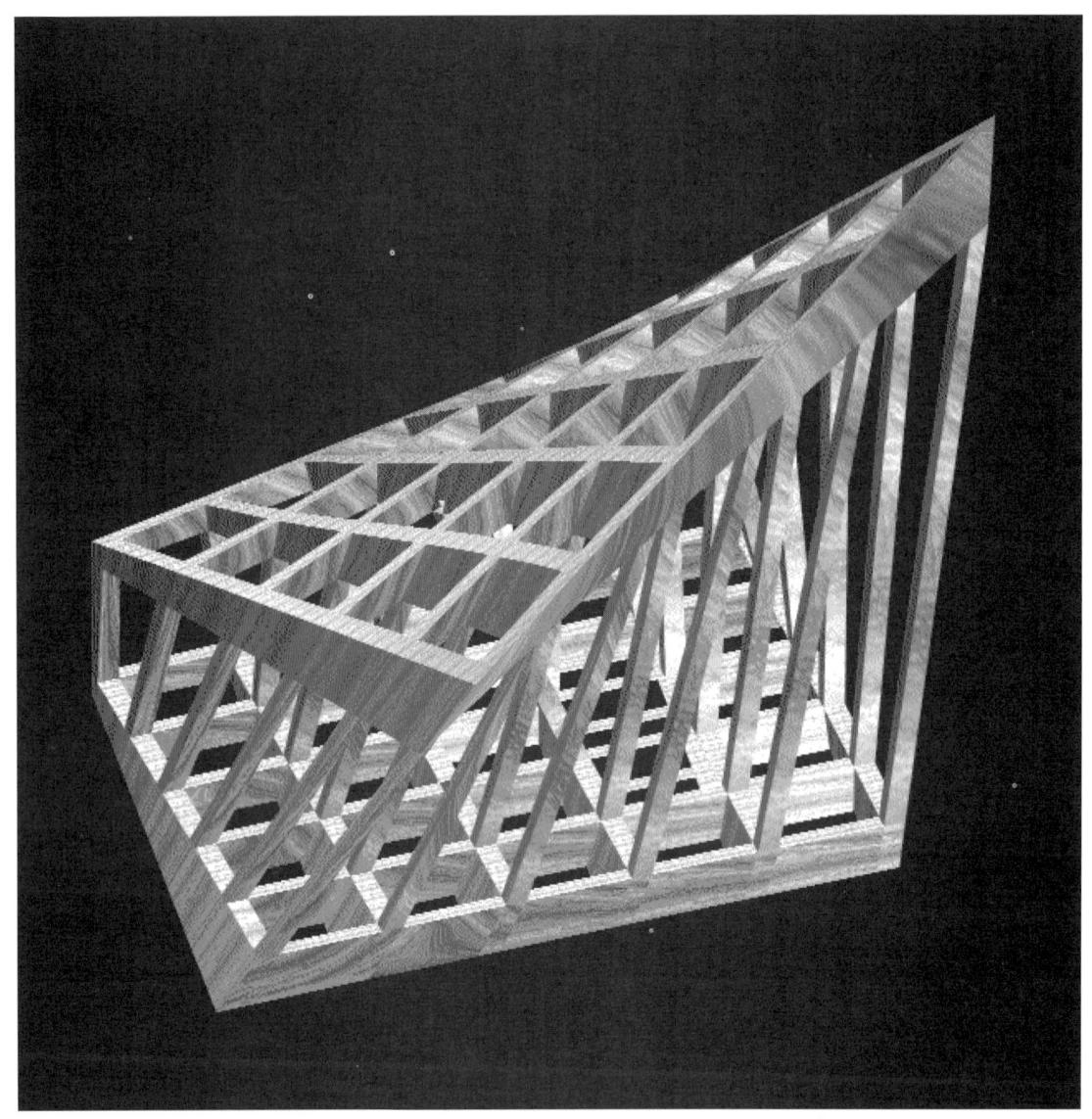

VI-28

VISUALIZING WITH CAD

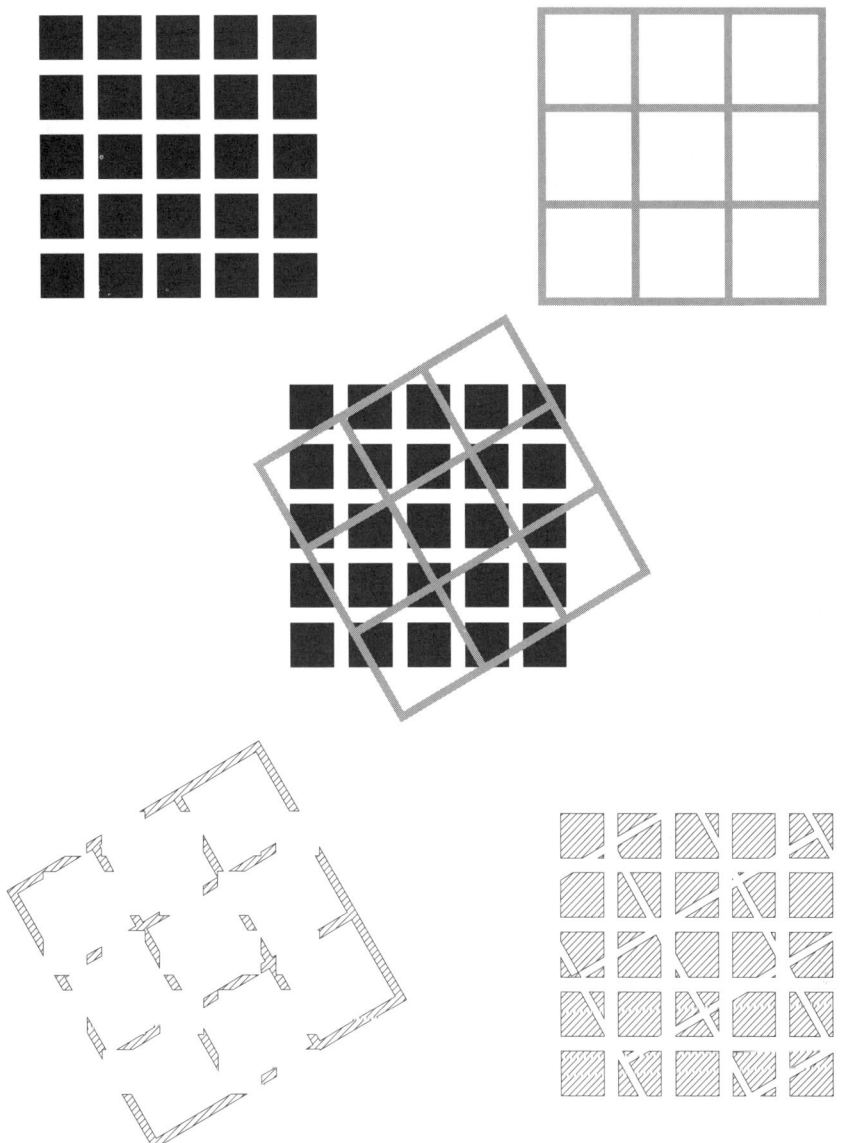

VI-29
"Deconstructivist" operations in the plane
Composition of two different grids.

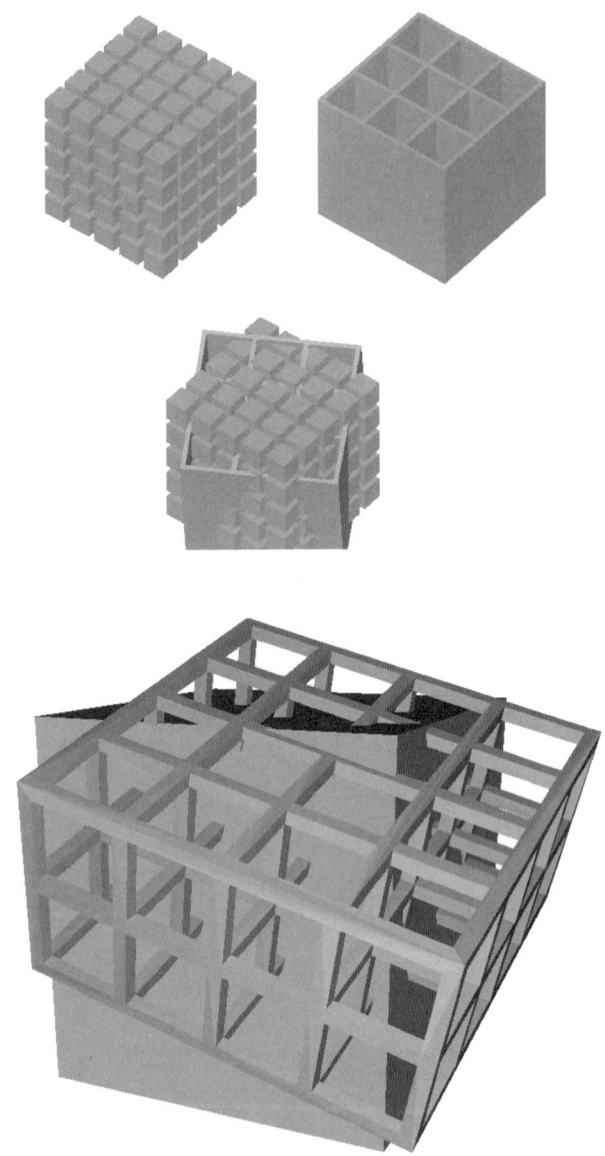

VI-30

"Deconstructivist" operations in three-dimensional space:
Composition of two different grids.

VISUALIZING WITH CAD

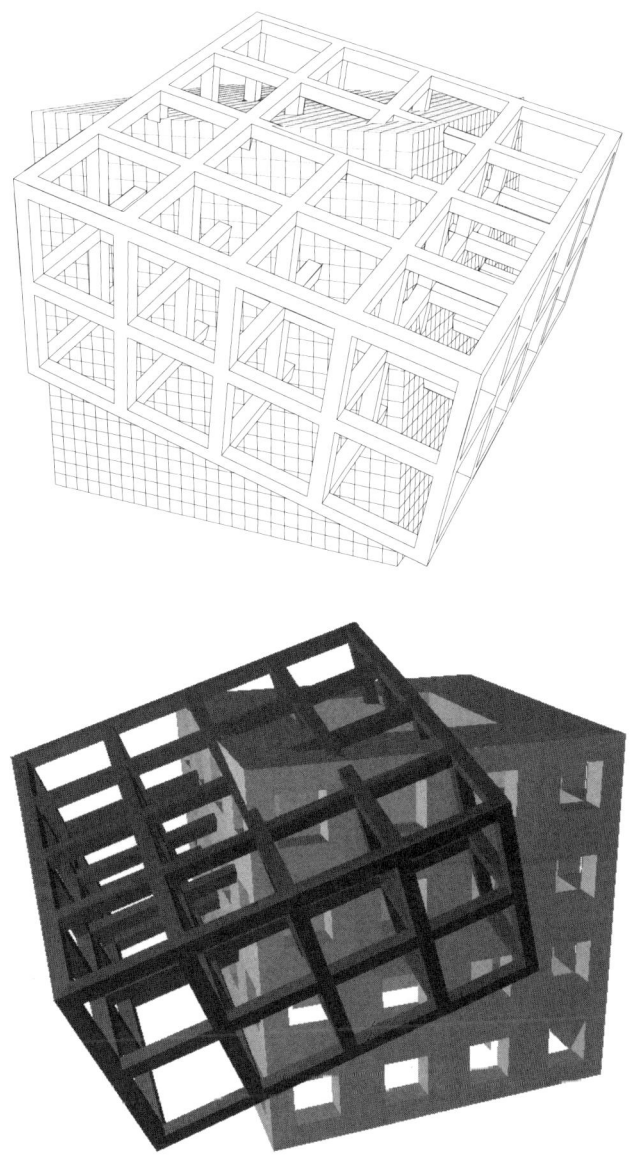

VI-31

"Deconstructivist" operations in three dimentional space:

Evolution from a composition of two different grids.

grid can again be an organizing principle for one of the parts. If after a series of operations, the two original parts are no longer related according to classic principles of geometric regularity, the geometric transformations still provide the syntactic structure. Furthermore, as with more conventional designs, the parts of the composition can still evolve as dynamic models (figure VI-31).

A CONCLUDING THOUGHT

As demonstrated in the course of this chapter, dynamic models offer the possibility of generating a composition from geometrically defined architectural elements and spatial relations. The geometric foundations and models in this chapter provide a basis for designers to experiment in developing complex spatial relations. Dynamic models are defined by a structure (in terms of spatial relations) that can expand in different semantic directions, as illustrated here by the expansion of the given geometric models into more complex compositions.

In this chapter we have explored formal compositions; another possible line of development would be the investigation of compositions in which function motivates design. The flexibility of dynamic models can also be applied to this type of approach. The spatial relations define a diagram of ordering principles which can evolve through the replacement of geometric shapes with functions. The reader is invited to experiment in this direction.

This book is intended to provide an approach to the interpretation, creation, and understanding of models of forms, in which the computer serves not only as a replacement for ink and paper, but offers a new methodology. Now it is up to you to use these techniques for your own needs.

BIBLIOGRAPHY

Alberti, Leon Battista, *The Ten Books of Architecture*, Dover, New York, 1986

Benevolo, Leonardo, *History of Modern Architecture*, MIT Press, Cambridge, 1977

Culler, Jonathan, *Theory and Criticism after Structuralism*, Cornell University Press, Ithaca, 1982

Derrida, Jacques, *Of Grammatology*, Johns Hopkins University Press, Baltimore and London, 1974

Hersey, G. L., *Pythagorean Palaces*, Cornell University Press, Ithaca and London, 1976

Huxtable, Ada Louise, *The Tall Building Artistically Reconsidered*, Pantheon, New York, 1984

Jencks, Charles, *Modern Movements in Architecture*, Anchor Books, Garden City, 1973

Le Corbusier, *Creation Is a Patient Search*, Frederick A. Praeger, New York, 1960

Norberg-Schulz, Christian, *The Concept of Dwelling*, Rizzoli, New York, 1985

Norberg-Schulz, Christian, *Genius Loci*, Rizzoli, New York, 1980

Norberg-Schulz, Christian, *Intentions in Architecture*, MIT Press, Cambridge 1965

Norberg-Schulz, Christian, *Meaning in Western Architecture*, Frederick A. Praeger, New York, 1975

Palladio, Andrea, *The Four Books of Architecture*, Dover, New York, 1965

Pampaloni, Guido, *Palazzo Strozzi*, Istituto Nazionale delle Assicurazioni, Roma, 1982

Serlio, Sebastiano, *The Five Books of Architecture*, Dover, New York, 1982

Sullivan, Louis, *The Public Papers*, University of Chicago Press, Chicago, 1988

Vitruvius, *The Ten Books on Architecture*, Dover, New York, 1960

Yessios, Chris, "The Computability of Void Architectural Modeling," *Computability of Design*, Yehuda E. Kalay (ed.), John Wiley and Sons, New York, 1986

Zevi, Bruno, *Architecture As Space*, Horizon, New York, 1957

NOTES

1   The term *deconstruction* began to be used in the mid-1960s in philosophy.

2   Examples of compositions based on an orthogonal grid in combination with polygonal shapes include the Ideal Cities designed by Vincenzo Scamozzi and Pietro Cataneo, both from the late sixteenth century.

ARCHITECTURE OF GEOMETRIES ▶

1.   A wall, a grid, and windows.

2.   A grid, a Moebius strip, and a mirror.

3.   A stellar dodecahedron and four mirrors.

4.   a. Grid based on spheres.
      b. Grid based on tetrahedra.

5.   a. A torus.
      b. Evolution of the torus into a colonnade.

6.   a. Walls.
      b. Portico.

7.   Paestum: one and two temples.

8.   Columns and rotations.

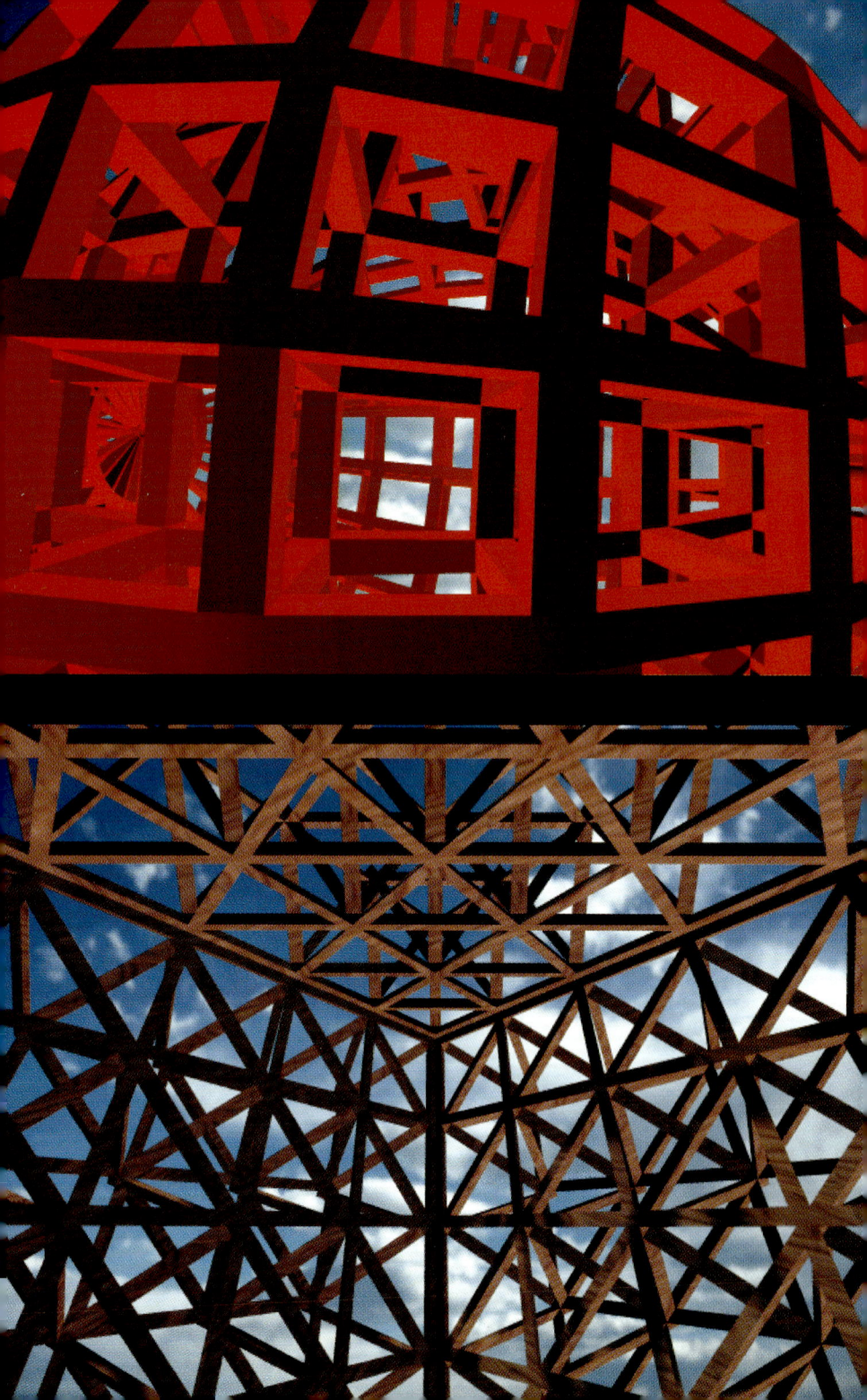

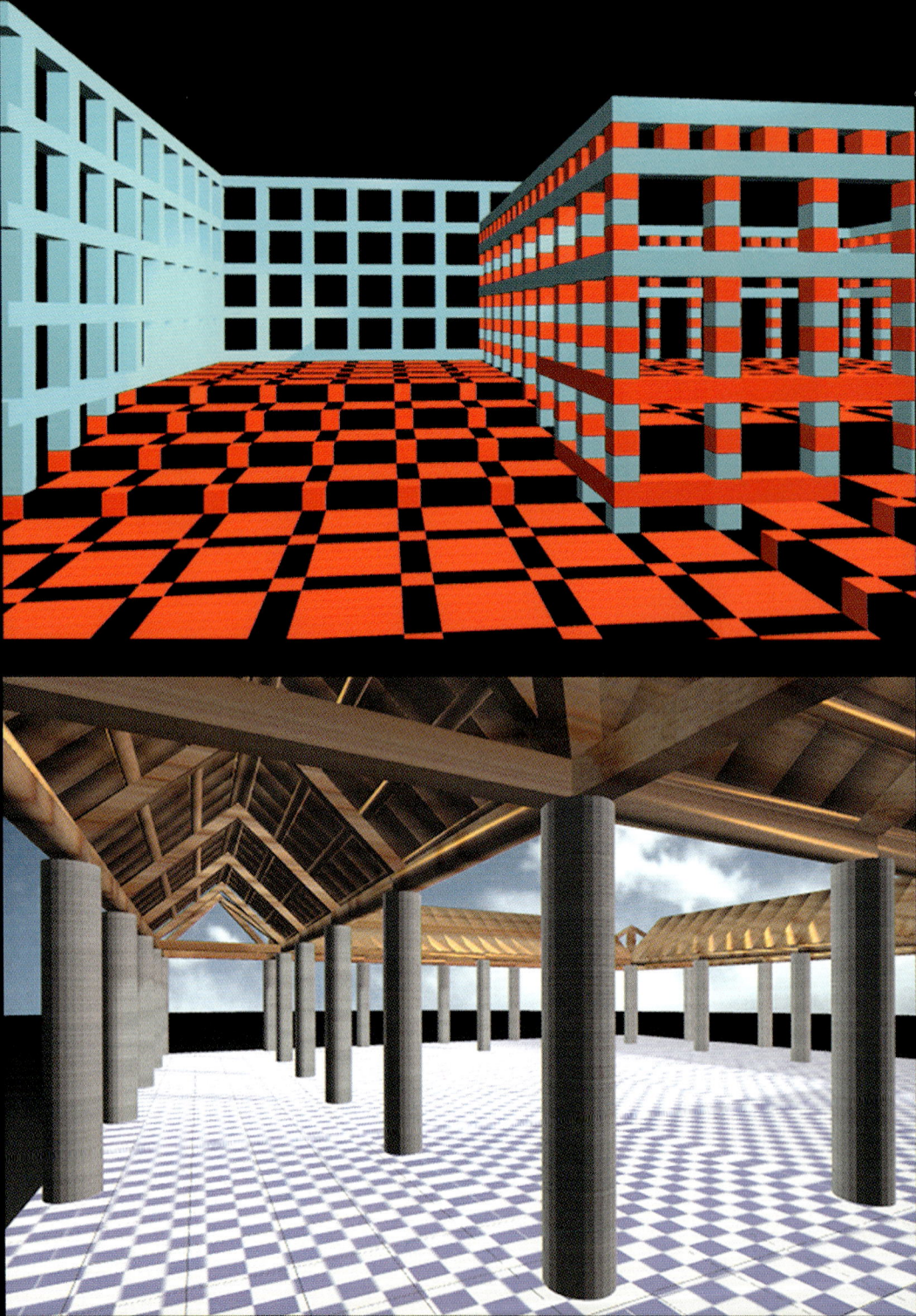

ABOUT THE AUTOLISP ROUTINES

The enclosed AutoLISP routines (*.lsp files) provide commands to generate three-dimensional models more complex than those created by the graphics primitives that are part of the AutoCAD software. The emphasis is on forms generated as solid-void compositions. The user can input several different parameters, such as the solid-void proportions and the number of modules in plan and elevation. Each of the enclosed AutoLISP routines represents a good application of the philosophy of this book, illustrating how several different compositions can be obtained from the same geometric relations by altering parameters such as solid-void proportions.

An AutoLISP routine can be loaded by invoking the "load" function or selecting the "applications" function from the "file" menu, and can be run with the command listed after the AutoLISProutine is loaded. A few suggestions are offered. The AutoLISP routine creates new layers and blocks, the names of which may conflict with those of existing blocks and layers. Therefore, it is recommended that the AutoLISP routine be run from a new drawing or after renaming existing blocks and layers, if the names are the same as those created by the AutoLISP routine. The user should also do a trial run using the parameters shown between parentheses at each prompt to get a feeling for the process involved. Even if a very large range of numbers, all the positive real numbers accepted by AutoCAD, is allowed, very large values can create a memory overflow or incoherent compositions.

The user is encouraged to run each AutoLISP routine several times with different values as parameters and, to experiment by changing the content of each block. An AutoLISP routine, as text file, can also be listed and edited, using the DOS command "edit."

ABOUT THE ELECTRONIC MODELS

All the AutoCAD drawing files contain a three-dimensional model, the images of which are seen in the book illustrations. In each drawing file there are defined views—plan, elevation, isometrics, axonometrics—that can be invoked through the AutoCAD "view" command. In some files additional perspective images have been created, which can be accessed through the AutoCAD "view" command.

The user is encouraged to explore the geometry of the enclosed models and to change the content of the blocks that constitute the hierarchical structure of the model.

347

DIRECTORY	FILE	DESCRIPTION	CHAPTER
Polyhedr (cont.)	tetrastl.dwg	tetrahedron with stellar faces	Chapter IV
	tetra_ev.dwg	tetrahedreon evolution	
	octa.dwg	octahedron	
	octahol.dwg	octahedron with void faces	
	octastl.dwg	octahedron with stellar faces	
	octa_ev.dwg	octahedron evolution	
	icosa.dwg	icosahedron	
	icosahol.dwg	icosahedron with void faces	
	icosastl.dwg	icosahedron with stellar faces	
	icosa_ev.dwg	icosahedron evolution	
	cube.dwg	cube	
	cubehol.dwg	cube with void faces	
	cubestl.dwg	cube with stellar faces	
	cube_ev.dwg	cube evolution	
	dode.dwg	dodecahedron	
	dodehol.dwg	dodecahedron with void faces	
	dodestl.dwg	dodecahedron with stellar faces	
	dode_ev.dwg	dodecahedron evolution	
Geometry	helix.dwg	helix	Chapter IV
	helix_ev.dwg	helix evolution	
	moebius.dwg	moebius strip	
	moebi_ev.dwg	moebius strip evolution	
	torus_ev.dwg	torus evolution	

anticlastic surface
> a saddle-like surface that, at any point, has both convex and concave curvature

atom
> the most basic element in a given universe of discourse, not divisible in simpler ones

axiality
> a spatial organization based on a line (axis)

axonometric projection
> a type of orthographic projection in which the picture plane is perpendicular to the direction of projection but not to the direction of any coordinate axis

Boolean operations
> the operations of union, subtractions, and intersection by which new forms are generated from initial elements

boundaries
> the elements by which regions of space are defined

bump map
> in rendering programs an image mapped on the surface to be rendered, in such a fashion that the surface assumes a bumped or dimpled appearance

CAD
> Computer Aided Design, area of computer graphics dealing with the drafting and design of architectural, structural, electrical, mechanical, and electronic artifacts

cyberspace
> term coined by the science fiction author William Gibson, denoting an information defined space

Cartesian coordinates
> a spatial coordinate system based on three mutually perpendicular planes; see also {polar coordinates}

centrality
> a spatial configuration based on rotational symmetry about a central point

continuous
> the property of a geometric figure, such as a line or surface, of being uninterrupted in

extension and infinitely divisible; in mathematics, a class of functions the graphs of which are uninterrupted; see also {discrete}

discrete
consisting or distinct or unconnected elements; noncontinuous; see also {continuous}

dynamic model
a CAD model defined in terms of hierarchical components which, when replaced by others, yields alternative designs

figure
a region defined, by its boundaries, with respect to a surrounding field (ground)

geometric primitives
the most elemental entities—points, lines, and surfaces—of which more complex geometric figures are composed

geometric transformation
a mathematical operation by which the coordinates of a geometric figure are replaced by new coordinates, resulting in the figure's distortion or displacement; see also {orthogonal transformation; reflection; rotation; scaling; translation}

grammar
a system of rules

graphics primitive
in CAD and solid modeling, a geometric entity (point, line, circle, ellipse, polyline, polygon, face, solid, box, sphere, torus, cylinder, and so forth) often corresponding to a command

ground
the surrounding field with respect to which a region is defined by its boundaries

(analogous, in a solid-void distinction, to the void with respect to which a solid (figure) is defined)

hidden line representation
line representation of a three-dimensional object, where the lines hidden from the view (according to viewer position) are removed

horizon
in perspective drawing, the converging line for all horizontal planes

isometric projection
a form of axonometric projection in which the lines of projection form equal angles with each of the three coordinate axes

knowledge base system
term mainly used in artificial intelligence to denote a set of facts (or knowledge base) and a set of formal logic rules

layer
in CAD, the structure by which information related to different aspects of an object is distinguished

model
representation of an object or phenomenon focusing on the aspects relevant to the field in which the model is used

orthographic projection
a type of parallel projection—exemplified by plan and elevation views—in which all the projection lines are perpendicular to the picture plane

orthogonal transformation
a transformation, such as translation, that changes the position of a geometric configuration without changing its shape or size

parallel projection
> a geometric projection in which the projection lines emanating from each point of the projected figure are all parallel (in contrast, for instance, to perspective projection)

perspective projection
> a geometric projection in which the projection lines emanating from each point of the projected object converge to a point, defined as center of projection, or point of view (POV)

pixel
> picture element; the smallest element in a computer image

platonic solid
> one of the five regular polyhedra: the tetrahedron, cube, octahedron, dodecahedron, and icosahedron

point of view
> in perspective drawing, the center of projection

polar coordinates
> a coordinate system in which points are located by reference to their distance from a central point and an angle of rotation with respect to a horizontal axis

projection
> the reproduction of a spatial object in which each point of the object is projected onto a plane or curved surface; see also {axonometric projection; isometric projection; orthographic projection}

projective geometry
> a branch of geometry that deals with properties of figures that are unaltered by projection

recursion
> in a sequential process, the determination of each successive element by means of operations on one or more preceding elements according to a given rule

reflection
> the operation by which of a geometric figure is transformed into its mirror image across a defined axis

rendering
> one of various graphic manipulations of the views obtained as geometric projections of a constructed computer model

ruled surface
> surfaces generated by a straight line transposed along a path

rotation
> the displacement of a geometric figure by means or rotation about a fixed point

scaling
> the geometric transformation by which the size of a figure is changed

semantics
> related to the meaning of signs

semiotics
> a theory of signs and symbols dealing with their function in artificial and natural languages

solid modeling
> area of computer graphics dealing with three-dimensional objects represented as solids

surface model
> model of a three-dimensional object defined by the surfaces bounding it

symbol
> in CAD a set of elements of any nature

(points, lines, faces, solids, text, other symbols, and so forth) grouped together under one name

syntax
:   arrangement of elements showing their relations

tessellation
:   the juxtaposition of geometric elements in a mosaic-like pattern

texture map
:   in rendering programs, image to be mapped on the surface to be rendered

topology
:   a branch of mathematics concerned with those properties of geometric configurations that are unaltered by elastic deformations

translation
:   the displacement of a geometric configuration to a new position without rotating it or changing its shape

vanishing point
:   in perspective drawing, the point at which all parallel lines not parallel to the picture place converge

virtual reality
:   a computer generated world in which the perceiver is totally immersed by wearing sensorial devices such as head mounted display and sensor-laced gloves

wireframe model
:   definition of a three-dimensional object by its edges